Active In-Wheel Suspension System Design for On-Road Vehicles

Active In-Wheel Suspension System Design for On-Road Vehicles presents a comprehensive and systematic design and control framework for an Active In-Wheel Suspension System (IWASS), a transformative technology for modern on-road vehicles. This book addresses the fundamental challenge of integrating a high-performance active suspension actuator directly within the confined space of a vehicle's wheel.

Taking a multidisciplinary approach which rigorously tackles the mechanical, electromagnetic, and thermal domains to create a viable and robust system, the work presents a complete trajectory from theoretical modeling and advanced control synthesis to high-fidelity simulation and rigorous robustness analysis. It therefore establishes a foundational roadmap for developing this disruptive automotive innovation. International case studies are integrated throughout, with summaries and problem sets at the end of each chapter to reinforce professional understanding of key engineering decisions. By consolidating key findings and charting essential future research directions, this book contributes significantly to the technologies that will underpin the next generation of vehicles, particularly as the industry evolves toward full electrification and intelligent, connected transportation systems.

Designed to serve as a vital resource for automotive engineers, control systems researchers, and graduate students, it bridges the gap between theoretical control science and practical automotive engineering, offering both the depth required for specialists and the foundational context for those new to the field.

Active In-Wheel Suspension System Design for On-Road Vehicles

Do Trong Tu

CRC Press
Taylor & Francis Group
Boca Raton London New York

CRC Press is an imprint of the
Taylor & Francis Group, an **informa** business

First edition published 2027
by CRC Press
2385 NW Executive Center Drive, Suite 320, Boca Raton FL 33431

and by CRC Press
4 Park Square, Milton Park, Abingdon, Oxon, OX14 4RN

CRC Press is an imprint of Taylor & Francis Group, LLC

ISBN: 978-1-041-31516-2 (hbk)
ISBN: 978-1-041-31748-7 (pbk)
ISBN: 978-1-003-78053-3 (ebk)

DOI: 10.1201/9781003780533

Typeset in TeXGyreTermesX font
by KnowledgeWorks Global Ltd.

Dedication

To my family for their unconditional support, my supervisors for their mentorship, my colleagues for collaboration, universities for opportunity, and students who inspire continued learning—this work is dedicated to all who contributed to its completion.

Contents

Foreword

The rapid evolution of electric and autonomous vehicles represents one of the most significant transformations in the history of the automotive industry. This technological revolution brings with it new challenges and opportunities, particularly in the domain of vehicle dynamics and suspension control systems. The integration of in-wheel motors, while offering numerous advantages in terms of vehicle design and control capabilities, introduces complex dynamic interactions that demand sophisticated solutions.

This monograph arrives at a crucial moment in this technological transition, presenting a comprehensive exploration of advanced control strategies that address the fundamental challenge of increased unsprung mass in electric vehicles with in-wheel motors. What makes this work particularly valuable is its systematic approach to bridging theoretical control theory with practical automotive applications, taking the reader from established control methodologies to cutting-edge optimization techniques, demonstrating a clear progression in complexity and performance. The inclusion of multi-objective optimization for controller tuning represents a significant contribution to the field, providing an automated framework for addressing the multi-objective design challenges inherent in suspension systems. Similarly, the treatment of robust control strategies acknowledges the practical realities of vehicle operation in uncertain environments.

As the automotive industry continues its transition toward electrification and automation, the insights and methodologies presented in this work will undoubtedly play an important role in shaping the future of vehicle suspension systems. This monograph serves as an invaluable resource for automotive engineers, control systems researchers, and graduate students working in the field of vehicle dynamics. The clarity of exposition, the depth of technical content, and the practical relevance of the presented material make it a significant contribution to the literature. This monograph stands as a testament to the innovative thinking and technical excellence driving this field forward, advancing our understanding of suspension control while providing practical tools for improving the ride comfort, safety, and performance of next-generation vehicles.

Hanoi, Viet Nam **Do Trong Tu**

Preface

The evolution of on-road vehicles has entered a transformative phase with the integration of advanced suspension technologies, particularly in the realm of electric and hybrid vehicles. The emergence of in-wheel motor systems presents both unprecedented opportunities and significant challenges in vehicle dynamics and suspension design. This monograph, ***Active In-Wheel Suspension System Design for On-Road Vehicles***, represents the culmination of extensive research dedicated to addressing the complex interplay between suspension performance, vehicle stability, and ride comfort in modern automotive systems.

The motivation for this work stems from the growing need to overcome the fundamental limitations of traditional suspension systems when applied to vehicles with in-wheel motors. The substantial increase in unsprung mass associated with wheel-mounted motors necessitates innovative approaches to suspension design and control strategies. This monograph systematically addresses these challenges through a comprehensive framework that combines theoretical foundations with practical implementation methodologies.

The organization of this monograph reflects a logical progression from fundamental concepts to advanced control strategies. Beginning with **Chapter 1** establishes the foundational context, framing the IWASS as a disruptive innovation within the historical evolution of automotive technology. Building on this, **Chapter 2** introduces a comprehensive, multidisciplinary design framework, addressing the critical integration of mechanical, electromagnetic, and thermal domains within severe packaging constraints. The theoretical core of the work is presented in **Chapter 3**, which constructs a rigorous mathematical foundation from first principles, and **Chapter 4**, which synthesizes advanced control strategies to manage the system's multi-objective dynamics. **Chapter 5** validates these control methodologies through high-fidelity simulation, demonstrating their efficacy in overcoming the unsprung mass challenge, while **Chapter 6** subjects the system to a rigorous parametric uncertainty analysis, establishing the necessity and performance of robust H_∞ control. Following that, **Chapter 7** comprehensively develops and validates a constrained Model Predictive Control framework for IWASS through systematic multi-objective optimization while explicitly handling physical constraints and parameter variations. Finally, **Chapter 8** provides a conclusive synthesis, consolidating the key findings and charting the essential future research directions to advance the IWASS from a high-performance prototype to an intelligent subsystem of future vehicles.

This monograph is designed to serve multiple audiences: automotive engineers seeking practical solutions for suspension system design, researchers exploring advanced control methodologies, and graduate students specializing in vehicle dynamics and control systems. The content has been structured to provide both depth for specialists and accessibility for those new to the field.

The methodologies, analyses, and design frameworks presented in this monograph will contribute significantly to the innovative technologies for on-road vehicles. As the automotive industry continues to evolve toward electrification and intelligent transportation systems, the insights provided herein will serve as valuable resources for developing next-generation systems that enhance both vehicle performance and passenger comfort.

Hanoi, Viet Nam **Do Trong Tu**

Acknowledgement

I would like to express my sincere gratitude to the many individuals, laboratories and institutions whose support and contributions have been essential to the completion of this monograph.

First and foremost, I extend my deep appreciation to my academic advisors and professors at the universities, including Electric Power University (EPU), the University of Transport and Communications (UTC), Grenoble INP, and Myongji University where this research was conducted. Their expert guidance, invaluable insights, and continuous encouragement have profoundly shaped the direction and quality of this work. Their mentorship has been a source of inspiration throughout this academic journey.

I am also profoundly thankful to my colleagues and research team members at the laboratories (Gipsa-lab, ITV-lab, NGV-lab) and academic departments (FMACE-EPU, FAME-UTC) involved in this project. The collaborative environment, stimulating discussions, and shared dedication to research excellence created a supportive atmosphere that greatly enriched this work. Their camaraderie and professional support were invaluable during the challenging phases of this research.

This work was made possible through the generous financial support provided by various research grants and funding agencies such as MOET and MOC, as well as institutional support from EPU. The resources and opportunities afforded by this funding were crucial for conducting the extensive simulations and experimental validations presented in this monograph.

I would like to acknowledge the laboratories and research centers at EPU, UTC, and Grenoble-INP that provided the facilities and research infrastructure necessary for this monograph. The access to advanced computational resources and research facilities significantly contributed to the depth and scope of this work.

To my family and loved ones, I owe my deepest gratitude for their unwavering support, understanding, and encouragement throughout this demanding yet rewarding journey. Their patience and belief in my work provided the foundation that sustained me through this endeavor.

Do Trong Tu
Hanoi, Viet Nam

Author

Do Trong Tu is a Lecturer specializing in advanced automotive engineering at Electric Power University, Vietnam. He began his academic journey with a Bachelor of Engineering from the University of Transport and Communications, which included an international research internship at Gipsa-Lab, Grenoble INP, France. Following initial industry experience as a research engineer in Hanoi, he advanced his expertise through a Master of Science in Mechanical Engineering at the Advanced Automotive Laboratory in South Korea. His research is dedicated to the core technologies enabling next-generation vehicles, with a specific focus on vehicle dynamics and control and energy management systems. His work encompasses a broad range of electric vehicles, including battery electric, hybrid electric, and fuel cell electric vehicles.

Acronyms

AFPM Axial Flux Permanent Magnet
ARE Algebraic Riccati Equation
CAD Computer-Aided Design
DAE Differential-Algebraic Equation
DOF Degree of Freedom
EMC Electromagnetic Compatibility
ER Electro-Rheological
EV Electric Vehicle
FEA Finite Element Analysis
FR Functional Requirement
GA Genetic Algorithm
GA-LQR Genetic Algorithm-Optimized Linear Quadratic Regulator
IC Instant Center
IWSS In-Wheel Suspension System
IWASS In-Wheel Active Suspension System
IWM In-Wheel Motor
LPV Linear Parameter-Varying
LQG Linear Quadratic Gaussian
LQR Linear Quadratic Regulator
MIMO Multi-Input, Multi-Output
MR Magneto-Rheological
MPC Model Predictive Control
MTTR Mean Time To Repair
NVH Noise, Vibration, and Harshness
OEM Original Equipment Manufacturer
PCM Phase Change Material
PID Proportional-Integral-Derivative
PMSM Permanent Magnet Synchronous Motor
PSD Power Spectral Density
R&D Research and Development
RFPM Radial Flux Permanent Magnet
RMS Root Mean Square
SMC Soft Magnetic Composite
SMC Sliding Mode Control
TIM Thermal Interface Material
UMP Unbalanced Magnetic Pull

1 Unsprung to Unbound: The Century-Long Journey of In-Wheel Suspension

1.1 INTRODUCTION: A PARADIGM SHIFT IN MOTION

For over a century, the fundamental architecture of the automobile has been governed by a set of seemingly immutable principles. An internal combustion engine provides power, a mechanical drivetrain transmits it, and a suspension system, comprising springs, dampers, and linkages, connects the wheels to the chassis, managing the dynamic dialogue between vehicle and road. This arrangement, while refined to a remarkable degree, carries inherent compromises rooted in the fundamental physics of vehicle dynamics [1]. The suspension components, along with the engine and transmission, constitute a significant portion of the vehicle's "sprung mass"—the part of the car isolated from road impacts by the springs. Wheels, tires, and brakes, by contrast, are the "unsprung mass," forced to directly follow every contour and imperfection of the tarmac.

This relationship between sprung and unsprung mass represents one of the most fundamental trade-offs in vehicle design [41]. The classical suspension theory, dating back to the early 20th century, demonstrates that the ratio of unsprung to sprung mass critically determines both ride comfort and road holding capability. The governing equations of motion reveal why this is so critical:

$$m_s\ddot{z}_s + k_s(z_s - z_u) + c_s(\dot{z}_s - \dot{z}_u) = 0 \tag{1.1}$$

$$m_u\ddot{z}_u + k_s(z_u - z_s) + c_s(\dot{z}_u - \dot{z}_s) + k_t(z_u - z_r) = 0 \tag{1.2}$$

where m_s and m_u represent sprung and unsprung mass, respectively, and their ratio dictates the system's natural frequencies and damping characteristics. A high unsprung mass not only degrades ride quality but also reduces tire contact patch consistency, fundamentally limiting both comfort and performance.

This chapter chronicles the story of a radical and persistent challenge to this century-old convention: the in-wheel suspension system (IWSS) [2]. It is a tale not of a single invention, but of a prolonged engineering evolution—a story of brilliant vision, formidable physics, false dawns, and a final, powerful convergence with technological trends that is now bringing this once-heretical idea to the brink of mainstream adoption. We will follow its path from a sketched idea in a patent office to a key technological enabler for the future of mobility, examining the key players, the critical hurdles, and the strategic decisions that are shaping its journey to market.

DOI: 10.1201/9781003780533-1

The in-wheel suspension, also known as wheel-hub suspension or, when combined with a motor, a hub motor system [4], proposes a fundamental inversion of conventional automotive architecture. Instead of having a large, sprung mass connected to a small, unsprung mass, why not miniaturize the suspension and place it directly inside or adjacent to the wheel itself? This concept promises to liberate vehicle design, redefine ride quality, and unlock unprecedented levels of control [3]. Its journey from idea to market is a masterclass in innovation, demonstrating that the path to progress is often long, winding, and dependent on the ecosystem of technology catching up with a pioneering vision.

The significance of this technological transition cannot be overstated. We are witnessing nothing less than the decomposition of the automobile into modular, self-contained corner units—a shift as profound as the move from body-on-frame to unibody construction that characterized mid-20th-century automotive design. This chapter will explore not only the technical evolution but also the broader implications for vehicle architecture, manufacturing processes, and the very nature of mobility itself.

This chapter introduces the conceptual and historical origins of the in-wheel suspension system, tracing how a century of engineering innovation and dynamic challenges converged toward the modern active in-wheel suspension architecture.

1.2 THE GENESIS: THE IDEA AND EARLY CONCEPTUALIZATION

The core idea of locating suspension components at the wheel is deceptively simple, and its appeal has captivated engineers for generations [1]. Long before the advent of modern materials and electronics, inventors recognized the potential benefits of re-thinking the relationship between the wheel and the chassis. What's remarkable is how early in automotive history these concepts emerged, suggesting that the limitations of conventional suspension architecture were apparent almost from the beginning.

1.2.1 KEY MOTIVATIONS BEHIND THE IDEA

The theoretical advantages that drove these early pioneers remain the central value proposition today [1], though our understanding of them has deepened considerably through advanced dynamics analysis and computational modeling.

1.2.1.1 The Mass Inversion Hypothesis

The primary theoretical benefit, now supported by extensive vehicle dynamics simulation, concerns the fundamental relationship between unsprung mass and vehicle behavior [41]. In a conventional suspension, a large unsprung mass (typically comprising brake calipers, wheel hubs, and partial suspension components) possesses significant inertia. When encountering road irregularities, this mass is subjected to substantial vertical accelerations governed by:

$$a_u = \frac{k_t(z_r - z_u) - F_{susp}}{m_u} \tag{1.3}$$

where a_u is the unsprung mass acceleration, k_t is tire stiffness, z_r and z_u are road and unsprung mass displacements, and F_{susp} is the suspension force. A heavier unsprung mass reduces the acceleration for a given force, meaning it responds more slowly to road inputs. This manifests in two critical ways:

First, when hitting a bump, the heavy unsprung mass is thrown upwards violently, losing contact with the road. The tire's ability to maintain consistent normal force F_z is compromised, reducing the available friction force according to the tire friction circle:

$$\sqrt{F_x^2 + F_y^2} \leq \mu F_z \tag{1.4}$$

Second, after the bump passes, the weight of the unsprung mass overpowers the spring, keeping the tire from quickly re-establishing contact. This leads to extended periods of reduced traction during transient events.

By moving the spring, damper, and linkage mass to the wheel, the unsprung mass (in an ideal scenario, now just the tire and rim) is drastically reduced. A lighter unsprung mass can react more quickly and with less force to road imperfections, leading to improved grip, enhanced comfort, and greater stability, especially at high speeds. The wheel stays "glued" to the road surface more effectively, maintaining optimal tire loading across a wider range of conditions.

Modern analysis shows that the benefits extend beyond simple vertical dynamics [2]. The reduced rotational inertia I_w of the wheel assembly improves both acceleration and braking performance:

$$T = I_w \dot{\omega} + F_x R \tag{1.5}$$

where lower I_w reduces the torque T required for a given angular acceleration $\dot{\omega}$, improving responsiveness and efficiency.

1.2.1.2 Unprecedented Packaging Freedom

The spatial implications of in-wheel suspension represent perhaps its most visually apparent advantage [34]. Removing the bulky shock towers, control arms, and anti-roll bars from the engine bay and chassis liberates vast amounts of space that can be reallocated according to new design priorities. This freedom manifests in several transformative ways:

The elimination of traditional suspension mounting points allows for completely flat vehicle floors, a feature particularly valuable in EVs, where battery packaging efficiency becomes paramount. The available height $h_{battery}$ for battery modules increases substantially:

$$h_{battery} = H_{floor} - h_{clearance} - h_{packaging} \tag{1.6}$$

where the reduction in $h_{packaging}$ requirements directly increases energy storage capacity within the same exterior dimensions.

Vehicle designers gain unprecedented freedom in optimizing aerodynamic shapes. Without the constraints of conventional suspension geometry, front and rear overhangs can be minimized, leading to improved drag coefficients and better aerodynamic efficiency. The fundamental relationship between drag force and vehicle geometry:

$$F_d = \frac{1}{2}\rho v^2 C_d A \tag{1.7}$$

can be optimized through both reduced frontal area A and improved drag coefficient C_d enabled by the packaging freedom.

The entire vehicle architecture can be rethought around occupant and cargo space rather than mechanical component accommodation. This represents a fundamental shift from function-follows-form to form-follows-function design philosophy.

1.2.1.3 Individual Wheel Control

The decentralization of suspension and drive components enables a revolution in vehicle dynamics control [3, 13, 14]. With each wheel's suspension and drive (if an electric motor is included) being a self-contained unit, each corner of the vehicle can be independently and precisely controlled by a central computer. This opens the door to revolutionary advancements in active safety and vehicle dynamics through several mechanisms.

The system can implement individual wheel force control through suspension actuation [10, 11], allowing for real-time compensation of road disturbances. The control law for each corner can be expressed as:

$$F_{a,i} = f(z_{s,i} - z_{u,i}, \dot{z}_{s,i} - \dot{z}_{u,i}, \ddot{z}_{s,i}, \phi, \theta, r, \ldots) \tag{1.8}$$

where the active force $F_{a,i}$ at each wheel is computed based on multiple vehicle state variables.

Torque vectoring becomes inherent rather than added functionality [38]. The distribution of drive torque between wheels can be optimized for both stability and agility:

$$\begin{bmatrix} T_{fl} \\ T_{fr} \\ T_{rl} \\ T_{rr} \end{bmatrix} = \mathbf{K} \begin{bmatrix} \delta \\ a_x \\ a_y \\ r \\ \beta \\ \dot{\beta} \end{bmatrix} \tag{1.9}$$

where the torque vector $\mathbf{T}$ is computed based on driver inputs and vehicle states through gain matrix $\mathbf{K}$.

The system can react to loss of traction, crosswinds, or evasive maneuvers on a millisecond basis, adjusting damping, torque, and even wheel position individually to maintain optimal stability and control [17]. This multi-input, multi-output (MIMO) control capability represents a fundamental advancement over conventional stability control systems.

1.2.2 EARLY PIONEERS AND PATENTS

The historical record reveals that the conceptual foundation for in-wheel suspension was established remarkably early in automotive history. Patent archives from the late 19th and early 20th centuries show numerous inventors grappling with the limitations of conventional suspension systems and proposing radical alternatives.

One of the earliest documented examples comes from American inventor William B. Biddle, who filed patents in the 1920s for rudimentary in-wheel mechanical spring and lever systems. Biddle's designs, while primitive by modern standards, demonstrated remarkable insight into the fundamental problem. His patents described "improvements in vehicle springing" that sought to "provide a more direct and responsive suspension mechanism" by locating spring elements within the wheel assembly itself.

Biddle's designs typically employed systems of levers and small coil springs contained within an enlarged wheel hub assembly. The mechanical advantage was carefully calculated to provide the necessary spring rates while keeping components compact enough to fit within the wheel envelope. While these systems were never commercialized, they established the basic architectural principle that would inspire future generations of engineers.

In Europe, similar concepts emerged independently during the same period. German and French inventors filed patents for "radnabenfederung" (wheel hub suspension) systems that used various mechanical arrangements to achieve the same basic goal. These European approaches often favored torsion-based systems rather than compression springs, reflecting different engineering traditions and manufacturing capabilities.

What's striking about these early designs is their recognition of the fundamental trade-offs involved. Patent documents from this era discuss the benefits of reduced unsprung mass, improved packaging, and better wheel control with surprising sophistication. However, these early systems were ultimately limited by the available technology of their time.

The materials science of the era—primarily heavy steel and cast iron—meant that any system placed in the wheel would be prohibitively heavy, thus negating the unsprung mass benefit [35]. The strength-to-weight ratios of available materials simply couldn't support the functional requirements without excessive mass penalty.

Precision manufacturing and sealing technologies were insufficient to create durable, reliable mechanisms that could survive the brutal, dirty environment of the wheel [35, 39]. Tolerances that are routine today were impossible to maintain with the manufacturing equipment of the 1920s and 1930s.

Furthermore, without electronic control, these were entirely passive systems, unable to adapt to changing conditions or provide the active control capabilities that make modern implementations so promising [18, 19]. The idea was sound in principle, but the technological foundation to execute it reliably and effectively did not yet exist. It was an idea ahead of its time, destined to lie dormant in patent files, waiting for the world to catch up.

The period between the 1930s and 1970s saw sporadic interest in the concept, primarily in specialized applications like military vehicles and racing cars where the

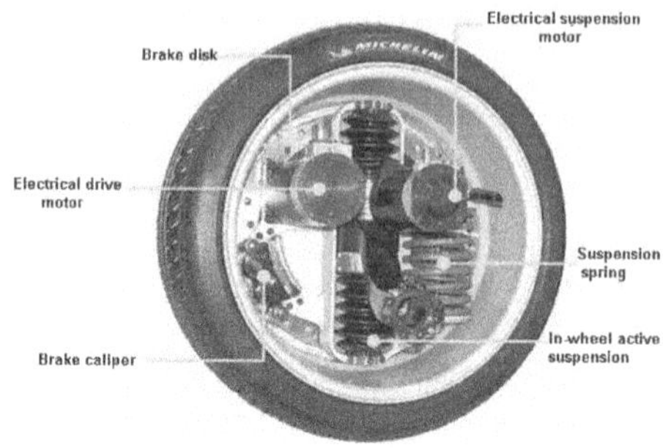

Figure 1.1: Michelin's active wheel technology [6]

performance benefits could justify the complexity and cost. However, these implementations remained niche applications, and the fundamental technological barriers prevented widespread adoption.

1.3 THE PIONEER: MICHELIN'S ACTIVE WHEEL AND THE MODERN REVIVAL

The modern era of in-wheel suspension began in earnest not with a car manufacturer, but with a tire company—a fact that reveals much about the interdisciplinary nature of true automotive innovation. In 2008, at the Challenge Bibendum, a prestigious forum for sustainable mobility, Michelin unveiled its Active Wheel system. This was a landmark moment that transformed the in-wheel concept from an engineering theory into a tangible, working prototype, capturing the imagination of the entire automotive industry.

Michelin's brilliance was in recognizing that the move towards electric propulsion created a unique opportunity to integrate multiple systems into a single, compact wheel unit [4]. The Active Wheel was not merely a suspension; it was a holistic "corner module" that redefined what was possible at each wheel of a vehicle. This systems-level thinking represented a fundamental shift from previous approaches that had treated suspension, braking, and propulsion as separate problems.

1.3.1 TECHNICAL DEEP DIVE: DECONSTRUCTING THE ACTIVE WHEEL

The system was an engineering marvel of integration, packing three critical functions into a standard-sized wheel rim. What made the Active Wheel particularly innovative was not just the individual components, but how they were engineered to work together as a unified system.

1.3.1.1 The Electric Motor Subsystem

At the heart of the system was a high-speed, lightweight electric motor providing power directly to the wheel [2]. This represented a radical implementation of the wheel-hub motor concept, eliminating the need for a central motor, transmission, driveshafts, and differentials. The motor design reflected sophisticated electromagnetic engineering [35]:

The motor utilized a permanent magnet synchronous (PMSM) architecture optimized for high torque density and efficiency. The electromagnetic torque production can be described by:

$$T_e = \frac{3}{2}\frac{P}{2}\left[\lambda_{pm} i_q + (L_d - L_q) i_d i_q\right] \tag{1.10}$$

where careful optimization of the permanent magnet flux λ_{pm} and dq-axis inductances L_d, L_q enabled high torque production within the constrained volume.

The motor was designed for direct drive operation, eliminating the need for reduction gears and their associated losses, backlash, and maintenance requirements. This simplified the mechanical design but placed greater demands on the motor's torque production capabilities.

Advanced cooling systems were integrated to manage heat generation, which followed the fundamental loss equation [8, 37]:

$$P_{loss} = P_{copper} + P_{iron} + P_{mech} = \sum I^2 R + k_h f B^{\alpha} + k_c (f B)^2 + k_f \omega^2 \tag{1.11}$$

The motor's compact design demonstrated that sufficient power and torque could be achieved within the spatial constraints of a conventional wheel.

1.3.1.2 The Active Suspension Subsystem

Perhaps the most revolutionary aspect was the suspension system [4]. Michelin replaced traditional passive springs and dampers with a miniature, electrically-actuated hydraulic suspension system. This "active" system represented a fundamental departure from conventional thinking in several ways [10, 11]:

The system used high-bandwidth hydraulic actuators capable of generating forces sufficient to control both high-frequency road inputs and low-frequency body motions. The actuator dynamics can be modeled as:

$$F_a = A_p P_L - B_{eq} \dot{x}_p - F_f \tag{1.12}$$

where the active force F_a is generated through precise control of the hydraulic pressure P_L acting on piston area A_p.

Sophisticated sensors and an electronic control unit constantly adjusted the suspension in real-time, implementing advanced control algorithms that could anticipate and react to road conditions and driver inputs [15, 16]. The control law typically took the form:

$$F_{a,cmd} = K_p e + K_i \int e dt + K_d \frac{de}{dt} + K_{skyhook} \dot{z}_s + K_{groundhook} \dot{z}_u \tag{1.13}$$

Incorporating both traditional PID terms and more advanced "skyhook" and "groundhook" damping strategies.

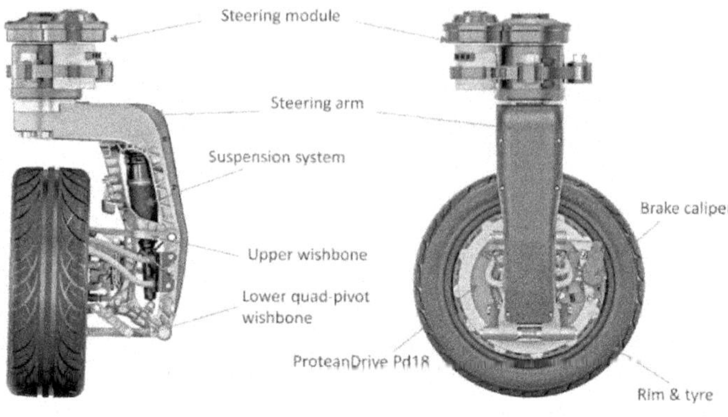

Figure 1.2: Motor, steering and suspension in a single module by Protean [5]

The system could counteract body roll in corners, dive under braking, and squat under acceleration, effectively creating a "magic carpet" ride that remained level regardless of driving conditions [20, 21]. It could also actively push the wheel down to maintain contact with the road over bumps, optimizing tire forces according to:

$$F_z = k_t(z_u - z_r) + F_{a,downforce} \tag{1.14}$$

This active control capability represented a quantum leap beyond what passive or even semi-active suspensions could achieve.

1.3.1.3 The Braking Subsystem

Integrated within the same compact housing was an electric disc brake system that demonstrated sophisticated systems integration [4]:

Michelin developed a brake-by-wire system that was fully decoupled from the hydraulic suspension, ensuring fail-safe operation and independent control of braking and suspension functions.

The braking system was designed for maximum energy recovery during deceleration, working in concert with the electric motor's regenerative braking capabilities [12, 32]. The total braking force distribution followed:

$$F_{brake,total} = F_{friction} + F_{regenerative} \tag{1.15}$$

With sophisticated control algorithms optimizing the blend between mechanical and electrical braking.

The entire assembly was designed for serviceability and reliability [8], with special attention paid to thermal management of the brake components in close proximity to the electric motor.

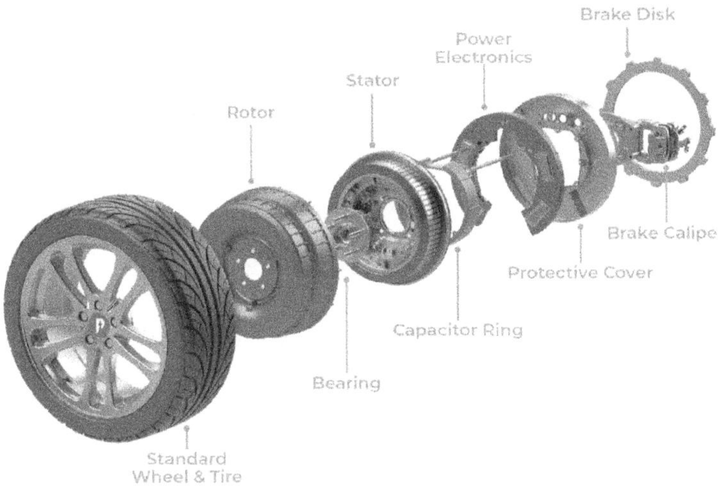

Figure 1.3: ProteanDrive system [7]

1.3.2 DEMONSTRATION AND IMPACT: PROVING THE CONCEPT

Michelin didn't just build a prototype; they demonstrated its application in working concept cars that showed the real-world benefits of the technology. The French coachbuilder Heuliez created the WILL, a sleek compact car, and Venturi unveiled the stunning Volage sports car. These vehicles served as rolling testaments to the benefits of the Active Wheel system.

The WILL demonstrated the packaging advantages with dramatic effect [34]. The vehicle featured a completely flat floor, as there was no transmission tunnel or suspension towers to accommodate. The interior space utilization was remarkable for the vehicle's exterior dimensions, with the wheelbase-to-length ratio approaching theoretical maximums. The mathematical relationship:

$$\text{Space Efficiency} = \frac{\text{Usable Interior Volume}}{\text{Exterior Envelope Volume}} \tag{1.16}$$

Reached unprecedented values, demonstrating the architectural advantages of the in-wheel approach.

The Venturi Volage showcased the performance potential [3, 23]. Journalists who experienced the prototypes reported an almost surreal ride quality—the vehicle glided over bumps and remained perfectly flat during aggressive maneuvers. The combination of active suspension and torque vectoring created handling characteristics that defied conventional expectations [14]. The vehicle's ability to maintain optimal tire loading during cornering followed the relationship:

$$F_{z,i} = F_{z,static} + \Delta F_{z,transfer} + \Delta F_{z,active} \tag{1.17}$$

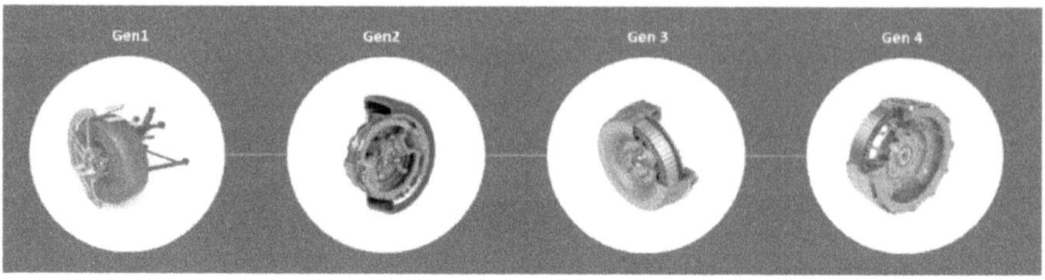

Figure 1.4: Generation of IWSS by Potent

where the active suspension component $\Delta F_{z,active}$ could counteract the weight transfer effects that normally degrade cornering performance.

The Active Wheel was the definitive "proof of concept" that the industry needed [1]. It tangibly demonstrated the three core benefits—mass inversion (in theory), packaging freedom (in practice), and individual active control (dramatically)—to a global audience. It proved that the immense challenges were not insurmountable and ignited a new wave of research and development across the industry.

Perhaps most importantly, the Active Wheel changed the conversation around in-wheel suspension from "is it possible?" to "how do we make it practical?" This shift in perspective was crucial for attracting the investment and engineering talent needed to advance the technology toward production readiness.

1.4 THE CRUCIBLE: TECHNICAL HURDLES AND ENGINEERING CHALLENGES

The brilliance of the Michelin Active Wheel, however, also cast a bright light on the immense practical challenges of bringing such a system to a production-ready, cost-effective, and reliable reality. For over a decade after its unveiling, these hurdles largely confined in-wheel suspension to the realms of concept cars and the ultra-exclusive world of high-end motorsports [1, 24]. This period was the crucible in which the concept was tested and refined, separating theoretical promise from practical reality.

1.4.1 THE UNSPRUNG MASS PARADOX

This was, and remains, the most significant and counterintuitive challenge [41]. The entire theoretical foundation of in-wheel suspension is predicated on reducing unsprung mass to improve ride and handling [9, 24]. However, the reality of packaging an electric motor, a braking system, and an active suspension mechanism inside the wheel inevitably adds a significant amount of mass to the very assembly you are trying to lighten.

This created a fundamental paradox that can be expressed mathematically. The total unsprung mass $m_{u,total}$ becomes:

$$m_{u,total} = m_{u,conventional} + \Delta m_{motor} + \Delta m_{suspension} + \Delta m_{brakes} - \Delta m_{removed} \tag{1.18}$$

where the additional components often outweighed the mass of the conventional components they replaced. This increased rotational inertia had several negative consequences that could be quantified precisely [2].

The impact on acceleration and braking performance followed the relationship:

$$\alpha = \frac{T - F_x R}{I_w} \tag{1.19}$$

where higher rotational inertia I_w required more torque T for the same angular acceleration α, negatively impacting acceleration, regenerative braking efficiency, and stopping distances.

The effect on ride comfort could be understood through the quarter-car model natural frequencies [41]:

$$\omega_{n,u} = \sqrt{\frac{k_t}{m_u}}, \quad \omega_{n,s} = \sqrt{\frac{k_s}{m_s}} \tag{1.20}$$

where increased unsprung mass m_u lowered the unsprung mass natural frequency $\omega_{n,u}$, potentially bringing it closer to typical road input frequencies and exacerbating wheel hop.

The engineering battle, therefore, shifted from simple integration to a relentless pursuit of lightweighting [35]. This demanded the use of advanced materials like forged aluminum, titanium, and carbon-fiber composites, as well as incredibly compact and power-dense motor designs. The material selection process involved sophisticated trade-off analysis considering:

$$\text{Merit Index} = \frac{E^\alpha}{\rho^\beta} \tag{1.21}$$

where different merit indices applied to structural components ($\alpha = 1, \beta = 1$), spring elements ($\alpha = 2, \beta = 1$), and rotating components ($\alpha = 2, \beta = 3$).

1.4.2 DURABILITY AND ENVIRONMENTAL SEALING

A traditional suspension is relatively protected within the vehicle's chassis. An in-wheel system is located in the worst possible environment [39]: it is constantly bombarded with road salt, abrasive grime, pressure-washer jets, and physical impacts from road debris and potholes. The engineering challenges here were monumental and multifaceted.

Creating a robust, long-lasting seal to protect the sensitive electric motor windings, high-precision hydraulic actuators, and bearings from this relentless assault required solving multiple problems simultaneously [17, 25, 27]:

The sealing system had to accommodate thermal expansion and contraction of different materials with coefficients of thermal expansion following:

$$\Delta L = L_0 \alpha \Delta T \tag{1.22}$$

where mismatched α values could lead to seal degradation over temperature cycles.

The system had to withstand extreme shock loads—the kind of impacts that would bend a traditional steel rim. The dynamic forces involved in pothole impacts could be estimated as:

$$F_{impact} = \frac{mv^2}{2\delta} \tag{1.23}$$

where even at modest speeds, the forces could reach tens of thousands of Newtons.

Electromagnetic Compatibility (EMC) presented another challenge, as the high-power electronics in close proximity to sensors and control systems created potential for interference [28, 33]. The near-field coupling could be modeled as:

$$V_{induced} = -j\omega M I_{source} \tag{1.24}$$

requiring sophisticated shielding and layout strategies.

The solution required not just better seals, but a complete rethinking of structural integrity and impact resistance for a vastly more complex assembly [31, 35]. Multi-material bonding techniques, advanced surface treatments, and redundant sealing architectures all played roles in addressing these challenges.

1.4.3 THE THERMAL MANAGEMENT NIGHTMARE

Electric motors and friction brakes are inherently inefficient [8]; they generate significant heat as a byproduct of their operation. Packaging them inside a sealed, compact wheel rim with minimal airflow created a thermal management challenge of the highest order [37]. The problem was fundamentally one of energy density and heat rejection capability.

Under heavy braking or aggressive acceleration, the heat generated could be immense. The power dissipation during braking followed:

$$P_{braking} = F_x v \tag{1.25}$$

which could easily reach hundreds of kilowatts in a performance vehicle. Similarly, motor losses followed:

$$P_{motor,loss} = I^2 R + k_h f B^\alpha + k_c (f B)^2 \tag{1.26}$$

creating substantial heat generation within the confined space.

Without an effective way to dissipate this heat, temperatures would soar, leading to multiple failure modes [8].

Motor demagnetization of permanent magnets occurs when temperature exceeds:

$$T > T_{curie} - \Delta T_{safety} \tag{1.27}$$

where neodymium magnets typically have Curie temperatures around $310°C$ but begin degrading at lower temperatures.

Insulation breakdown in motor windings follows Arrhenius behavior [37]:

$$\text{Lifetime} \propto e^{\frac{E_a}{kT}} \tag{1.28}$$

where every $10°C$ increase roughly halves the insulation life.

Brake fade occurs as the friction coefficient decreases with temperature according to complex thermal-wear relationships.

Engineers had to develop sophisticated and compact liquid-cooling systems, often using oil that could circulate through the motor and brakes, and then to a heat exchanger [8, 37]. The heat transfer capability was governed by:

$$q = hA\Delta T \tag{1.29}$$

where maximizing the heat transfer coefficient h and surface area A within the constrained volume became critical design challenges.

They also experimented with heat-dissipating materials and designs that could passively radiate heat away from the critical components. Phase change materials (PCMs) were investigated for their ability to absorb large amounts of heat isothermally [37]:

$$Q = mL \tag{1.30}$$

where the latent heat L provided significant thermal buffering capacity.

1.4.4 HIGH COST, COMPLEXITY, AND SERVICEABILITY

The precision engineering, exotic materials, intricate assembly, and sophisticated electronic control required made early in-wheel systems prohibitively expensive [42]. They were works of art, not products for mass consumption. A single Active Wheel unit was estimated to cost many times more than a conventional suspension corner, placing it firmly in the realm of exotic supercars rather than volume production vehicles.

This high cost was compounded by complexity that grew non-linearly with integration level [4]. The number of potential failure modes $N_{failures}$ could be modeled as:

$$N_{failures} = \sum_{i=1}^{n} \binom{n}{i} \tag{1.31}$$

where integrating multiple systems dramatically increased the combinatorial explosion of potential issues.

Furthermore, from a service perspective, the system was a nightmare [39]. A simple wheel bearing replacement on a conventional car is a straightforward job. On an in-wheel system, it could require disassembling the entire integrated unit, a highly specialized and time-consuming procedure. The mean time to repair (MTTR) increased substantially:

$$\text{MTTR} = \sum_{i=1}^{n} t_{operation,i} \tag{1.32}$$

where the numerous disassembly steps $t_{operation,i}$ made servicing impractical for conventional repair facilities.

This made the technology completely incompatible with the high-volume, cost-sensitive, and service-friendly requirements of the mainstream automotive industry [42]. The business case simply didn't close when all costs were considered:

$$\text{Business Case} = \frac{\text{Customer Value}}{\text{Manufacturing Cost} + \text{Warranty Cost} + \text{Service Cost}} \tag{1.33}$$

where all terms in the denominator were significantly higher than conventional solutions.

1.5 THE CONVERGENCE: MARKET FORCES AND TECHNOLOGICAL ENABLERS

The tide began to turn for in-wheel suspension not through a single, miraculous breakthrough, but through a powerful and synergistic convergence of external market forces and internal technological advancements [1]. Over the 2010s and into the 2020s, the stars began to align, systematically addressing the core challenges that had previously kept the technology on the sidelines. This convergence represented the classic pattern of disruptive innovation finally reaching its enabling moment.

1.5.1 THE ELECTRIC VEHICLE REVOLUTION: THE ULTIMATE CATALYST

The single greatest enabler for in-wheel suspension has been the rapid and global shift toward electric propulsion [1, 42]. Electric vehicles (EV) fundamentally dismantle the traditional automotive architecture, creating a natural demand for new solutions and making the inherent advantages of in-wheel systems suddenly far more compelling. Several aspects of this revolution were particularly significant:

The natural synergy with hub motors became apparent as EV designers sought to optimize packaging and efficiency [2]. The elimination of the internal combustion engine and mechanical drivetrain removed the traditional barriers to wheel-hub motor adoption. The system-level benefits could now be properly evaluated using whole-vehicle analysis:

$$\text{System Efficiency} = \eta_{motor} \cdot \eta_{inverter} \cdot \eta_{battery} \cdot \eta_{regeneration} \tag{1.34}$$

where direct drive hub motors could achieve excellent overall efficiency despite potentially lower peak motor efficiency.

The "skateboard" platform architecture of most modern EVs provided the perfect foundation for in-wheel suspension integration. The flat battery pack structure naturally complemented the packaging advantages of in-wheel systems. The vehicle architecture optimization could now be expressed as:

$$\text{Packaging Efficiency} = \frac{\text{Battery Volume} + \text{Passenger Volume}}{\text{Total Envelope Volume}} \tag{1.35}$$

where in-wheel suspension maximized this ratio by minimizing mechanical intrusion.

Figure 1.5: Phenikaa-X vehicle with IWSS [22]

Torque vectoring became a native capability rather than an added feature [3, 38]. The individual control of wheel torques enabled sophisticated vehicle dynamics control strategies:

$$\begin{bmatrix} T_{fl} \\ T_{fr} \\ T_{rl} \\ T_{rr} \end{bmatrix} = f(\delta, a_x, a_y, r, \beta, \dot{\beta}, \mu, \ldots) \tag{1.36}$$

where the multi-variable control law could optimize stability, responsiveness, and efficiency simultaneously.

1.5.2 THE MATERIALS SCIENCE RENAISSANCE

The development of new high-strength, lightweight materials has been critical in tackling the unsprung mass paradox [35]. Advances across multiple material classes provided the necessary properties to make in-wheel systems technically feasible:

Advanced alloys and composites enabled dramatic weight reduction without sacrificing strength or durability. The specific strength and specific stiffness improvements followed:

$$\text{Specific Strength} = \frac{\sigma_{uts}}{\rho}, \quad \text{Specific Stiffness} = \frac{E}{\rho} \tag{1.37}$$

where new materials like carbon fiber composites and advanced aluminum alloys provided values an order of magnitude better than conventional steels.

High-flux magnets based on neodymium and other rare-earth elements enabled the creation of motors that were significantly smaller and lighter for the same power output [2, 35, 37]. The energy product improvement:

$$\text{Energy Product} = (BH)_{max} \tag{1.38}$$

Allowed for more compact magnetic circuits, directly reducing motor mass and volume.

Advanced thermal materials with high thermal conductivity helped address the thermal management challenges. Materials like thermal pyrolytic graphite (TPG) offered in-plane thermal conductivity up to:

$$k_{TPG} \approx 1700 \text{ W/mK} \tag{1.39}$$

Enabling efficient heat spreading within the constrained volume.

1.5.3 THE RISE OF SOPHISTICATED ACTIVE AND SEMI-ACTIVE DAMPING

While Michelin's fully active hydraulic system was brilliant, it was also complex and expensive. In the interim, the industry widely adopted fast-processing, semi-active dampers like MagneRide (from GM/Delphi) and similar systems from other suppliers [18]. These technologies provided a crucial stepping stone by demonstrating that advanced suspension control could be practical and reliable.

These systems used magneto-rheological (MR) or electro-rheological (ER) fluids that could change their viscosity in milliseconds when exposed to a magnetic or electric field [18]. The fundamental relationship:

$$\tau = \tau_y(H) + \eta\dot{\gamma} \tag{1.40}$$

where the yield stress τ_y could be controlled by the applied field H, enabling rapid damping adjustment.

The success of these systems proved that complex suspension control could be achieved in a relatively compact, reliable, and increasingly cost-effective package [10, 11]. This technology, and the control software behind it, provided a stepping stone and a core technology that could be miniaturized and adapted for in-wheel applications.

1.5.4 THE ERA OF THE SOFTWARE-DEFINED VEHICLE

Modern vehicles are increasingly "computers on wheels", controlled by sophisticated software and networks of high-speed sensors [36]. This paradigm is a perfect match for in-wheel systems [16]. The individual wheel control they offer is a powerful hardware platform that can be fully exploited by software.

Features like ultra-stable ride quality, active roll cancellation, and advanced torque vectoring become a matter of software updates [15], allowing for new features and performance enhancements to be delivered to the vehicle long after it has left the factory. The control sophistication can be continuously improved through:

$$\text{Control Performance} = f(\text{Sensor Quality}, \text{Processor Speed}, \text{Algorithm Sophistication}) \tag{1.41}$$

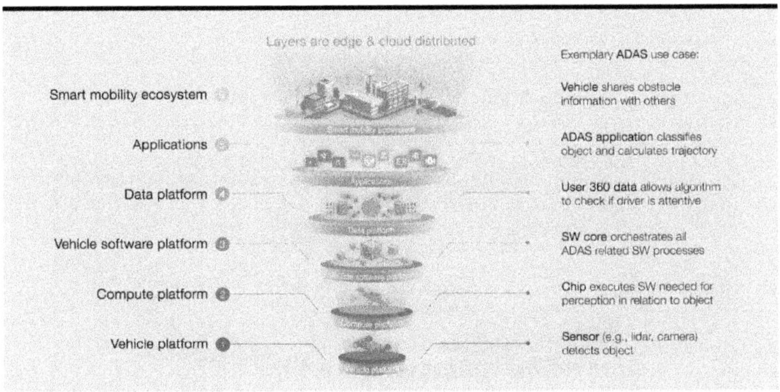

Figure 1.6: Layers of the Software-defined vehicle by world economic forum [26]

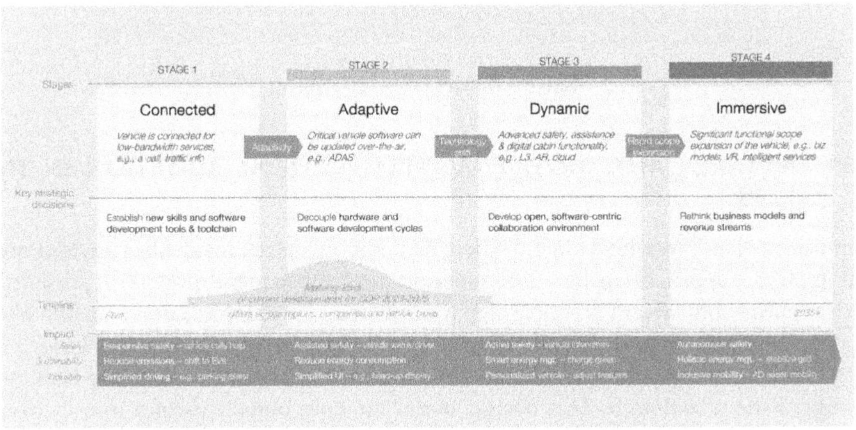

Figure 1.7: Journey towards the vehicle's software [26]

where all three factors have been improving exponentially according to Moore's Law and related trends.

The ability to implement sophisticated MIMO control strategies enables performance that was previously theoretical [14, 40]. The general control architecture:

$$\mathbf{u} = \mathbf{K}\mathbf{x} + \mathbf{K}_f\mathbf{r} + \mathbf{K}_d\mathbf{d} \tag{1.42}$$

1.6 CASE STUDIES IN PRODUCTION AND PROTOTYPING

The journey from laboratory to market is best illustrated by examining real-world applications, from high-performance supercars that test the limits to startups focused on perfecting the core technology. These case studies reveal both the progress made and the challenges that remain.

Figure 1.8: McLaren 720s Spider

1.6.1 MCLAREN AUTOMOTIVE: A PHILOSOPHICAL ADOPTION IN THE 720S

While not a "true" in-wheel system where all components are inside the rim, McLaren Automotive's approach in its 720S supercar (and subsequent models) represents a critical philosophical and practical step towards the concept. McLaren's unique front suspension design features a "double-wishbone" layout where the top wishbone is so high that the damper and spring units are mounted horizontally, extending back into the top of the wheel arch. This design shares the core objectives of a true in-wheel system [41] and demonstrates several important principles.

The mass centralization benefits are significant. By moving the suspension components inwards and lower, it lowers the vehicle's center of gravity and reduces the moment of inertia around vehicle axes:

$$I_{xx} = \sum m_i(y_i^2 + z_i^2), \quad I_{yy} = \sum m_i(x_i^2 + z_i^2) \tag{1.43}$$

where moving mass components inward reduces these inertia terms, improving agility.

The packaging benefits are substantial. It allows for a dramatically lower hood line, improving aerodynamics and forward visibility. The aerodynamic advantages can be quantified through:

$$C_d A = C_d \cdot A \tag{1.44}$$

where both the drag coefficient C_d and frontal area A benefit from the lower hood line.

The McLaren 720S is lauded for its sublime ride quality and razor-sharp handling—a combination often thought to be mutually exclusive [11]. This demonstrated to a skeptical industry and enthusiast community that rethinking suspension geometry

Figure 1.9: Autonomous passenger shuttle by Olli [29]

and packaging around the principles of mass centralization could deliver tangible, dramatic performance benefits in a production vehicle. It served to legitimize the core philosophy behind the more radical in-wheel concept.

1.6.2 ELAPHE PROPULSION TECHNOLOGIES AND LORDSTOWN MOTORS: THE HUB MOTOR IN THE REAL WORLD

While the full integrated suspension-motor unit remains rare, the in-wheel motor has progressed significantly [2]. Companies like Elaphe Propulsion Technologies, a Slovenian company, have focused exclusively on developing robust, high-torque, and lightweight in-wheel motors. Their technology represents the state of the art in hub motor design.

Elaphe's motors demonstrate several key advancements [35, 37]:

The power density has reached levels that make automotive applications feasible. The specific power:

$$\text{Specific Power} = \frac{P_{max}}{m_{motor}} \tag{1.45}$$

has reached values competitive with conventional central motors when system-level benefits are considered.

The thermal management systems have been refined to handle continuous high power operation [8]. The thermal time constant:

$$\tau_{thermal} = \frac{mc_p}{hA} \tag{1.46}$$

has been optimized through advanced cooling strategies and thermal mass management.

Figure 1.10: Endurance EV pickup by Lordstown Motors [30]

The Lordstown case is particularly instructive. The company's plan was to use four Elaphe motors, one in each wheel, to provide all-wheel drive and simplified manufacturing. While Lordstown itself faced well-documented challenges, the selection of Elaphe's motors for a vehicle intended for volume production was a significant milestone.

It proved that a major automotive player (for a time) believed the durability and performance of in-wheel motors had reached a viable level for a production vehicle [39], tackling head-on the challenges of sealing and thermal management. The validation process included:

$$\text{Reliability} = e^{-\lambda t} \tag{1.47}$$

where the failure rate λ had to meet automotive standards for the business case to close.

1.6.3 THE CHINESE EV ECOSYSTEM: A FERTILE GROUND FOR INNOVATION

The Chinese automotive market, particularly its vibrant EV sector, has become a fertile testing ground for advanced technologies [42]. Several Chinese startups and established OEMs have unveiled concepts and pre-production models featuring in-wheel motors, often developed in partnership with domestic tech companies.

This high-paced, innovation-driven environment is accelerating the real-world testing and refinement of the technology through several mechanisms [1]:

The rapid iteration cycle typical of Chinese EV development allows for quick learning and improvement. The development timeline is significantly compressed compared to traditional automotive development:

$$T_{development} = f(\text{Testing Intensity}, \text{Iteration Speed}) \tag{1.48}$$

The large market size provides the volume needed to achieve manufacturing scale and cost reduction [42]. The learning curve relationship:

$$C(N) = C_1 N^{-b} \tag{1.49}$$

where cost decreases with cumulative production N, benefits from the massive Chinese market.

The integration with autonomous driving development creates natural synergies [40]. The control architecture for autonomous vehicles which naturally benefits from the individual wheel control capabilities of in-wheel systems.

$$\mathbf{u}_{vehicle} = f(\mathbf{x}_{env}, \mathbf{x}_{vehicle}, \mathbf{x}_{mission}) \tag{1.50}$$

1.7 THE MARKET LAUNCH: PHASED STRATEGIES AND FUTURE OUTLOOK

The market introduction of in-wheel suspension is not a single "big bang" event where it appears on every new car simultaneously. Instead, it is following a classic technology adoption curve, with a phased rollout across different vehicle segments, each with its own business case and go-to-market strategy [42]. This pattern follows the classic technology adoption life cycle but with some unique characteristics specific to automotive technology.

1.7.1 PHASE 1: NICHE AND PREMIUM SEGMENTS (PRESENT - NEAR FUTURE)

Phase 1: Niche and Premium Segments (Present - Near Future)

The technology is first finding commercial viability in applications where its high cost can be justified by a compelling, unique value proposition [42]. These initial applications serve as technology demonstrators and learning platforms while the technology matures and costs decrease.

The high-performance EV segment represents a natural early adopter [3]. The ultra-high-end electric hypercar and supercar segment is a perfect first home, where the benefits of instantaneous torque vectoring, superior handling from reduced unsprung mass (if achieved), and the "wow factor" of cutting-edge technology easily justify a price premium. The business case in this segment follows:

$$\text{Value Proposition} = \frac{\text{Performance Benefit} + \text{Exclusivity Premium}}{\text{Technology Cost}} \tag{1.51}$$

where both numerator terms are substantial in this segment.

Last-mile autonomous people movers represent another promising early application [40]. Companies developing low-speed, self-driving pods for urban ride-sharing and shuttle services prize the packaging freedom above all else. A flat, spacious interior with no driver's controls is the ultimate goal, and in-wheel suspension/motors are the key enabler. The value equation here emphasizes:

$$\text{Revenue per Trip} = f(\text{Passenger Capacity}, \text{Comfort Level}, \text{Reliability}) \tag{1.52}$$

where in-wheel systems can improve all three factors.

Specialty commercial vehicles, particularly for low-speed urban delivery applications, benefit from the operational advantages [25, 34]. The ability to have a completely flat, low load floor maximizes cargo space and simplifies loading/unloading. The operational efficiency gain can be calculated as:

$$\text{Efficiency Gain} = \frac{\text{Cargo Volume}_{new} - \text{Cargo Volume}_{conventional}}{\text{Cargo Volume}_{conventional}} \tag{1.53}$$

which can provide a clear return on investment for the higher initial cost of the technology.

The go-to-market strategy for Phase 1 positions the technology as a key differentiator and premium feature. Marketing focuses on the direct benefits: "unparalleled handling control," "a revolutionary magic carpet ride," or "the most spacious interior in its class." It is sold as a headline innovation that justifies the vehicle's price and position at the apex of the market.

1.7.2 PHASE 2: MAINSTREAM ADOPTION IN THE NEXT DECADE

As costs fall through economies of scale, continued engineering refinement, and competition between suppliers like Bosch, ZF, and dedicated startups, in-wheel suspension will begin to trickle down to premium family sedans and SUVs, and eventually to volume segments [42]. This transition follows the classic technology diffusion pattern but accelerated by the EV transformation.

The cost reduction follows predictable learning and scale curves:

$$C(N) = C_1 N^{-b} \tag{1.54}$$

where the learning coefficient b typically ranges from 0.1 to 0.3 for automotive technologies, leading to 10-30% cost reduction with each doubling of cumulative production.

The performance and reliability improvements follow similar learning patterns:

$$\text{Reliability} = 1 - e^{-\lambda(t)t} \tag{1.55}$$

where the failure rate $\lambda(t)$ decreases with time and experience.

In this phase, the technology transitions from a differentiator to an invisible enabler. The marketing shifts away from "look at our amazing suspension" to the user-experience benefits it enables. The selling points become:

Space and comfort: "The most interior space in its class, with a serene, silent cabin" [34]. The quantitative benefits can be expressed as:

$$\text{Space Advantage} = \frac{V_{interior}}{L_{vehicle} \cdot W_{vehicle}} \tag{1.56}$$

Safety and stability: "An uncannily stable and safe ride, thanks to our advanced vehicle dynamics platform" [17]. The safety improvement can be quantified through:

$$\text{Stability Margin} = \frac{\text{Actual Performance}}{\text{Limit Performance}} \tag{1.57}$$

Design and aerodynamics: "A sleek, aerodynamic design that maximizes efficiency and range" [34]. The aerodynamic benefit follows:

$$\text{Range Improvement} = \frac{C_d A_{conventional} - C_d A_{new}}{C_d A_{conventional}} \tag{1.58}$$

The complexity of the system is hidden from the consumer, who simply enjoys a better, safer, and more spacious vehicle.

1.7.3 THE LONG-TERM VISION: REDEFINING THE VEHICLE

Looking further ahead, the true potential of in-wheel systems lies in enabling truly revolutionary vehicle designs that transcend current automotive paradigms [40]. In a fully autonomous vehicle, with no need for a steering wheel, pedals, or a driver-oriented cockpit, the entire cabin can be a flexible living space.

In-wheel suspension and steering (another technology in development, often called "e-Corner" modules) would allow the vehicle's cabin to be a fixed, stable module while the wheel pods move independently, enabling crab-walking and zero-radius turning for unparalleled maneuverability in urban environments [1, 34, 40]. The kinematic possibilities can be described by:

$$\begin{bmatrix} x \\ y \\ \theta \end{bmatrix} = f(\delta_{fl}, \delta_{fr}, \delta_{rl}, \delta_{rr}, \omega_{fl}, \omega_{fr}, \omega_{rl}, \omega_{rr}) \tag{1.59}$$

where all wheel angles and speeds can be controlled independently.

This capability transforms urban mobility by enabling to approach theoretical maximums:

$$\text{Maneuverability} = \frac{\text{Turning Capability}}{\text{Vehicle Length}} \tag{1.60}$$

The vehicle architecture becomes completely modular, with the cabin, battery, and corner modules being largely independent. This enables to increase dramatically, supporting everything from delivery pods to mobile offices to luxury lounges on the same basic platform:

$$\text{Design Freedom} = \frac{\text{Possible Configurations}}{\text{Conventional Configurations}} \tag{1.61}$$

In this future, the in-wheel system is not just a component; it is the foundation of a new mobility experience [1, 40]. It enables vehicle types and usage models that are impossible with conventional architecture, truly fulfilling the promise of the original vision that began a century ago.

1.8 CONCLUSION: THE LONG ROAD TO A REVOLUTION

The journey of the IWSS is a classic, almost textbook, story of disruptive innovation. It began not with a bang, but with a whisper—a sketch in a patent office, an idea that challenged a century of convention. For decades, it was an answer in search of a question, a solution hampered by the limitations of its time.

The historical pattern followed the typical technology maturation curve:

$$\text{Technology Readiness} = f(\text{Time}, \text{Investment}, \text{Enabling Technologies}) \tag{1.62}$$

where all three factors had to reach critical mass simultaneously.

It was kept alive by the vision and deep pockets of pioneers like Michelin, who had the courage to build a working prototype and show the world what was possible. The Active Wheel was the crucial catalyst that moved the concept from the fringes to the forefront of automotive R&D. The subsequent years were a necessary crucible, a period of intense engineering struggle where the concept's flaws were exposed and painstakingly addressed.

The challenges of unsprung mass, durability, heat, and cost were formidable, but they served to refine and focus the development efforts. Each challenge spawned multiple innovation streams:

$$\text{Innovation} = \sum \text{Materials} + \sum \text{Manufacturing} + \sum \text{Control} + \sum \text{Integration} \tag{1.63}$$

where progress in each domain gradually accumulated to overcome the barriers.

The breakthrough came not from a single company, but from a powerful convergence: the existential shift to electric propulsion created the perfect ecosystem, while parallel revolutions in materials science, software, and active damping provided the necessary tools. The convergence can be modeled as:

$$\text{Adoption} = \text{EV Revolution} \times \text{Materials Science} \times \text{Digital Control} \tag{1.64}$$

where the multiplicative nature meant that progress in all areas was necessary.

The case studies of McLaren and Elaphe show the concept being tested, validated, and refined in the real world, building the confidence needed for mass-market investment. Each success, and even each failure, contributed to the collective learning:

$$\text{Knowledge} = \int (\text{Successes} + \text{Failures})\,dt \tag{1.65}$$

where both positive and negative results advanced the technology.

Today, the in-wheel suspension is no longer a question of "if" but "when" and "where first." Its path to market is clear, logical, and already underway. It stands as a powerful testament to the enduring power of a simple, elegant idea and its long, arduous, but ultimately triumphant journey from the unsprung fringes of automotive design to a central role in shaping the future of how we move.

The story reminds us that the wheels of progress, though they may sometimes turn slowly, are always moving forward. More importantly, it demonstrates that true innovation often requires not just new ideas, but the patience and persistence to wait for the world to become ready for them. As we stand on the brink of widespread adoption, we can appreciate that the century-long journey was not a delay but a necessary incubation period for a technology whose time has finally come.

The historical trajectory outlined in this chapter establishes the technological basis for the modeling and control approaches developed in the subsequent chapters. Chapter 2 builds upon this foundation by constructing the structural and mathematical framework of the IWSS, translating conceptual vision into tangible engineering design.

PROBLEMS

Problem 1: Explain the "Mass Inversion Hypothesis" that is central to the theoretical benefit of IWSS. In your answer, describe the problem with high unsprung mass in a conventional suspension and how the in-wheel concept aims to solve it.

Problem 2: (a) Why were the early in-wheel suspension concepts from inventors like William B. Biddle in the 1920s not commercially successful? (b) Identify three specific technological enablers that emerged decades later that were necessary to make the in-wheel suspension concept viable.

Problem 3: The Michelin Active Wheel was a landmark "corner module." List the three primary functions it integrated into a single wheel unit. For one of these functions, describe a significant engineering challenge Michelin had to overcome to make the integration work.

Problem 4: The chapter discusses the "Unsprung Mass Paradox." What is this paradox? Explain why the initial implementation of in-wheel systems often resulted in a performance degradation that was the exact opposite of its intended goal.

Problem 5: Compare the operating environment of a traditional suspension component (e.g., a shock absorber) with that of a component within an in-wheel system. Why does this environmental difference present a monumental challenge for durability and sealing, and what are the potential consequences of seal failure?

Problem 6: An in-wheel system packages a motor, brakes, and potentially active suspension actuators in a sealed, compact space with minimal airflow. (a) List two key components that generate significant heat. (b) For each component, describe one potential failure mode that could occur if the heat is not managed effectively.

Problem 7: Explain why the rise of EV is described in the chapter as the "ultimate catalyst" for in-wheel suspension technology. Provide two distinct reasons based on the architectural differences between EVs and internal combustion engine vehicles.

Problem 8: In-wheel motors enable precise "torque vectoring." Define what torque vectoring is and explain how it can enhance vehicle stability and agility beyond the capabilities of a conventional differential and stability control system.

Problem 9: The development of new materials was crucial for tackling the unsprung mass paradox. Name two types of advanced materials mentioned in the chapter and state the primary property of each that is beneficial for in-wheel system design (e.g., high strength-to-weight ratio, high thermal conductivity).

Problem 10: The McLaren 720S does not use a true in-wheel suspension, but its front suspension design is cited as a philosophical step towards the concept. (a) Describe the key design feature of the 720S's front suspension. (b) What two benefits does this design share with a true in-wheel system?

Problem 11: The case of Lordstown Motors and Elaphe Propulsion Technologies is described as "instructive." What was the significant milestone demonstrated by Lordstown's plan to use Elaphe's in-wheel motors, regardless of the company's eventual fate?

Problem 12: The chapter proposes a two-phase market launch strategy for IWSS. (a) Name two vehicle segments identified for **Phase 1** (niche and premium segments). (b) For one of these segments, explain why the high cost of the technology is justifiable.

Problem 13: Contrast the "go-to-market strategy" for in-wheel systems in Phase 1 (Niche/Premium) versus Phase 2 (Mainstream Adoption). How does the marketing message and the perceived value proposition change between these phases?

Problem 14: Looking at the long-term vision, how could in-wheel suspension and steering systems (e.g., "e-Corner" modules) enable a fundamental rethinking of vehicle design for a fully autonomous future? Describe one specific capability that would be enabled.

Problem 15: The chapter concludes that the breakthrough for in-wheel suspension came from a "powerful convergence," not a single breakthrough. Describe the convergence of at least three distinct technological or market trends that were necessary to bring the concept from a radical idea to the brink of mainstream adoption.

REFERENCES

1. Aloeyi, Essossinam Frederic, Nasir Ali, and Qingsong Wang. "A review of in-wheel motors for electric vehicle propulsion." 2022 IEEE Transportation Electrification Conference and Expo, Asia-Pacific (ITEC Asia-Pacific). IEEE, 2022.
2. Li, Xiao Hua, and Hong Qian. "The present status and future trends of in-wheel motors for electric vehicles." Advanced Materials Research 433 (2012): 6943-6950.
3. Deepak, Kritika, et al. "In-wheel motor drive systems for electric vehicles: State of the art, challenges, and future trends." Energies 16.7 (2023): 3121.
4. Katsuyama, Etsuo, Makoto Yamakado, and Masato Abe. "A state-of-the-art review: toward a novel vehicle dynamics control concept taking the driveline of electric vehicles into account as promising control actuators." Vehicle System Dynamics 59.7 (2021): 976-1025.
5. Protean Electric. "360-Degree Corner Module Accelerates Revolution in Urban Mobility." Protean Electric, 18 July 2019, www.proteanelectric.com/360-degree-corner-module-accelerates-revolution-in-urban-mobility/ . Accessed 22 Aug 2025.
6. "Michelin Active Wheel to Hit Road at 2010." OnElectricCars.com, www.onelectriccars.com/tag/michelin-active-wheel/ . Accessed 09 May 2025.
7. Protean Electric. "Protean360+—Technology." Protean Electric, www.proteanelectric.com/technology/protean360plus . Accessed 19 Nov 2025.
8. Zhao, W., & Wang, C. "Thermal management of in-wheel permanent magnet synchronous motors for electric vehicles: A review." Applied Thermal Engineering 201 (2022): 117793.
9. Fakhfakh, Mohamed Amine, et al. "Thermal analysis of a permanent magnet synchronous motor for electric vehicles." Journal of Asian Electric Vehicles 6.2 (2008): 1145-1151.

10. Wu, Hang, et al. "Robust control for active suspension of hub-driven electric vehicles subject to in-wheel motor magnetic force oscillation." Applied Sciences 10.11 (2020): 3929.
11. Li, Zhe, et al. "Multi-objective optimization of active suspension system in electric vehicle with In-Wheel-Motor against the negative electromechanical coupling effects." Mechanical Systems and Signal Processing 116 (2019): 545-565.
12. Yan, Kehan, et al. "A critical review of radial field in-wheel motors: technical progress and future trends." eTransportation 22 (2024): 100353.
13. Yoon, Dal-Seong, and Seung-Bok Choi. "Adaptive control for suspension system of in-wheel motor vehicle with magnetorheological damper." Machines 12.7 (2024): 433.
14. Li, Liang, et al. "Critical Review of Direct-Drive In-Wheel Motors in Electric Vehicles." Energies 18.6 (2025): 1521.
15. Wang, Y., & Li, Y. Long, Guimin, et al. "Regenerative active suspension system with residual energy for in-wheel motor driven electric vehicle." Applied Energy 260 (2020): 114180.
16. Fan, Xiaobin, et al. "A review of integrated control technologies for four in-wheel motor drive electric vehicle chassis." Journal of the Brazilian Society of Mechanical Sciences and Engineering 47.2 (2025): 89.
17. Wu, Hang, Ling Zheng, and Yinong Li. "Coupling effects in hub motor and optimization for active suspension system to improve the vehicle and the motor performance." Journal of Sound and Vibration 482 (2020): 115426.
18. Lee, Daekyun, Sunwoo Jin, and Chibum Lee. "Deep reinforcement learning of semi-active suspension controller for vehicle ride comfort." IEEE Transactions on Vehicular Technology 72.1 (2022): 327-339.
19. Liang, Qian-Kun, et al. "State-observer-based asymptotic tracking control for electro-hydraulic actuator systems with uncertainties and unmeasurable velocity." Journal of the Franklin Institute 360.18 (2023): 14660-14675.
20. Gao, Huijun, James Lam, and Changhong Wang. "Multi-objective control of vehicle active suspension systems via load-dependent controllers." Journal of sound and vibration 290.3-5 (2006): 654-675.
21. Wang, Rongrong, et al. "Robust fault-tolerant H_∞ control of active suspension systems with finite-frequency constraint." Mechanical Systems and Signal Processing 62 (2015): 341-355.
22. Phenikaa-X. Autonomous Vehicle: Technology of the Future. Phenikaa-X, https://phenikaa-x.com/autocar . Accessed 21 Jun. 2025.
23. Sun, Weichao, Huijun Gao, and Peng Shi. "Active Suspension Control with the Unideal Actuators." Advanced Control for Vehicle Active Suspension Systems. Cham: Springer International Publishing, 2019. 143-187.
24. Huang, Yingbo, et al. "Robust adaptive parameter estimation and control for vehicle active suspension systems." Vibration Control and Actuation of Large-Scale Systems. Academic Press, 2020. 225-252.
25. Sun, Jinwei, and Kai Zhao. "Adaptive neural network sliding mode control for active suspension systems with electrohydraulic actuator dynamics." International Journal of Advanced Robotic Systems 17.4 (2020): 1729881420941986.
26. World Economic Forum. Automotive in the Software-Driven Era. World Economic Forum – DRIVE-A Initiative, https://initiatives.weforum.org/drive-a/sdv. Accessed 12 Sep. 2025.

27. Yin, Yunbao, et al. "Robust control design for active suspension system with uncertain dynamics and actuator time delay." Journal of Mechanical Science and Technology 36.12 (2022): 6319-6327.
28. Yeneneh, Kumlachew, et al. "Optimizing Active Suspension Systems with Robust H_∞ Control and Adaptive Techniques under Uncertainties." Applications in Engineering Science (2025): 100225.
29. "Self-Driving Bus in Whitby to Accept Passengers Starting Tuesday." Durham Radio News, 29 Oct. 2021, www.durhamradionews.com/archives/146661 . Accessed 10 Aug. 2025.
30. Merchants Fleet. Merchants Fleet Launches New National Fleet Electrification Initiative. Merchants Fleet, 1 Oct. 2024, https://www.merchantsfleet.com/merchants-fleet-launches-new-national-fleet-electrification-initiative/ . Accessed 22 Jul. 2025.
31. Liu, Yan-Jun, et al. "An adaptive neural network controller for active suspension systems with hydraulic actuator." IEEE Transactions on Systems, Man, and Cybernetics: Systems 50.12 (2018): 5351-5360.
32. Ichikawa, Haru, et al. "Model building and experimental validation for regeneration of perpendicular vibration by in-wheel motor." 2024 IEEE International Conference on Advanced Intelligent Mechatronics (AIM). IEEE, 2024.
33. Haseen, S. Fahira, P. Lakshmi, and S. Madhupreetha. "Efficient Vibration Control: Model Predictive Controller Design for In-Wheel Motor-Driven Electric Vehicle Suspension Systems." Journal of Vibration Engineering & Technologies 13.7 (2025).
34. Liu, Pengtao, et al. "A novel hybrid electromagnetic-hydraulically interconnected actuator suspension: Dynamic modeling, bench test and ride analysis." Mechanical Systems and Signal Processing 236 (2025): 113028.
35. Lei, Fei, et al. "Optimization of an implicit constrained multi-physics system for motor wheels of electric vehicle." Energy 113 (2016): 980-990.
36. Li, Boyuan, Haiping Du, and Weihua Li. "Fault-tolerant control of electric vehicles with in-wheel motors using actuator-grouping sliding mode controllers." Mechanical Systems and Signal Processing 72 (2016): 462-485.
37. Li, Bo, et al. "Thermal management of electrified propulsion system for low-carbon vehicles." Automotive Innovation 3.4 (2020): 299-316.
38. Xiong, Lu, et al. "Vehicle dynamics control of four in-wheel motor drive electric vehicle using gain scheduling based on tyre cornering stiffness estimation." Vehicle system dynamics 50.6 (2012): 831-846.
39. Kashyzadeh, Kazem Reza, Mohammad Jafar Ostad-Ahmad-Ghorabi, and Alireza Arghavan. "Fatigue life prediction of package of suspension automotive under random vibration based on road roughness." Mediterranean Journal of Modeling and Simulation 4.1 (2015): 037-050.
40. Feng, Jiwei, et al. "An Integrated Control Framework for Torque Vectoring and Active Suspension System." Chinese Journal of Mechanical Engineering 37.1 (2024): 10.
41. Valdivieso-Soto, Andrew, et al. "Lyapunov-Based Pitch Control for Electric Vehicles Using In-Wheel Motors." Vehicles 7.2 (2025): 37.
42. Pasha, Junayed, et al. "Electric vehicle scheduling: State of the art, critical challenges, and future research opportunities." Journal of industrial information integration 38 (2024): 100561.

2 The Architectural Blueprint: Structural Design and Mathematical Modeling of In-Wheel Suspension System

2.1 INTRODUCTION: THE ARCHITECTURAL PARADIGM SHIFT

The transition from conventional suspension systems to in-wheel architectures represents one of the most profound transformations in automotive engineering since the advent of independent suspension. This chapter provides an exhaustive examination of the structural design principles and mathematical foundations that underpin modern IWSS. We will systematically deconstruct these complex mechatronic systems, progressing from fundamental mechanical configurations through advanced multi-physics integration to sophisticated control-oriented mathematical representations [7, 25].

The design challenge for IWSS is fundamentally multi-objective and constrained: engineers must achieve superior dynamic performance while simultaneously overcoming severe spatial limitations, managing unprecedented thermal loads, and maintaining structural integrity under extreme operating conditions. This requires a holistic approach that integrates traditionally separate engineering domains—structural mechanics, electromagnetics, thermal dynamics, and control theory—into a unified design framework [3, 23].

The architectural significance of in-wheel systems extends beyond mere component relocation. These systems enable a complete rethinking of vehicle architecture, transforming the traditional relationship between chassis and running gear [14, 17]. By consolidating suspension, propulsion, and braking functions into self-contained corner modules, in-wheel technology facilitates the development of truly modular vehicle platforms with unprecedented design flexibility and performance potential [16].

This chapter is structured to provide both theoretical depth and practical engineering insight. We begin with a detailed analysis of mechanical suspension topologies, progress through electromagnetic motor design, explore thermal management strategies, and culminate in comprehensive mathematical models that capture the full complexity of system dynamics [9, 19]. Throughout this exploration, we maintain a focus on the practical engineering trade-offs that define successful in-wheel suspension implementation.

DOI: 10.1201/9781003780533-2

2.2 STRUCTURAL CONFIGURATION OF IN-WHEEL SUSPENSION SYSTEM

The mechanical suspension component of an in-wheel system represents a remarkable feat of engineering miniaturization and integration. Unlike conventional suspensions that operate within the relatively spacious confines of a vehicle's chassis, in-wheel suspensions must deliver equivalent or superior performance while operating within the severely constrained volume of a wheel assembly.

2.2.1 FUNCTIONAL REQUIREMENTS AND DESIGN CONSTRAINTS

The design of any IWSS must satisfy six fundamental functional requirements while operating within four primary constraint domains. Understanding these requirements and constraints is essential for evaluating different architectural approaches [2, 21].

2.2.1.1 Functional Requirements

1. **Kinematic Guidance:** The system must constrain the wheel to follow a predetermined path relative to the vehicle body, controlling critical parameters such as camber angle, toe angle, and scrub radius throughout the suspension travel [6]. The kinematic function can be mathematically described by the transformation:

$$\begin{bmatrix} x_w \\ y_w \\ z_w \\ \theta_w \\ \phi_w \\ \psi_w \end{bmatrix} = \mathbf{T}(\mathbf{q}, t) \tag{2.1}$$

   where $\mathbf{q}$ represents the suspension joint coordinates and $\mathbf{T}$ defines the kinematic transformation [19].
2. **Force Management:** The suspension must manage forces in all six degrees of freedom—three translational and three rotational—while providing appropriate compliance for ride comfort and sufficient stiffness for handling precision [29]. The force-displacement relationship follows:

$$\mathbf{F} = \mathbf{Kx} + \mathbf{C}\dot{\mathbf{x}} + \mathbf{F}_{active} \tag{2.2}$$

   where $\mathbf{K}$ and $\mathbf{C}$ represent stiffness and damping matrices [3].
3. **Mass Optimization:** The system must minimize unsprung mass while maintaining structural integrity, creating a fundamental tension between lightweight design and durability requirements [15]. The mass optimization problem can be formulated as:

$$\min_{\mathbf{x}} m(\mathbf{x}) \quad \text{subject to} \quad \sigma(\mathbf{x}) \le \sigma_{yield}, \quad \delta(\mathbf{x}) \le \delta_{max} \tag{2.3}$$

4. **Packaging Efficiency:** All components must fit within the limited volume defined by the wheel rim, brake assembly, and tire envelope, requiring innovative spatial organization strategies [1]. The packaging constraint can be expressed as:

$$\bigcup_{i=1}^{n} V_i \subseteq V_{available} \quad \text{where} \quad V_{available} = V_{wheel} - V_{tire} - V_{brake} \tag{2.4}$$

5. **Service Accessibility:** Despite high integration density, the system must allow for maintenance and replacement of wear components such as bearings, seals, and friction elements.
6. **Environmental Protection:** The system must provide robust sealing against water, dust, salt, and mechanical impacts while managing internal heat generation and dissipation [4].

2.2.1.2 Design Constraints

The successful realization of the system's functional requirements is a complex engineering challenge that must be navigated within a tightly bounded and often conflicting set of design constraints. These limitations are not merely secondary considerations but are fundamental drivers that shape the architecture, material selection, and manufacturing processes of every component [7]. The design effort is therefore a continuous process of optimization and compromise across four primary domains:

Spatial and Geometric Constraints: The physical envelope available for the in-wheel assembly is exceptionally restrictive [1]. Components must be packaged within the confines of the wheel itself, a volume dictated by the standard rim diameter and width. This typically translates to a cylindrical envelope with a diameter of 400-500 mm and a width (axial length) of only 150-200 mm. This spatial limitation has profound implications [25]:

- **Component Integration:** The motor, braking system, suspension linkages, and power electronics must all be co-located within this fixed volume, leading to complex and highly integrated designs where every cubic millimeter is contested.
- **Clearance Issues:** Adequate clearance must be maintained for full steering articulation, suspension travel through jounce and rebound, and tire deflection under load without interference with the vehicle's chassis or bodywork.
- **Modularity and Serviceability:** The dense packaging can complicate assembly, disassembly, and maintenance procedures. Replacing a single component might require the removal of several others, potentially increasing lifecycle costs and repair times.

Mass and Inertial Constraints: The system's mass is a critical performance parameter, as it contributes directly to the vehicle's unsprung mass—the mass not supported by the suspension springs (i.e., wheels, tires, brakes, and outboard suspension components) [15]. Adhering to a target of 40–50 kg per corner for passenger vehicles is paramount for several reasons [14]:

- **Vehicle Dynamics:** Excessive unsprung mass severely degrades ride quality, handling, and road-holding. A heavier wheel assembly has more inertia, making it slower to react to road surface imperfections. This results in a loss of tire contact with the road, compromising traction and stability, especially on uneven surfaces.
- **Energy Efficiency:** A lower mass reduces the rotational inertia of the drivetrain, which directly improves acceleration, regenerative braking efficiency, and overall vehicle range.
- **Suspension Loads:** Higher unsprung mass imposes greater dynamic loads and shock forces on suspension components like springs, dampers, and control arms. This can necessitate the use of heavier and more robust parts elsewhere in the suspension system, creating a detrimental mass-compounding effect.

Thermal Management Constraints: The in-wheel system is a significant source of heat generation, with peak power dissipation from friction brakes and electric motor losses capable of exceeding 10 kW per corner during aggressive driving, braking, or sustained hill descent [4]. Managing this substantial thermal load is a major challenge, as [18], due to:

- **Concentrated Heat Sources:** Brake discs and motor windings generate intense heat in very confined areas with limited thermal mass, leading to rapid temperature spikes.
- **Poor Convective Cooling:** The enclosed nature of the wheel well offers restricted airflow, hindering the primary method of passive cooling. Active cooling solutions, such as liquid cooling loops, add mass, complexity, cost, and potential failure points.
- **Material Degradation:** Prolonged or repeated thermal cycling can degrade lubricants, damage bearing seals, demagnetize permanent magnets in the motor, and reduce the performance and lifespan of sensitive power electronics, ultimately threatening system reliability.

Economic and Manufacturing Constraints: Beyond pure performance, the design must be grounded in commercial reality. The pursuit of technical excellence must be balanced against stringent cost targets to ensure the system is viable for mass production and competitive in the marketplace [24].

- **Material Selection:** While advanced materials like carbon composites or high-strength aerospace alloys may offer superior performance benefits in terms of strength-to-weight ratio or thermal properties, their high cost often precludes their use. Engineers must often select cost-effective materials like standardized steel and aluminum alloys and optimize their application through sophisticated design and analysis.
- **Manufacturing Complexity:** Designs that are difficult to machine, cast, or assemble will have a high per-unit cost. Designs must be simplified for

manufacturability (DFM) and leverage scalable, high-volume production techniques like casting, forging, and automated assembly to keep costs down.

- **Reliability and Durability:** The system must be designed for a long service life under harsh operating conditions, including exposure to moisture, road salt, and mechanical impact. Achieving this required reliability without resorting to over-engineering or excessively expensive components and validation processes, which is a key economic challenge.

2.2.2 SUSPENSION TOPOLOGIES AND KINEMATIC ARCHITECTURES

Several distinct suspension topologies have emerged as viable solutions for in-wheel applications, each offering different trade-offs between performance, packaging, and complexity [2, 21]. We examine these topologies in detail, analyzing their kinematic characteristics and practical implementation challenges.

2.2.2.1 Double Wishbone In-Wheel Configuration

The double wishbone architecture represents a direct adaptation of one of the most respected conventional suspension designs, modified for in-wheel packaging constraints [1]. This topology provides excellent kinematic control and has been implemented in several high-performance applications [19].

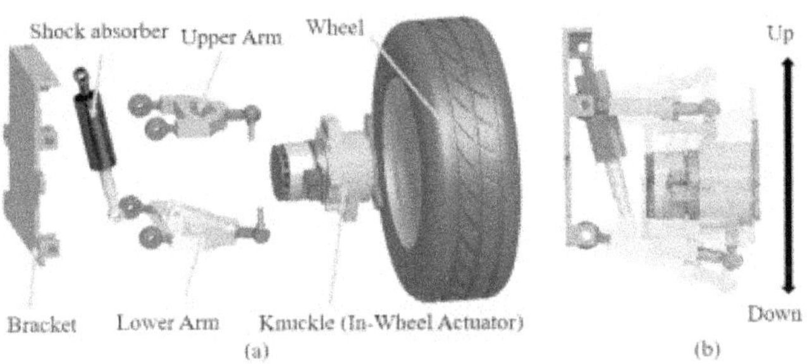

Figure 2.1: Double Wishbone In-Wheel suspension system in six-wheel mobile robot [1]

The kinematic behavior of a double wishbone suspension can be analyzed using screw theory and homogeneous transformations [6]. The position and orientation of the wheel carrier relative to the chassis is determined by the constraint equations:

$$\mathbf{r}_{wc} = \mathbf{r}_{chassis} + \mathbf{R}(\phi)\mathbf{r}_{upper} + \mathbf{R}(\theta)\mathbf{r}_{lower} \tag{2.5}$$

where $\mathbf{r}_{upper}$ and $\mathbf{r}_{lower}$ represent the positions of the upper and lower wishbone connections, and $\mathbf{R}(\phi)$, $\mathbf{R}(\theta)$ are rotation matrices describing the suspension articulation.

The instant center of rotation for the wheel assembly can be determined through the intersection of the upper and lower wishbone planes [6]:

$$\mathbf{r}_{IC} = \frac{(\mathbf{r}_{u1} \times \mathbf{r}_{u2}) \times (\mathbf{r}_{l1} \times \mathbf{r}_{l2})}{||(\mathbf{r}_{u1} \times \mathbf{r}_{u2}) \times (\mathbf{r}_{l1} \times \mathbf{r}_{l2})||} \tag{2.6}$$

Figure 2.2: Double Wishbone In-Wheel suspension in real six-wheel mobile robot [1]

- **Roll Center Height:** The instant center's vertical position relative to the ground plane projects laterally to define the vehicle's roll center at that corner [6]. A higher instant center elevates the roll center, which reduces the body roll moment for a given lateral acceleration. However, an excessively high roll center can lead to undesirable "jacking" forces, effectively lifting the vehicle's body during cornering. Conversely, a lower instant center increases the roll moment but provides a more consistent and predictable contact patch load transfer [14].
- **Camber Gain Characteristics:** As the suspension articulates, the wheel's camber angle changes [19]. The rate and direction of this change are dictated by the instant center's longitudinal location relative to the wheel center. A properly positioned instant center induces negative camber gain as the body rolls in a corner. This is critical because it helps to maintain a more perpendicular orientation of the tire relative to the road surface, thereby maximizing the contact patch area and available lateral grip during cornering maneuvers [21].

The double wishbone design offers a high degree of freedom in positioning this instant center, making it a preferred choice for performance applications where fine-tuned handling is paramount [2].

Advantages of the Double Wishbone Configuration:

- **Superior Kinematic Control:** The independent mounting points for the upper and lower arms allow engineers to precisely tailor the arc through which the wheel travels [2]. This enables optimal control over camber variation throughout the entire suspension travel. In a well-tuned double wishbone system, typical camber gain ranges from $0.5°$ to $0.8°$ per degree of body roll. This dynamic negative camber gain actively compensates for body roll, ensuring the tire remains flush with the road surface to maximize grip and cornering stability [19].
- **High Structural Stiffness and Precision:** The triangulated A-arm configuration forms a highly rigid structure that provides exceptional resistance

to forces acting in all directions [1]. It effectively manages longitudinal forces from acceleration and braking, lateral forces during cornering, and vertical impacts from road irregularities. This inherent stiffness results in minimal deflection at the wheel hub, with typical measured hub stiffness values exceeding $20\,\mathrm{N\,\mu m^{-1}}$. This precision ensures that the kinematic design is maintained under load, providing predictable and consistent handling feedback to the driver [21].

- **Proven Reliability and Maturity:** With an extensive operational history spanning decades across automotive, racing, and aerospace applications, the double wishbone suspension has a substantial and well-documented knowledge base [7]. This maturity means that failure modes, load paths, and design best practices are thoroughly understood. This historical data provides a robust foundation for design refinement, fatigue analysis, and the validation of new iterations, thereby reducing development risk and time [25].

Disadvantages and Design Challenges of the Double Wishbone Configuration:

- **Lateral Space Consumption and Packaging Conflict:** The upper and lower wishbone arms, along with their mounting structures, protrude into the space immediately adjacent to the wheel rim and tire sidewal [1]. This creates a significant packaging challenge, particularly in applications with wide tires or limited space within the wheel well. It necessitates careful three-dimensional optimization of the arm shapes (often resulting in complex, kinked, or forged profiles) and the placement of other components like brake calipers, steering tie-rods, and drive shafts to avoid interference throughout the full range of suspension and steering travel.
- **High Component Count and Associated Complexity:** A single corner utilizing a double wishbone suspension requires a multitude of components: two control arms (upper and lower), two ball joints, one or two suspension knuckles/spindles, and a tie-rod [2]. This high part count increases the overall system complexity, mass, and number of potential failure points (e.g., ball joint wear, bushing degradation). Furthermore, it impacts manufacturing cost, assembly time, and the complexity of service procedures, as replacing a single component often requires realignment of the entire corner [7].
- **Elevated and Complex Bearing Loads:** The geometry of the double wishbone suspension often results in the wheel bearing being subjected to significant overturning moments and offset loads, rather than pure radial loads [21]. This is due to the forces being transmitted through the control arms at points offset from the bearing's centerline. If not meticulously addressed during the design phase through proper bearing selection (e.g., using angular contact bearings), precise hub design, and rigorous load analysis, these conditions can lead to premature bearing wear, reduced service life, and potential catastrophic failure under high-load conditions.

2.2.2.2 MacPherson Strut In-Wheel Configuration

The MacPherson strut adaptation for in-wheel applications represents a pragmatic approach that prioritizes packaging efficiency over kinematic perfection [25]. This topology is particularly well-suited for applications where lateral space is at a premium.

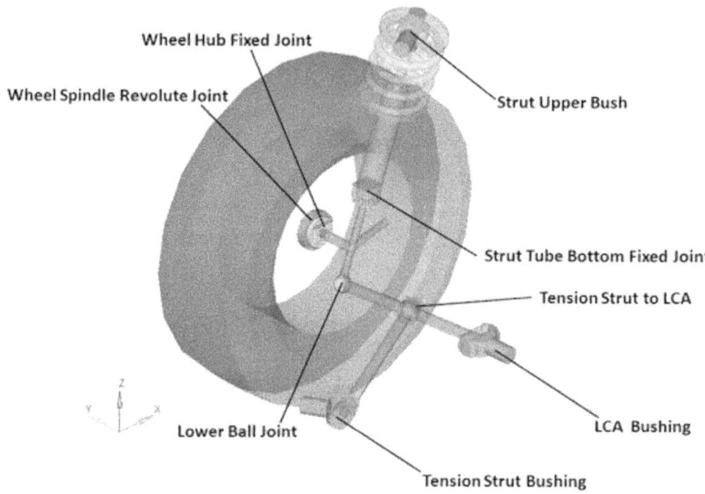

Figure 2.3: MacPherson strut In-Wheel configuration [11]

The kinematic analysis of MacPherson strut systems requires consideration of the strut's sliding joint constraint combined with the lower control arm's rotational freedom [21]. The system can be modeled as a serial chain with one translational and two rotational degrees of freedom:

$$\mathbf{q} = [z, \theta, \phi]^T \tag{2.7}$$

where z represents the strut compression, θ the steering angle, and ϕ the camber angle.

The camber variation with suspension travel follows an approximately linear relationship [6]:

$$\frac{d\phi}{dz} \approx \frac{r_{lower}}{l_{strut}} \sin(\alpha) \tag{2.8}$$

where r_{lower} is the lower control arm length, l_{strut} is the strut length, and α is the control arm angle.

The double wishbone suspension, while more complex, offers distinct advantages and disadvantages when evaluated against a strut-type design [2]. Its multi-link architecture fundamentally differs in how it manages space, forces, and kinematics [19].

Advantages of the Double Wishbone Configuration:

- **Superior Kinematic Control and Camber Management:** The independent upper and lower arms provide unparalleled control over the wheel's path of motion [2]. This allows engineers to precisely tune the camber gain curve, typically achieving optimal negative camber gain of 0.5° to 0.8° per degree of body roll. This active camber compensation maintains a larger, more effective tire contact patch during cornering, directly translating to higher ultimate lateral acceleration and superior grip compared to strut designs [19].
- **Reduced Friction and Enhanced Small-Bump Sensitivity:** Unlike a MacPherson strut, the double wishbone design separates the damper's damping function from its role as a structural link and sliding element [27, 29]. The loads are channeled through ball joints and bushings, which are designed to handle them with minimal stiction. This results in significantly lower internal friction, allowing the suspension to react more sensitively to high-frequency, low-amplitude road imperfections (e.g., pavement ripples), thereby improving ride quality and tire contact on rough surfaces [3].

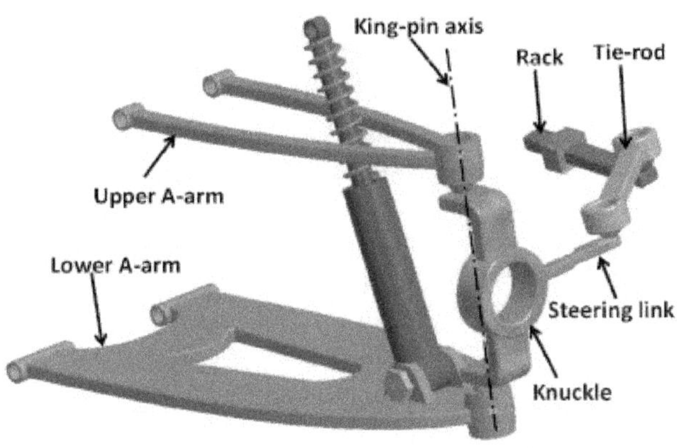

Figure 2.4: Double wishbone configuration [12]

- **Optimized Load Paths and Reduced Bending Moments:** The triangulated A-arm structure transmits lateral and longitudinal forces directly into the chassis through robust, widely spaced pick-up points [1]. This prevents the damper shaft and tube from being subjected to large bending moments, a critical weakness of the strut design. This leads to reduced seal and damper wear, more consistent damping performance over the system's lifetime, and a higher tolerance for aggressive driving conditions [21].

Disadvantages of the Double Wishbone Configuration:

- **Significant Lateral Space Consumption:** The presence of an upper control arm and its associated mounting structure intrudes heavily into the lateral space adjacent to the wheel [1]. This creates a major packaging challenge, competing for critical volume required for the wheel rim, tire sidewall, and—especially in in-wheel motor applications—the electric motor and braking components. This often necessitates complex, forged, or kinked arm designs to create clearance, adding cost and complexity [25].

- **Increased Component Count and Structural Complexity:** A double wishbone system inherently requires more parts: two control arms (each with two bushings and a ball joint), a knuckle, and a separate damper and spring assembly [2]. This higher part count increases vehicle mass, manufacturing cost, assembly time, and the number of potential wear points (e.g., bushing degradation, ball joint slack). Service procedures are often more involved, as replacing one component can necessitate a full wheel alignment [7].
- **Non-Coaxial Spring Placement and Packaging Challenges:** The spring and damper are typically not mounted coaxially and must be positioned separately, often on one of the control arms [21]. This "remote" spring location requires a more complex design to transmit forces efficiently, can lead to less direct packaging options, and may require additional components like a separate spring perch. This contrasts with the strut's compact, integrated, and highly efficient coaxial spring-damper unit [25].

2.2.2.3 Multi-Link In-Wheel Configuration

Multi-link suspensions represent the pinnacle of kinematic refinement, offering unparalleled design freedom to optimize suspension behavior across multiple performance criteria. These systems use three or more carefully positioned links to control wheel motion with exceptional precision [21].

The general form of the constraint equations for an n-link suspension system can be expressed as:

$$\mathbf{J}(\mathbf{q})\dot{\mathbf{q}} = \mathbf{0} \tag{2.9}$$

where $\mathbf{J}(\mathbf{q})$ is the Jacobian matrix representing the kinematic constraints imposed by the suspension links.

For a typical five-link configuration, the system provides precise control over the instant center migration path [6]:

$$\mathbf{r}_{IC}(z) = \mathbf{f}(\mathbf{l}_1, \mathbf{l}_2, \mathbf{l}_3, \mathbf{l}_4, \mathbf{l}_5, z) \tag{2.10}$$

where $\mathbf{l}_i$ represent the link vectors and z is the suspension travel.

The multi-link suspension represents the pinnacle of independent suspension design, offering unparalleled tuning flexibility at the cost of significant complexity [19]. It can be conceptually understood as an evolution of the double wishbone, where each traditional control arm is replaced by two or more separate links, each with their own dedicated mounting points [21]. This decomposition of the suspension's degrees of freedom allows for an exceptionally precise and independent control over its kinematic and compliance characteristics.

Advantages of the Multi-Link Configuration:

- **Optimal Kinematic Synthesis and Parameter Isolation:** The primary advantage of a multi-link design is the ability to independently optimize specific kinematic parameters that are coupled in simpler designs [19]. Engineers can use virtual or physical kinematic synthesis tools to define the exact length and mounting points of each link to precisely control camber gain, toe variation, and roll center movement throughout the suspension travel. For instance, it is possible to design a system that provides aggressive negative camber gain for cornering grip while maintaining minimal toe change to ensure straight-line stability and reduce tire wear [21].

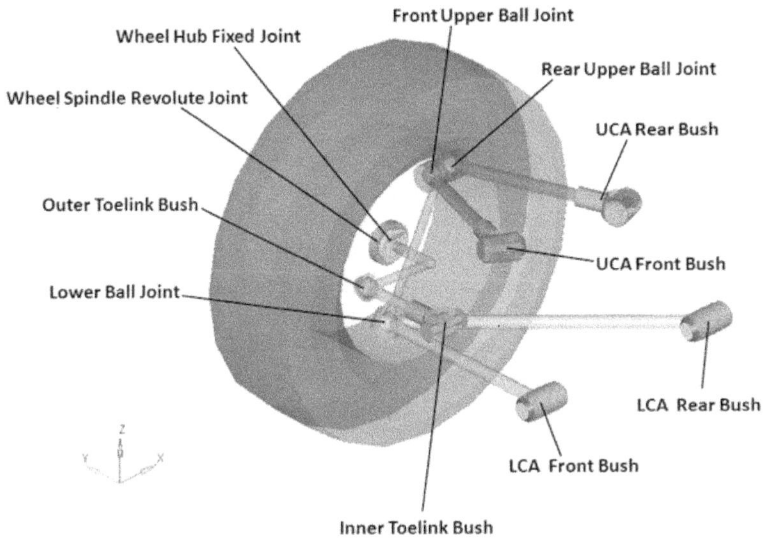

Figure 2.5: Multi-Link In-Wheel configuration [11]

- **Superior Force Isolation and Refined Vibration Damping:** By utilizing multiple, strategically oriented links, the multi-link suspension can effectively decouple different force paths entering the chassis [29]. Longitudinal forces from braking and acceleration can be channeled through specific links, while lateral cornering forces are managed by others. This isolation prevents these forces from interacting in a way that excites unwanted vibrations, leading to a significant reduction in noise, harshness, and vibration (NVH) transmitted to the vehicle's cabin. The result is a ride quality that feels both controlled and exceptionally refined [3].
- **Precise Compliance Control and Elastokinematic Tuning:** Beyond pure rigid-body kinematics, the multi-link design offers exquisite control over the suspension's elastokinematic behavior—the way it moves under load due to bushing deflection [21]. By carefully selecting the durometer (stiffness) and geometry of the numerous bushings, engineers can design in beneficial

compliance. For example, the suspension can be tuned to induce a slight toe-in under lateral cornering loads to enhance stability, or a specific compliance under longitudinal forces to improve traction over bumps. This allows the vehicle to achieve a superior blend of sharp handling and ride comfort [19].

Disadvantages of the Multi-Link Configuration:

- **Extreme Design, Manufacturing, and Maintenance Complexity:** The multi-link suspension has the highest component count of any common independent design [2]. A single corner may consist of three, four, or even five separate links, each with two bushings or ball joints, a complex knuckle, and a subframe. This proliferation of parts drastically increases design analysis time, tooling costs, assembly time, and the number of potential wear points. Furthermore, wheel alignment is a complex procedure involving the adjustment of multiple links, making it more time-consuming and expensive for service technicians [7].
- **Significant Packaging Challenges and Spatial Intrusion:** Accommodating the multitude of links and their respective mounting points within the confined volume of a wheel well, especially one already constrained by an in-wheel motor system, is a formidable challenge [1]. The links compete fiercely for space with the motor, brake components, steering system, and the wheel itself. This often necessitates complex, contoured subframes and links that are expensive to fabricate, and can limit the maximum possible steering angle or tire size [25].
- **Premium Cost and Weight Penalties:** The high part count, complex cast or forged components (such as the knuckle), and extensive use of high-grade bearings and bushings make the multi-link suspension typically the most expensive solution in both initial manufacturing and long-term maintenance [7]. The cumulative mass of all the individual links, bolts, and subframes also makes it the heaviest option, directly conflicting with the critical unsprung mass targets for a high-performance in-wheel application [15]. This creates a direct trade-off between achieving the ultimate handling refinement and meeting stringent mass and cost constraints.

2.2.2.4 Trailing Arm and Link-Lever Mechanisms

Trailing arm configurations, often used in conjunction with sophisticated link-lever systems, represent a specialized approach particularly well-suited for active suspension applications [29]. These systems separate the spring and damper elements from the primary structural members, allowing for flexible packaging and actuation integration.

The kinematic analysis of trailing arm systems focuses on the rotational motion about the arm pivot [21]:

$$\theta_{arm} = \tan^{-1}\left(\frac{z}{r_{arm}}\right) \tag{2.11}$$

where r_{arm} is the effective arm length and z is the wheel displacement.

The force transmission through the lever mechanism follows the principle of virtual work [19]:

$$F_{wheel} = \frac{r_{lever}}{r_{arm}} F_{spring} \eta \tag{2.12}$$

where η represents the mechanism efficiency.

The trailing arm suspension with a link-lever mechanism, often termed a "bell crank" or "rocker" suspension, decouples the spring and damper unit from the direct path of the wheel's motion [29]. In this architecture, the wheel is mounted to a trailing arm that pivots about an axis parallel to the vehicle's transverse axis. The vertical motion of this arm is then transmitted to a remotely mounted spring-damper unit via a connecting link and a pivoting lever [3]. This separation of functions introduces a distinct set of trade-offs, making it particularly interesting for packaging-critical applications [1].

Advantages of the Trailing Arm and Link-Lever Configuration:

- **Flexible Actuator Placement and Packaging Optimization:** The most significant advantage of this system is the complete decoupling of the spring and damper ("actuator") from the wheel carrier [29]. This allows them to be positioned in areas of the chassis with abundant free space, away from the cramped and thermally challenging environment of the wheel well. This is crucial for in-wheel motor systems, as it frees up the critical lateral and radial volume for the motor, brake caliper, and steering mechanism. The actuator can be mounted horizontally, vertically, or at an angle, whichever best suits the available package [1].
- **Inherently Progressive Rate Characteristics:** The pivoting lever acts as a mechanical transformer, with its effective ratio changing throughout the suspension's travel [21]. This creates a non-linear relationship between wheel travel and spring compression. Typically, this is designed to provide a progressively higher effective spring rate as the suspension is compressed. This characteristic offers a soft initial ride for comfort over small bumps, which automatically stiffens significantly to manage large impacts and control body roll during aggressive cornering, all without the need for complex and expensive progressive-rate springs [19].
- **Compact Envelope and Efficient Volume Utilization:** By separating the structural arm (which handles high loads) from the force-generating actuator (the spring-damper), the system achieves a very efficient use of space [25]. The trailing arm itself can be a relatively compact, high-strength forging, while the spring-damper unit can be tucked into an unused cavity within the chassis or subframe. This component separation prevents the need for a large, vertically-oriented strut tower, leading to a more compact overall envelope that is easier to integrate within a tight wheel diameter [1].

Disadvantages of the Trailing Arm and Link-Lever Configuration:

- **Increased Joint Count and System Complexity:** While conceptually simple, the mechanism introduces several additional components compared to a

Figure 2.6: Semi-trailing arm suspension [26]

strut: a pivoting lever (rocker), at least one additional link (pushrod or pull-rod), and the associated high-strength bearings or bushings at each pivot [2]. This higher joint count increases the mass of the unsprung and semi-sprung components, adds more potential points of wear and failure (e.g., rod-end joints), and complicates the assembly process. Friction in these multiple joints can also detract from the suspension's sensitivity [29].

- **Compounded Compliance Effects Under High Loads:** Every joint and link in the force path introduces a small amount of elastic deflection under load [21]. In a system with multiple pivots, these small compliances can stack up, leading to a perceptible loss of wheel position precision under high cornering, braking, or acceleration forces. This elastokinematic "wind-up" can manifest as a slight change in toe or camber under load, reducing the predictability and sharpness of the handling response compared to a more direct, stiffly-mounted damper system [19].
- **Limited Anti-Dive and Anti-Squat Geometry Control:** The trailing arm's single pivot axis provides limited ability to control the vehicle's pitch behavior during braking (dive) and acceleration (squat) [6]. The geometry for these characteristics is largely fixed by the height of the arm's pivot and the location of the virtual swing center. It is difficult to achieve the same level of independent tuning for anti-dive/anti-squat as is possible with a double wishbone or multi-link suspension, where the instant center can be precisely positioned in both the lateral and longitudinal planes [19]. This can result in more pronounced body pitch during aggressive driving, which can impact aerodynamic stability and driver confidence [14].

2.2.3 SPRING AND DAMPER INTEGRATION STRATEGIES

The integration of spring and damper elements within the constrained volume of an in-wheel suspension presents unique challenges that have inspired innovative engineering solutions across multiple domains [29].

2.2.3.1 Spring Technologies

The selection of spring technology for IWSS represents a critical engineering compromise between mass efficiency, spatial constraints, and dynamic performance requirements [21]. As quantified in Table 2.1, the specific energy density and volumetric efficiency parameters reveal fundamental limitations and opportunities for each technology class.

Conventional metallic spring solutions, including steel coil springs and torsion bars, demonstrate limited specific energy capacity (0.5–1.0 kJ/kg), establishing a significant mass penalty that directly contravenes unsprung mass minimization objectives [15]. While torsion bars achieve superior volume efficiency through their structural configuration, both technologies exhibit fundamental material limitations in energy storage per unit mass [7]. This mass inefficiency becomes particularly problematic in in-wheel applications, where each additional kilogram of unsprung mass adversely affects high-frequency road holding capability and vehicle dynamics stability.

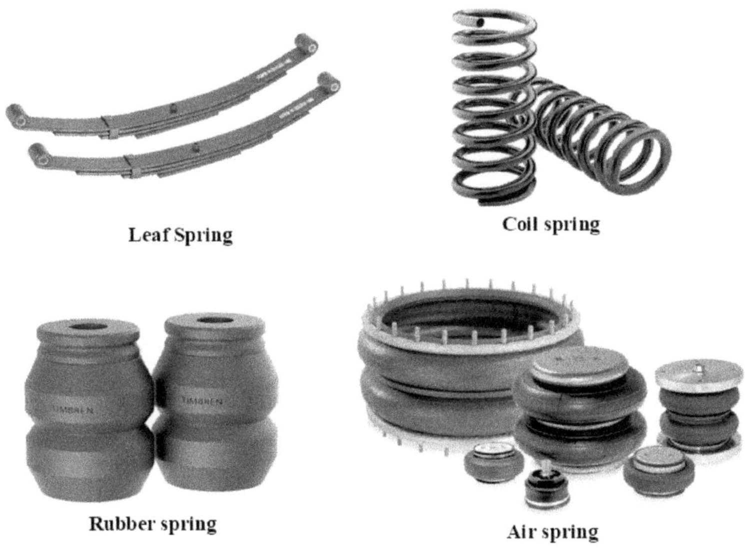

Figure 2.7: Spring categories on suspension

Advanced composite and fluid-based technologies offer substantially improved mass characteristics [29]. Composite leaf springs leverage anisotropic material properties of fiber-reinforced polymers to achieve specific energy values of 1.2–1.8 kJ/kg, representing a 140–260% improvement over conventional steel springs while maintaining high volumetric efficiency. Hydraulic gas springs demonstrate the ultimate specific energy performance (2.0–3.0 kJ/kg) through compressible fluid

Table 2.1
Comparison of Spring Technologies for In-Wheel Applications

Technology	Specific Energy [kJ/kg]	Volume Efficiency	Application Suitability
Steel Coil Springs	0.5–0.8	Moderate	General purpose, cost-sensitive
Torsion Bars	0.6–1.0	High	Space-constrained, high-load
Air Springs	1.5–2.5	Variable	Luxury/comfort, active systems
Composite Leaf Springs	1.2–1.8	High	Performance, lightweight
Hydraulic Gas Springs	2.0–3.0	Moderate	High-performance, compact

thermodynamics, though this comes with increased system complexity and potential thermal management challenges in confined wheel arch environments [4].

The spatial limitations inherent to in-wheel packaging necessitate careful evaluation of volume efficiency [1]. Torsion bars and composite leaf springs provide optimal solutions for scenarios where radial and axial clearances are severely constrained by brake components, electric motor assemblies, and steering linkages [25]. The variable volume efficiency of air springs presents both an opportunity and challenge; while enabling adaptable packaging solutions, the required compressor and reservoir systems introduce additional complexity and failure modes that must be accommodated within overall vehicle architecture.

The selection of spring technology involves careful consideration of multiple factors [21]:

$$\text{Spring Merit Index} = \frac{U_s}{\rho V} \cdot \frac{f_{fatigue}}{C_{manufacturing}} \tag{2.13}$$

where U_s is the stored energy capacity, ρ is density, V is volume, $f_{fatigue}$ is fatigue strength, and $C_{manufacturing}$ is manufacturing cost.

The technological selection matrix reveals clear application correlations [29]. High-performance scenarios demanding minimal unsprung mass benefit substantially from composite leaf springs or hydraulic gas springs, despite their premium cost structure and manufacturing complexity [15]. Luxury applications prioritizing ride quality and adaptive capability find justification for air spring implementation, accepting the mass and reliability compromises for superior NVH characteristics and active leveling functionality [3].

Cost-sensitive implementations remain best served by conventional steel springs, where the maturity of manufacturing processes and predictable failure modes provide economic advantages, albeit with permanent mass and performance compromises [7]. The emerging application space for in-wheel EV particularly favors composite solutions, where the combined benefits of mass reduction, corrosion resistance, and packaging flexibility align with the fundamental design constraints of integrated motor-suspension systems [1].

It meant that no single spring technology currently dominates all performance metrics [21]. The selection process must therefore be driven by specific application weightings of mass, volume, cost, and dynamic performance requirements [29]. Future development trajectories suggest increasing adoption of composite materials and hybrid spring-damper systems that can provide adaptive characteristics while meeting the stringent mass and packaging constraints of next-generation in-wheel suspension architectures.

2.2.3.2 Damper Technologies for IWSS

The integration of advanced damper technologies represents a critical enabling factor for IWSS, where the constraints of unsprung mass and packaging volume must be balanced against increasingly demanding performance requirements for ride comfort, vehicle stability, and road holding [3]. Modern implementations have evolved beyond conventional passive designs to encompass semi-active and fully active systems, each offering distinct trade-offs in complexity, energy consumption, and performance capability [29]. This evolution is particularly significant for in-wheel configurations, where the close coupling between suspension dynamics and electric motor operation necessitates sophisticated vibration management strategies [15].

Passive Dampers represent the conventional approach, employing fixed-parameter hydraulic systems optimized for a specific compromise between ride quality and handling precision [21]. These dampers operate on fundamental fluid dynamics principles where hydraulic oil is forced through precisely calibrated orifices and valves, converting kinetic energy into thermal energy that is subsequently dissipated to the environment [7]. The force-velocity characteristics of passive dampers typically exhibit nonlinear behavior described by:

$$F_d = c_1 v + c_2 v^2 + F_{friction} + F_{gas} + F_{temperature} \tag{2.14}$$

where c_1 represents the linear damping coefficient, c_2 characterizes the quadratic damping component arising from high-velocity fluid dynamics, $F_{friction}$ accounts for mechanical friction in seals and bearings, F_{gas} represents the gas spring force from the nitrogen charge, and $F_{temperature}$ captures the significant viscosity-dependent behavior that affects performance consistency [29]. While passive dampers benefit from mechanical simplicity, proven reliability, and minimal energy requirements, their fixed characteristics represent an inherent compromise that cannot adapt to varying road conditions, vehicle loading states, or driver preferences [3]. This limitation becomes particularly pronounced in in-wheel applications where the unsprung mass characteristics differ significantly from conventional suspensions [15].

Semi-Active Dampers bridge the gap between passive and fully active systems by providing real-time adjustable damping characteristics without significant energy injection into the suspension system [29]. These systems employ sophisticated control algorithms that continuously modulate damping forces based on sensor inputs measuring body acceleration, wheel velocity, and vehicle dynamics states [3]. The predominant technologies in this category include:

- **Magnetorheological (MR) Dampers** utilize a sophisticated fluid containing micron-sized ferromagnetic particles suspended in a synthetic hydrocarbon base oil [29]. When subjected to a magnetic field generated by an electromagnetic coil integrated into the piston assembly, these particles form chain-like structures aligned with the field lines, dramatically increasing the fluid's apparent viscosity and yield stress within milliseconds. The constitutive behavior of MR fluids is governed by:

$$\tau = \tau_y(H) \cdot \text{sgn}(\dot{\gamma}) + \eta\dot{\gamma} + \rho\frac{\partial v}{\partial t} \tag{2.15}$$

where $\tau_y(H)$ is the field-dependent yield stress that follows a sigmoidal relationship with magnetic field intensity H, η represents the field-independent plastic viscosity, $\dot{\gamma}$ is the shear rate, and the final term accounts for fluid inertia effects that become significant at high frequencies. Modern MR dampers can achieve response times of less than 10 milliseconds and provide continuously variable force ranges exceeding 1000 N, making them particularly suitable for handling the high-frequency content characteristic of in-wheel suspension dynamics [3].

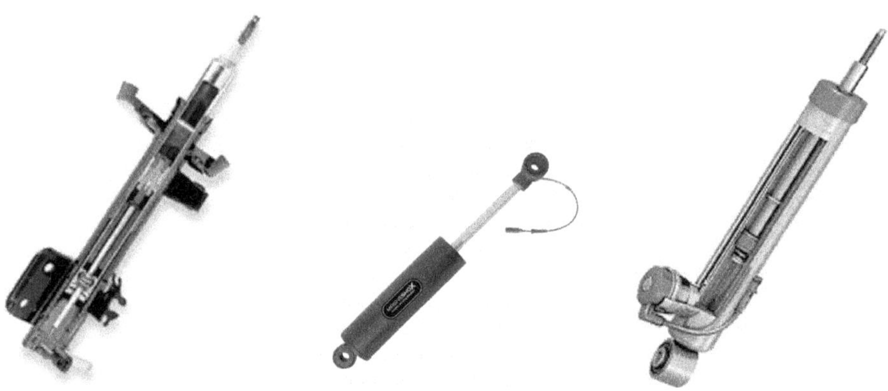

(a) Magnetorheological damper (b) Electrorheological damper (c) Electrohydraulic damper

Figure 2.8: Controllable damper categories on suspension

- **Electrorheological (ER) Dampers** operate on a similar field-responsive principle but employ fluids whose viscosity is controlled by applied electric fields rather than magnetic fields. While ER fluids can achieve even faster response times (typically 1-2 milliseconds) and require less complex power electronics, they generally produce lower maximum yield stresses and are more sensitive to temperature variations and contamination [3]. These limitations have restricted their widespread adoption in automotive applications despite their theoretical advantages in response speed and energy efficiency.

Active Damper Systems represent the pinnacle of suspension technology, employing hydraulic actuators or electric linear motors to inject and remove energy from the suspension system independently of the relative motion between wheel and body [29]. These systems typically incorporate high-bandwidth hydraulic servovalves or precision ball-screw mechanisms driven by brushless motors, coupled with sophisticated multi-variable control algorithms [3]. The general control law for active systems extends beyond simple PID formulation to include optimal control strategies:

$$F_a = K_p e + K_i \int e dt + K_d \frac{de}{dt} + K_{LQR}\mathbf{x} + F_{feedforward} \tag{2.16}$$

where e represents the error between desired and actual suspension states, K_{LQR} is the optimal feedback gain matrix derived from Linear Quadratic Regulator theory, $\mathbf{x}$ is the state vector encompassing body position, velocity, wheel position, and velocity, and $F_{feedforward}$ incorporates predictive control actions based on road preview information when available [23]. Active systems can theoretically achieve perfect vibration isolation and eliminate body roll during cornering, but they impose substantial energy requirements (typically 1-5 kW per corner), introduce significant mass penalties, and present formidable challenges in packaging within the constrained volume of an in-wheel application [1]. Furthermore, stability assurance requires careful attention to control system design, as time delays or sensor failures can lead to dangerous oscillatory behavior or complete performance degradation [3].

The selection between these damper technologies for in-wheel applications involves careful consideration of the trade-space encompassing performance requirements, energy availability from the vehicle's electrical system, mass constraints, packaging limitations, and cost targets [29]. While passive systems remain relevant for cost-sensitive implementations, the trend toward electrified and automated vehicles is driving increased adoption of semi-active and active technologies that can provide the adaptive characteristics necessary for optimal performance across diverse operating conditions [3].

2.2.4 HUB CARRIER AND STRUCTURAL INTEGRATION

The hub carrier serves as the structural nucleus of the IWSS, integrating multiple functions and load paths into a single compact component [1]. Its design represents one of the most challenging aspects of in-wheel suspension engineering [7].

2.2.4.1 Structural Requirements

The hub carrier represents a safety-critical structural component that must simultaneously satisfy multiple demanding mechanical requirements under complex, multi-axial loading conditions [2]. Its primary function is to transfer forces from the wheel-road interface through the suspension linkages to the vehicle chassis while maintaining precise wheel alignment and providing reliable mounting for the brake system and wheel bearings [21]. The structural integrity must be verified across three fundamental domains: static strength, service stiffness, and long-term durability [7].

Static Strength Verification The hub carrier must withstand extreme load cases encountered during severe maneuvers such as emergency braking, cornering at the limits of adhesion, and impact with road obstacles [19]. These scenarios generate complex stress states within the component, necessitating evaluation using the von Mises yield criterion to account for multiaxial stress conditions:

$$\sigma_{von\ Mises} = \sqrt{\frac{(\sigma_{11}-\sigma_{22})^2+(\sigma_{22}-\sigma_{33})^2+(\sigma_{33}-\sigma_{11})^2+6(\sigma_{12}^2+\sigma_{23}^2+\sigma_{31}^2)}{2}} \leq \frac{\sigma_y}{SF} \tag{2.17}$$

where σ_{ij} represents the components of the stress tensor at critical locations, σ_y is the material yield strength, and SF is an appropriate safety factor [7]. For automotive applications, safety factors typically range from 1.5 to 2.5 for normal operating conditions, increasing to 3.0-4.0 for extreme or crash scenarios [25]. The most critical stress concentrations typically occur at geometric discontinuities such as fillet radii, bolt holes, and section changes, requiring careful attention during the design phase to avoid premature yielding or fracture [2].

Stiffness and Deflection Constraints Beyond ultimate strength, the hub carrier must maintain sufficient structural stiffness to ensure proper vehicle dynamics and handling characteristics [21]. Excessive deformation under load alters suspension kinematics, degrades steering precision, and compromises bearing alignment, leading to premature wear and potential failure [19]. The stiffness requirement is mathematically expressed as:

$$\delta_{max} = \max\left(\frac{F_i}{K_i}\right) \leq \delta_{allowable} \quad \text{for } i = x, y, z \text{ directions} \tag{2.18}$$

where δ_{max} is the maximum displacement at critical locations under design loads, F_i represents forces in different coordinate directions, K_i is the corresponding structural stiffness, and $\delta_{allowable}$ is the maximum permissible deflection [21]. Typical allowable displacements are on the order of 0.1-0.5 mm under normal operating loads, with even stricter requirements for high-performance applications [19]. The hub carrier's compliance directly influences the wheel center stiffness, which modern vehicles typically target in the range of 20-40 N/µm to balance handling precision with ride comfort [14].

Fatigue Life and Durability As a component subjected to cyclic loading throughout its service life, the hub carrier must demonstrate adequate fatigue resistance to withstand millions of load cycles from road irregularities, braking, and cornering forces [7]. The fatigue life requirement is quantified using:

$$N_{fatigue} = \int_0^{N_{design\ life}} \frac{dn}{N(S_i)} \geq N_{design\ life} \quad \text{with } N(S_i) = C \cdot S_i^{-m} \tag{2.19}$$

where $N_{fatigue}$ is the calculated fatigue life, $N_{design\ life}$ is the required service life (typically 10,000-100,000 hours of operation), S_i is the stress amplitude for each load cycle, and $N(S_i)$ represents the number of cycles to failure at stress amplitude S_i according to the material's S-N curve parameters C and m [25]. Modern design approaches employ strain-life (ϵ-N) methods and fracture mechanics to account for localized plasticity and crack propagation, particularly at stress concentrations. For cast and forged components, the fatigue strength is typically 30-50% of the ultimate tensile strength after accounting for surface finish, size, and loading condition factors [2, 7].

Beyond these primary requirements, the hub carrier design must address several secondary structural considerations including: resonance avoidance by ensuring natural frequencies exceed excitation frequencies from wheel imbalance and road inputs (typically >150 Hz); corrosion resistance to withstand road salt and environmental exposure; and thermal stability to maintain mechanical properties under braking heat soak conditions that can elevate local temperatures to 200-300°C [4, 7]. These multi-faceted requirements necessitate a systems engineering approach that balances competing objectives through iterative design refinement and validation using both analytical methods and physical testing.

2.2.4.2 Material Selection

Hub carrier materials must balance strength, stiffness, weight, and manufacturing considerations [7, 25]. The data presented in Table 2.2 reveals a clear trade-space where no single material dominates all performance metrics.

Cast iron serves as the traditional baseline, offering high stiffness and low cost, but its severe mass penalty (7200 kg/m^3) makes it generally unsuitable for in-wheel applications where minimizing unsprung mass is critical [15]. Aluminum alloys present the most common compromise; cast variants offer a favorable weight reduction at a moderate cost premium, while forged aluminum provides significantly improved yield strength (up to 450 MPa) for high-performance applications, retaining the density advantage but at increased expense [2]. Magnesium alloys offer the ultimate weight savings due to their exceptionally low density (1800 kg/m^3), however, this comes with significant compromises in lower yield strength, the lowest stiffness of the group, and higher cost coupled with corrosion and flammability concerns during processing. For the most extreme performance applications where strength-to-weight ratio is paramount and cost is secondary, titanium alloys are unmatched, offering yield strengths exceeding 1 GPa [19, 25, 27]. However, their high density relative to aluminum and magnesium, combined with an order-of-magnitude higher cost, severely limits their practicality for all but niche motorsport or aerospace applications.

The selection is therefore highly application-dependent. For mass-produced vehicles, cast aluminum often represents the optimal balance. For performance-focused or in-wheel motor systems where unsprung mass is a primary design constraint, forged aluminum provides the best combination of strength, weight, and manufacturability, despite its cost. Magnesium remains a specialist choice for maximum weight reduction where its mechanical limitations can be managed, while titanium is reserved for cases where ultimate strength and fatigue life are required regardless of cost [15, 19].

Table 2.2
Hub Carrier Material Comparison

Material	Yield Strength [MPa]	Young's Modulus [GPa]	Density [kg/m^3]	Cost Factor
Cast Iron	250–350	170–190	7100–7300	1.0
Cast Aluminum	200–300	70–80	2700–2800	1.5–2.0
Forged Aluminum	300–450	70–80	2700–2800	2.0–3.0
Magnesium Alloy	160–240	42–45	1800–1850	3.0–4.0
Titanium Alloy	800–1100	110–120	4400–4500	10–15

2.2.4.3 Manufacturing Processes

The manufacturing process selection involves trade-offs between cost, performance, and design flexibility [7]:

$$\text{Manufacturing Merit} = \frac{\text{Design Freedom} \times \text{Material Utilization}}{\text{Tooling Cost} \times \text{Production Time}} \tag{2.20}$$

common processes include casting (for complex geometries), forging (for high strength), and additive manufacturing (for prototype and low-volume production).

2.3 HUB MOTOR DESIGN AND ELECTROMAGNETIC INTEGRATION

The electric motor represents both the greatest opportunity and most significant challenge in in-wheel suspension design [4, 5, 16]. Successful integration requires careful balancing of electromagnetic performance, thermal management, and structural requirements within severe spatial constraints.

2.3.1 MOTOR TOPOLOGIES AND CONFIGURATION SELECTION

The selection of motor topology represents a fundamental design decision that influences all aspects of system performance. We examine the primary candidates in detail.

2.3.1.1 Outer Rotor Permanent Magnet Synchronous Motors

Outer rotor configurations have emerged as the dominant topology for in-wheel applications due to their superior torque density and natural integration with the wheel structure.

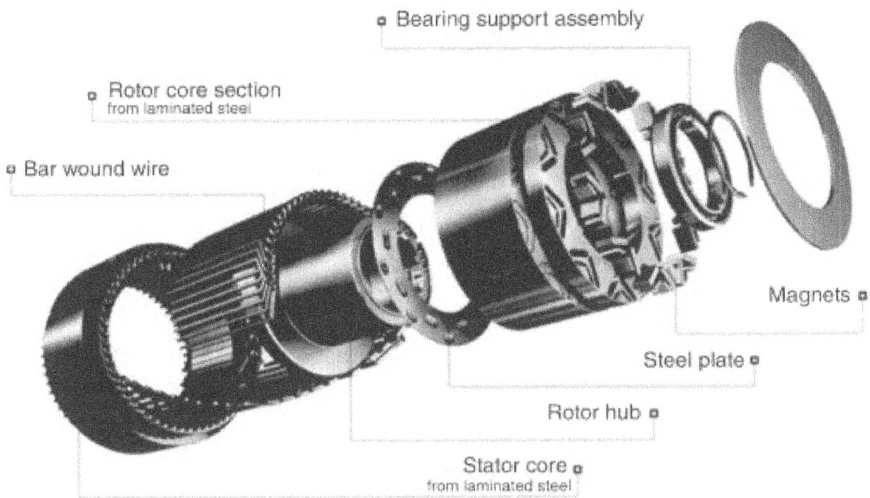

Figure 2.9: Outer rotor permanent magnet synchronous motor [28]

The electromagnetic torque production in a PMSM follows [10]:

$$T_e = \frac{3}{2}\frac{P}{2}\left[\lambda_{pm} i_q + (L_d - L_q) i_d i_q\right] \tag{2.21}$$

where P is the number of poles, λ_{pm} is the permanent magnet flux linkage, L_d and L_q are dq-axis inductances, and i_d, i_q are dq-axis currents [24].

The key advantage of outer rotor configurations lies in their torque production capability [5]:

$$T_{max} \propto D^2 L \quad \text{for constant current density} \tag{2.22}$$

where D is the rotor diameter and L is the stack length. The larger diameter of outer rotor designs provides superior torque for a given electromagnetic loading [16].

Design Considerations:

- **Magnet Retention:** The centrifugal force on magnets at maximum operating speed follows:

$$F_c = \rho_m \omega^2 r_m t_m L_m \tag{2.23}$$

  requiring robust mechanical retention systems [24].
- **Cogging Torque Mitigation:** Cogging torque minimization requires careful optimization of magnet shaping and stator slot geometry [22]:

$$T_{cog} = -\frac{\partial W}{\partial \theta} \approx \sum_{n=1}^{\infty} T_n \sin(n N_{slot} \theta) \tag{2.24}$$

- **Thermal Management:** The buried stator presents challenges for heat extraction, requiring innovative cooling strategies [22].

2.3.1.2 Inner Rotor Configurations

Inner rotor motors represent the conventional motor architecture adapted for in-wheel applications [5]. While less common than outer rotor designs, they offer specific advantages in certain applications [16].

$$\omega_{max} = \frac{\sigma_{yield}}{\rho r^2} \quad \text{for stress-limited operation} \tag{2.25}$$

The smaller diameter of inner rotor designs enables higher maximum speeds but reduces torque production capability for a given electromagnetic loading [5].

2.3.1.3 Axial Flux Motors for In-Wheel Propulsion Systems

Axial flux permanent magnet (AFPM) motors, characterized by their compact disc-like geometry with magnetic flux parallel to the axis of rotation, present a fundamentally different electromagnetic architecture compared to conventional radial flux machines [8]. This topology offers particularly compelling advantages for direct in-wheel propulsion applications, where the axial space between the inner and outer wheel rims is often less constrained than the radial envelope, enabling a more efficient utilization of the available wheel volume [13, 16]. The pancake-shaped profile allows for shorter end-windings and a larger active torque-producing diameter relative to their overall package size.

The fundamental scaling laws governing axial flux machines reveal their inherent suitability for high-torque, low-speed applications characteristic of in-wheel drives. Unlike radial flux motors where torque typically scales with the square of the rotor diameter and active length ($T \propto D^2 L$), the torque production in axial flux machines follows a more favorable relationship for constant current density and magnetic loading:

$$T = k \cdot B_g \cdot A \cdot D \cdot N_r \propto D^3 \tag{2.26}$$

where k is a machine constant, B_g is the airgap flux density, A is the electrical loading, D is the active diameter, and N_r is the number of rotor discs [8, 13, 15]. This cubic relationship with diameter enables axial flux motors to achieve exceptional torque density—typically ranging from 25 to 45 Nm/L—making them capable of delivering the high wheel torques required for vehicle acceleration and hill climbing without requiring mechanical gear reduction. The short axial length furthermore minimizes the overhang moment on the suspension components, improving packaging and handling dynamics.

Critical Design Challenges and Mitigation Strategies:

- **Magnetic Attraction Forces and Structural Design:** The opposing magnetic faces in axial flux configurations generate substantial normal forces that can reach several kilonewtons in automotive-scale machines [8]. This attractive force between rotor and stator follows:

$$F_{magnetic} \approx \frac{B_g^2 A_{pole}}{2\mu_0} \cdot \cos(\theta_m) \tag{2.27}$$

where B_g is the airgap flux density, A_{pole} is the pole area, μ_0 is the permeability of free space, and θ_m is the mechanical angle. These forces impose significant demands on the bearing system, requiring careful selection of angular contact bearings or specialized magnetic bearing solutions capable of handling both radial and axial loads. Additionally, the stator backing structure must possess sufficient mechanical stiffness to prevent deflection that would alter the critical airgap dimension, potentially leading to magnetic rubbing and catastrophic failure.

- **Manufacturing Complexity and Thermal Management:** The disc-shaped components, particularly the stator with its typically non-standard core geometry, present significant manufacturing challenges. Stator construction often requires specialized segmented lamination techniques or advanced soft magnetic composite (SMC) materials that enable three-dimensional flux paths. Cooling represents another critical challenge; the compact geometry limits surface area for heat dissipation while the dual airgap configuration complicates the implementation of conventional cooling jackets. Advanced thermal management strategies such as direct oil cooling, phase-change materials, or integrated heat pipes are often necessary to maintain operational temperatures below the insulation class limits, particularly during sustained high-torque operation such as hill climbing or repeated acceleration events.
- **Rotor Dynamics and Vibration Considerations:** The large diameter-to-length ratio creates unique rotor dynamic behavior, with critical speeds often falling within the operating range. Careful balancing and potentially active magnetic bearing systems may be required to suppress whirling modes. Furthermore, the unbalanced magnetic pull (UMP) resulting from manufacturing tolerances or eccentric operation can excite structural resonances in the wheel assembly, necessitating comprehensive rotodynamic analysis and robust control strategies to ensure smooth operation across the entire speed-torque envelope.

Despite these challenges, ongoing advancements in materials, manufacturing processes, and thermal management techniques continue to improve the viability of axial flux motors for in-wheel applications, particularly where maximizing torque density within strict geometric constraints is the primary design objective [8, 13].

2.3.2 ELECTROMAGNETIC DESIGN AND OPTIMIZATION

The electromagnetic design process for in-wheel motors involves simultaneous optimization of multiple competing objectives within severely constrained design spaces, requiring sophisticated multi-physics analysis and trade-off management between torque density, efficiency, manufacturability, and thermal performance [10, 24].

2.3.2.1 Magnetic Circuit Design

The magnetic circuit design begins with the fundamental magnetic Ohm's law, which governs flux distribution throughout the motor:

$$\mathcal{F} = \Phi\mathcal{R} = NI = \Phi(\mathcal{R}_g + \mathcal{R}_{core} + \mathcal{R}_{leakage}) \tag{2.28}$$

where $\mathcal{F}$ is the magnetomotive force, Φ is the magnetic flux, $\mathcal{R}$ is the total reluctance comprising airgap reluctance $\mathcal{R}_g$, core reluctance $\mathcal{R}_{core}$, and leakage reluctance $\mathcal{R}_{leakage}$, N is the number of turns per phase, and I is the phase current [16].

The air gap flux density represents a critical design parameter that fundamentally limits torque production:

$$B_g = \frac{\lambda_{pm}}{k_w N A_g} = \frac{B_r A_m}{A_g(1+k_{leakage})} \cdot \frac{1}{1+\frac{\mathcal{R}_g}{\mathcal{R}_m}} \tag{2.29}$$

where k_w is the winding factor, A_g is the effective air gap area, λ_{pm} is the permanent magnet flux linkage, B_r is the magnet remanence, A_m is the magnet area, $k_{leakage}$ is the leakage factor, and $\mathcal{R}_m$ is the magnet reluctance [10, 24]. Modern designs typically target airgap flux densities of 0.8-1.0 T for optimal utilization of magnetic materials while avoiding excessive saturation in the stator teeth and back iron.

2.3.2.2 Winding Design and Optimization

The winding design involves complex trade-offs between copper loss minimization, thermal performance optimization, manufacturability constraints, and high-frequency effects. The slot fill factor represents a critical manufacturing metric:

$$k_{fill} = \frac{A_{copper}}{A_{slot}} \leq 0.4 - 0.6 \quad \text{for random windings, } 0.6 - 0.7 \quad \text{for pre-formed coils} \tag{2.30}$$

where A_{copper} is the total cross-sectional area of copper conductors within the slot and A_{slot} is the available slot area. Higher fill factors reduce DC resistance but increase manufacturing complexity and AC losses due to proximity effects.

The winding configuration selection (distributed vs. concentrated) significantly affects both electromagnetic performance and manufacturability:

$$k_w = k_p k_d k_{skew} \quad \text{with} \quad k_p = \sin\left(\frac{\pi}{2} \cdot \frac{W_{pitch}}{\tau_p}\right), \quad k_d = \frac{\sin\left(\frac{q\alpha}{2}\right)}{q \sin\left(\frac{\alpha}{2}\right)} \tag{2.31}$$

where k_p is the pitch factor, k_d is the distribution factor, k_{skew} is the skew factor, W_{pitch} is the coil pitch, τ_p is the pole pitch, q is the number of slots per pole per phase, and α is the electrical angle between slots [24]. Distributed windings typically achieve higher winding factors (0.85-0.95) and lower harmonic content, while concentrated windings offer shorter end-windings, higher slot fill factors, and simpler manufacturing.

2.3.2.3 Loss Modeling and Minimization

Accurate motor loss modeling is essential for efficiency optimization and thermal management system design [4, 18]:

$$P_{loss} = P_{cu} + P_{fe} + P_{pm} + P_{str} + P_{windage} + P_{bearing} \tag{2.32}$$

Copper Losses: Comprising both DC and AC components:

$$P_{cu} = I^2 R_{ac} + I_h^2 R_{ac,h} \quad \text{where} \quad R_{ac} = k_{skin} k_{proximity} k_{circulation} R_{dc} \tag{2.33}$$

where I_h represents harmonic currents, $k_{skin} = \frac{\delta}{d}\left[\frac{\sinh(\frac{2d}{\delta})+\sin(\frac{2d}{\delta})}{\cosh(\frac{2d}{\delta})-\cos(\frac{2d}{\delta})}\right]$ accounts for skin effect, $k_{proximity}$ models adjacent conductor influence, $k_{circulation}$ accounts for circulating currents in parallel strands, d is conductor diameter, and $\delta = \sqrt{\frac{2\rho}{\omega\mu}}$ is the skin depth.

Iron Losses: Modeled using improved Steinmetz equations with separation of hysteresis and eddy current components [18]:

$$P_{fe} = k_h f B^\alpha + k_c (fB)^2 + k_e (fB)^{1.5} + k_{excess} f^{1.5} B^{1.5} \tag{2.34}$$

where k_h, k_c, k_e, and k_{excess} are material-dependent loss coefficients, f is the frequency, B is the flux density, and α is the Steinmetz exponent (typically 1.5-2.0) [16, 24]. The excess loss term becomes particularly significant in modern thin-gauge electrical steels operating at high frequencies.

Magnet Losses: Particularly important in high-speed applications with significant harmonic content [20]:

$$P_{pm} = \int_V \frac{J_{eddy}^2}{\sigma} dV \quad \text{where} \quad J_{eddy} \approx \frac{\pi d_m B_h}{\rho} f \tag{2.35}$$

where d_m is the magnet segmentation size, B_h is the harmonic flux density, and σ is the electrical conductivity [22].

2.3.3 STRUCTURAL DESIGN FOR ROTATING COMPONENTS

The structural design of rotating components must ensure mechanical integrity under extreme operating conditions while minimizing mass and maximizing electromagnetic performance through optimal material utilization [5, 15].

2.3.3.1 Rotor Stress Analysis

The centrifugal stress distribution in rotating components follows the general equation for axisymmetric bodies:

$$\sigma_c(r) = \int_r^{r_o} \rho\omega^2 r dr + C = \frac{\rho\omega^2}{8}(3+\nu)(r_o^2 - r^2) \quad \text{for thin rotors} \tag{2.36}$$

where ν is Poisson's ratio and C is the integration constant determined by boundary conditions [24].

For a solid rotor, the maximum stress occurs at the center:

$$\sigma_{c,max} = \frac{\rho\omega^2}{8}(3+\nu)r_o^2 \tag{2.37}$$

The safety factor against yield must consider combined stress states including centrifugal, interference fit, thermal, and magnetic stresses [4]:

$$SF = \frac{\sigma_{yield}}{\sqrt{\sigma^2_{c,max} + \sigma^2_{interference} + \sigma^2_{thermal} - \sigma_{c,max}\sigma_{thermal}}} \tag{2.38}$$

Modern designs typically require minimum safety factors of 1.5-2.0 at maximum operating speed, increasing to 3.0-4.0 for overspeed conditions.

2.3.3.2 Critical Speed Analysis

The rotor critical speeds must be thoroughly analyzed and designed outside the operating range with sufficient margin [15]:

$$\omega_{critical} = \sqrt{\frac{k_{bearing} + k_{shaft}}{m_{rotor} + m_{added}}} \cdot \frac{1}{1 - \left(\frac{\omega}{\omega_{critical}}\right)^2} \tag{2.39}$$

where $k_{bearing}$ is the composite bearing stiffness, k_{shaft} is the shaft stiffness, m_{rotor} is the rotor mass, and m_{added} is the added mass effect from the surrounding fluid. For in-wheel applications, the first critical speed should typically exceed 1.3 times the maximum operating speed to avoid resonance amplification.

2.3.4 ADVANCED MATERIALS AND MANUFACTURING

The performance of hub motors is critically dependent on advanced material selection and precision manufacturing processes that enable the realization of complex electromagnetic and thermal designs [10, 24].

2.3.4.1 Magnetic Materials for High-Performance In-Wheel Motors

The selection and optimization of permanent magnet materials constitutes a fundamental aspect of electromagnetic design for in-wheel motors, directly influencing torque density, efficiency, operational stability, and overall system cost. The magnetic properties of commercially relevant permanent magnet materials are quantitatively compared in Table 2.3, highlighting the critical trade-offs between magnetic strength, thermal stability, and economic feasibility [22].

Modern high-performance in-wheel motors predominantly employ sintered Neodymium-Iron-Boron (NdFeB) magnets due to their exceptional maximum energy product $(BH)_{max}$, which reaches 440 kJ/m^3 for premium grades [22]. This parameter, representing the maximum energy density stored in the magnetic field, directly correlates with the motor's torque production capability per unit volume according to the relationship:

$$T \propto (BH)_{max} \cdot D^2 L \cdot k_w \cdot A \tag{2.40}$$

Table 2.3
Comparative Analysis of Permanent Magnet Materials for Automotive Traction Applications

Material	Remanence [T]	Coercivity [kA/m]	Energy Product [kJ/m³]	Max Temp [°C]	Cost Index
Ferrite (Ceramic)	0.38–0.42	250–300	25–35	450	1.0
Alnico	1.20–1.30	50–60	40–60	540	3.5–5.0
Samarium-Cobalt (SmCo)	0.90–1.10	600–800	150–240	350	12–18
Neodymium-Iron-Boron (NdFeB)	1.10–1.45	800–2000	200–440	150–220	8–15

where D is the rotor diameter, L is the active length, k_w is the winding factor, and A is the electrical loading. The superior remanence flux density B_r of NdFeB magnets (1.10-1.45 T) enables higher airgap flux densities, typically in the range of 0.8-1.1 T, which directly enhances torque production while potentially allowing for reduction in motor active volume.

However, the implementation of NdFeB magnets necessitates careful attention to their intrinsic temperature sensitivity, characterized by reversible temperature coefficients of remanence ($\alpha_{B_r} \approx -0.10$ to $-0.13\%/C$) and coercivity ($\beta_{H_{cJ}} \approx -0.50$ to $-0.70\%/C$). This temperature dependence follows the relationship [4]:

$$B_r(T) = B_{r,20}\left[1 + \frac{\alpha_{B_r}}{100}(T-20)\right], \quad H_{cJ}(T) = H_{cJ,20}\left[1 + \frac{\beta_{H_{cJ}}}{100}(T-20)\right] \tag{2.41}$$

where $B_{r,20}$ and $H_{cJ,20}$ represent room-temperature values. The particularly strong negative temperature coefficient of coercivity creates vulnerability to irreversible demagnetization at elevated temperatures, especially under high armature reaction fields during peak current operation or fault conditions. To mitigate this limitation, high-coercivity grades incorporating strategic additions of dysprosium (Dy) or terbium (Tb) are routinely employed, though these heavy rare-earth elements significantly increase material cost and introduce supply chain vulnerabilities [16, 18, 22]. The coercivity enhancement mechanism operates through the formation of a $(Nd,Dy/Tb)_2Fe_{14}B$ phase with higher crystal anisotropy field, effectively raising the intrinsic coercivity H_{cJ} by 200-500 kA/m while typically reducing remanence by 0.05-0.15 T due to the antiferromagnetic coupling of the heavy rare-earth moments.

Alternative magnet technologies present distinct advantages for specific operational scenarios. Samarium-Cobalt (SmCo) magnets offer superior thermal stability with maximum operating temperatures up to 350°C and lower temperature coefficients ($\alpha_{B_r} \approx -0.03$ to $-0.05\%/C, \beta_{H_{cJ}} \approx -0.15$ to $-0.30\%/C$), making them suitable for applications where consistent performance across wide temperature ranges is paramount, despite their higher cost and lower maximum energy product compared to premium NdFeB grades. Ferrite magnets remain economically attractive for cost-sensitive applications where moderate performance is acceptable, though their low remanence (0.38-0.42 T) necessitates larger magnetic circuits and increased motor volume to achieve equivalent torque output [10, 18, 24]. Alnico magnets exhibit exceptional thermal stability with maximum operating temperatures exceeding 500°C, but their low coercivity (50-60 kA/m) renders them highly susceptible to demagnetization from armature reaction fields, limiting their applicability in modern high-current density traction motors.

Recent developments in grain boundary diffusion processes for NdFeB magnets have enabled more efficient utilization of heavy rare-earth elements by concentrating them preferentially in the grain boundary phases where they most effectively enhance coercivity, reducing overall dysprosium consumption by 30-60% while maintaining equivalent high-temperature performance. Concurrently, intensive research focuses on reduced-heavy-rare-earth and rare-earth-free magnet technologies, including Mn-Al-C, Fe-Ni, and exchange-spring nanocomposite magnets, though these alternatives currently achieve energy products below 100 kJ/m^3, insufficient for premium traction applications [16, 18, 22]. The optimal magnet selection therefore represents a multi-dimensional optimization considering not only electromagnetic performance but also thermal management requirements, cost constraints, supply chain stability, and environmental sustainability considerations throughout the product lifecycle.

2.3.4.2 Lamination Materials

Electrical steel selection involves sophisticated trade-offs between core loss, cost, saturation flux density, and mechanical properties:

$$\text{Core Loss Merit} = \frac{B_{sat}^2 \cdot f_{max}}{P_{fe} \cdot Cost \cdot \rho} \cdot k_{stacking} \tag{2.42}$$

where f_{max} is the maximum operating frequency, ρ is the material density, and $k_{stacking}$ is the stacking factor. Modern non-grain oriented (NGO) electrical steels with thicknesses of 0.15-0.35 mm and silicon content of 2.0-3.5% provide optimal balance for most in-wheel applications, while advanced amorphous and nanocrystalline materials offer superior high-frequency performance at premium cost.

2.4 THERMAL MANAGEMENT SYSTEMS

Effective thermal management represents arguably the most critical challenge in in-wheel motor system design [4, 18]. The concentrated heat generation within a sealed, compact volume, combined with limited heat rejection pathways and severe transient operating conditions, demands sophisticated multi-physics thermal engineering solutions.

2.4.1 HEAT GENERATION MECHANISMS

Accurate quantification and characterization of heat generation mechanisms throughout the operational envelope provides the essential foundation for effective thermal management system design.

2.4.1.1 Motor Heat Generation

The primary heat sources in the motor exhibit complex interdependencies and operational variations:

$$q_{motor} = \frac{P_{cu}(I,T) + P_{fe}(f,B,T) + P_{pm}(f,B_h) + P_{str}(f,\sigma)}{V_{active}} \cdot k_{load} \cdot k_{duty} \quad (2.43)$$

where V_{active} is the active volume of the motor, k_{load} accounts for load variation, and k_{duty} represents the duty cycle effects.

The copper losses exhibit strong positive temperature dependence due to resistivity variation [16]:

$$R(T) = R_{20}[1 + \alpha_{20}(T - 20) + \beta_{20}(T - 20)^2] \quad (2.44)$$

where α_{20} is the linear temperature coefficient (0.00393/°C for copper) and β_{20} is the quadratic coefficient, creating a thermally regenerative feedback loop that must be carefully managed through derating or enhanced cooling.

2.4.1.2 Braking Heat Generation

During regenerative and friction braking, the kinetic and potential energy conversion follows:

$$E_{braking} = \frac{1}{2}m(v_1^2 - v_2^2) + mg(h_1 - h_2) + \int F_{roll}ds + \int F_{aero}ds \quad (2.45)$$

For repeated aggressive braking events, such as during mountain descent or track driving, the average power dissipation reaches extreme levels:

$$P_{brake,avg} = \frac{\sum E_{brake}}{t_{cycle}} \eta_{thermal} + P_{convection} + P_{radiation} \quad (2.46)$$

where $\eta_{thermal}$ represents the fraction of braking energy converted to heat in the wheel assembly, which can approach 90-95% for repeated heavy braking scenarios.

2.4.2 HEAT TRANSFER MECHANISMS AND ANALYSIS

Multiple heat transfer mechanisms operate simultaneously within the constrained in-wheel environment, each with distinct characteristics, limitations, and design implications [4].

2.4.2.1 Conduction Heat Transfer

Conduction through solid components and interfaces follows the generalized Fourier's law:

$$\mathbf{q}'' = -k\nabla T \quad \text{with} \quad \nabla\cdot(k\nabla T) + \dot{q} = \rho c_p \frac{\partial T}{\partial t} \tag{2.47}$$

For complex geometries typical of in-wheel assemblies, the thermal resistance network approach provides valuable system-level insights:

$$R_{th,cond} = \sum \frac{L_i}{k_i A_i} + \sum R_{th,contact} + \sum R_{th,spreading} \tag{2.48}$$

where $R_{th,spreading} = \frac{1-\frac{A_{small}}{A_{large}}}{\pi^{1/2} k D_{small}}$ accounts for heat spreading in enlarged sections.

2.4.2.2 Convection Heat Transfer

Convective cooling, whether natural or forced, follows Newton's law of cooling with strongly geometry-dependent coefficients [4]:

$$q'' = h(T_s - T_\infty) + \epsilon\sigma(T_s^4 - T_\infty^4) \tag{2.49}$$

The convection coefficient h exhibits complex dependence on flow regime, fluid properties, and surface geometry.

For forced convection in complex internal channels, modified Dittus-Boelter correlations provide practical estimates:

$$Nu = 0.023 Re^{0.8} Pr^{0.4} \left(\frac{\mu_b}{\mu_w}\right)^{0.14} \left(1 + \frac{6D_h}{L}\right) \quad \text{for turbulent flow} \tag{2.50}$$

where $Nu = \frac{hD_h}{k_f}$ is the Nusselt number, $Re = \frac{\rho v D_h}{\mu}$ is the Reynolds number, $Pr = \frac{\mu c_p}{k_f}$ is the Prandtl number, and the viscosity ratio accounts for temperature-dependent property variations.

2.4.2.3 Radiation Heat Transfer

Radiation becomes increasingly significant at higher temperatures and in vacuum-like environments, following the Stefan-Boltzmann law with view factor considerations [4]:

$$q''_{1-2} = \frac{\sigma(T_1^4 - T_2^4)}{\frac{1-\epsilon_1}{\epsilon_1 A_1} + \frac{1}{A_1 F_{1-2}} + \frac{1-\epsilon_2}{\epsilon_2 A_2}} \tag{2.51}$$

where F_{1-2} is the view factor between surfaces 1 and 2, and ϵ_1, ϵ_2 are the surface emissivities. Strategic surface treatments can enhance radiative heat rejection by 30-50%.

2.4.3 COOLING SYSTEM ARCHITECTURES FOR IN-WHEEL MOTOR APPLICATIONS

The thermal management of in-wheel motor systems presents one of the most formidable engineering challenges in electric vehicle development, owing to the extreme power densities encountered, severe packaging constraints within the wheel envelope, and the complex multi-physics nature of heat transfer in rotating systems. Various cooling system architectures have been developed to address these extreme thermal challenges, each exhibiting distinct advantages, implementation complexities, and performance limitations that must be carefully evaluated against specific application requirements [4, 18].

2.4.3.1 Liquid Cooling Systems

Liquid cooling represents the most technologically mature and effective approach for high-power density applications, offering superior heat transfer coefficients and the ability to reject heat to remote locations. The fundamental heat removal capacity of liquid cooling systems is governed by:

$$P_{cooling} = \dot{m} c_p \Delta T + \varepsilon \dot{m} \Delta P / \rho \quad \text{with parasitic pumping penalty} \tag{2.52}$$

where $\dot{m}$ is the coolant mass flow rate, ΔT is the temperature rise across the motor, c_p is the specific heat capacity of the coolant, ΔP is the system pressure drop, ρ is the coolant density, and ε represents the pump efficiency factor typically ranging from 0.3 to 0.6. The second term accounts for the parasitic power consumption of the circulation system, which can become significant in high-flow-rate applications and must be minimized through optimal hydraulic design.

Advanced Cooling Jacket Designs:

- **Multi-Pass Axial Channels:** These configurations employ multiple parallel or serial flow paths along the motor axis, creating enhanced heat transfer through increased flow velocity and turbulence. The heat transfer coefficient follows the correlation $h \approx 0.023 \frac{k}{D_h} Re^{0.8} Pr^{0.4}$ for turbulent flow, with typical values of 2000-5000 $W/m^2 \cdot K$ for water-glycol mixtures [4, 18]. However, the pressure drop increases proportionally to $\Delta P \propto f \frac{L}{D_h} \rho v^2$, where the friction factor f ranges from 0.02 to 0.05 for smooth channels, necessitating careful optimization to balance thermal performance against pumping power requirements.
- **Helical Radial Channels:** This architecture utilizes spiral or helical passages that follow the curvature of the stator outer diameter, providing optimal space utilization and improved heat transfer coefficients through secondary flow patterns and Dean vortices. The helical geometry induces centrifugal forces that enhance fluid mixing, particularly in the curved regions, with heat transfer enhancements of 15-30% compared to straight channels at equivalent Reynolds numbers. The compact nature of these designs makes them particularly suitable for in-wheel applications where radial space is less constrained than axial dimensions.

- **Direct Stator Cooling:** The most thermally efficient approach eliminates thermal interface resistances by integrating cooling channels directly into the stator core or housing in intimate contact with the heat sources. This configuration can achieve thermal resistances as low as 0.01-0.03 K/W, enabling heat flux dissipation up to 150 W/cm^2 [18]. Implementation challenges include increased manufacturing complexity, potential reduction in stator mechanical integrity, and the requirement for excellent dielectric properties in the coolant to prevent electrical shorting in the event of leakage.
- **Phase-Enhanced Microchannels:** Leveraging flow boiling phenomena in micro-scale channels (typically 100-500 μm hydraulic diameter) provides ultra-high heat flux capability through the latent heat of vaporization. These systems can achieve heat transfer coefficients exceeding 10,000 W/m^2·K and critical heat fluxes up to 500 W/cm^2 [4]. The two-phase flow behavior is characterized by the Boiling number $Bo = \frac{q''}{Gh_{fg}}$ and the convection number $Co = \left(\frac{1-x}{x}\right)^{0.8}\left(\frac{\rho_g}{\rho_f}\right)^{0.5}$, where design must carefully manage flow instability, dryout conditions, and system pressure control.

2.4.3.2 Phase Change Cooling Systems

Phase change cooling systems offer exceptional heat absorption capacity through latent heat utilization, making them ideally suited for handling extreme transient thermal loads encountered during aggressive driving maneuvers such as repeated acceleration, regenerative braking, and hill climbing. The total energy absorption capacity for a complete phase transition cycle is given by:

$$Q = m\left[\int_{T_i}^{T_m} c_{p,s}(T)\,dT + L + \int_{T_m}^{T_f} c_{p,l}(T)\,dT + \int_{T_f}^{T_v} c_{p,v}(T)\,dT\right] + Q_{\text{pressure-volume work}} \tag{2.53}$$

where m is the mass of the working fluid, $c_{p,s}$, $c_{p,l}$, and $c_{p,v}$ represent the temperature-dependent specific heats of the solid, liquid, and vapor phases respectively, L is the latent heat of vaporization, T_i, T_m, T_f, and T_v are the initial, melting, final, and vaporization temperatures, and the pressure-volume work term accounts for expansion/compression energy. These systems demonstrate heat absorption capacities 5-10 times greater than single-phase systems per unit mass, with typical values of 200-500 kJ/kg for refrigerants such as R134a or R1234yf compared to 20-50 kJ/kg for single-phase liquid cooling with a 10°C temperature rise.

The implementation of phase change cooling in in-wheel motors presents unique challenges including two-phase flow management under centrifugal forces, refrigerant charge optimization for varying orientation, and the design of compact, lightweight condensers that can be integrated within the limited wheel envelope. Advanced control strategies must manage the complex thermo-fluid dynamics to prevent critical heat flux conditions, flow instabilities, and maldistribution that could lead to localized overheating and performance degradation.

2.4.3.3 Thermal Storage Systems

Advanced thermal storage using engineered phase change materials (PCMs) provides effective buffering of peak thermal loads, enabling temporary absorption of heat generated during high-power transients that exceed the continuous cooling capacity of the primary heat rejection system. The comprehensive energy storage capacity accounts for all thermodynamic phenomena:

$$\begin{aligned} Q_{storage} = &\int_{T_i}^{T_m} mc_{p,s}(T)dT + mL + \int_{T_m}^{T_f} mc_{p,l}(T)dT \\ &+ Q_{supercooling} + Q_{hysteresis} + Q_{subcooling} \end{aligned} \tag{2.54}$$

where $Q_{supercooling}$ represents the energy associated with temperature depression below the equilibrium phase change temperature, $Q_{hysteresis}$ accounts for the difference between melting and solidification energies, and $Q_{subcooling}$ includes energy effects from temperature variations in the solidified phase. Modern composite PCMs incorporate thermally conductive fillers such as expanded graphite, carbon nanotubes, or metallic foams to achieve enhanced thermal conductivities of 5-20 W/m·K, representing 10-50 fold improvements over pure paraffin-based materials, while maintaining high latent heat capacities of 150-250 kJ/kg.

The strategic integration of PCM thermal storage enables 2-3 times longer operation at peak power before thermal limiting occurs, effectively decoupling the motor's transient thermal performance from the continuous capacity of the primary cooling system [18]. Advanced applications employ multiple PCMs with carefully sequenced phase change temperatures (typically 50-80°C for motor applications) to create cascaded thermal storage systems that maintain more stable temperatures during extended high-power operation. The design must carefully consider the cyclability, long-term stability, and packaging efficiency of PCM systems, with typical volumetric energy densities of 150-300 MJ/m³ representing a compromise between performance and practical implementation constraints.

2.4.4 THERMAL INTERFACE MATERIALS AND MANAGEMENT

The thermal interfaces between components often represent the dominant thermal resistance in high-performance in-wheel motor systems, frequently accounting for 40-60% of the total thermal impedance between heat-generating components and the ultimate heat sink [4]. This necessitates sophisticated materials science and mechanical engineering approaches to minimize interfacial resistances while maintaining structural integrity, electrical isolation, and long-term reliability under severe thermal cycling conditions.

The comprehensive thermal resistance of an interface is characterized by:

$$\begin{aligned} R_{th,interface} = &\frac{t_{TIM}}{k_{TIM} A_{contact}} + R_{th,contact} + R_{th,spreading} \\ &+ R_{th,constriction} + R_{th,oxidation} \end{aligned} \tag{2.55}$$

where t_{TIM} is the bond line thickness, k_{TIM} is the thermal conductivity of the interface material, $A_{contact}$ is the nominal contact area, $R_{th,contact} = \frac{1}{A_{nominal} h_c}$

with $h_c \approx \frac{k_s}{\sigma_{rms}+\frac{m}{\rho_s}+\frac{\sigma_{oxide}}{k_{oxide}}}$ representing the contact conductance dependent on surface roughness σ_{rms}, contact pressure through the microhardness parameter m, surface material density ρ_s, and any oxide layer contributions. The spreading resistance $R_{th,spreading}$ accounts for heat flow expansion at discrete contact points, while constriction resistance $R_{th,constriction}$ addresses the convergence of heat flow lines into actual contact areas, and $R_{th,oxidation}$ captures the degradation effects from surface oxidation over the operational lifetime.

Advanced thermal interface materials (TIMs) have evolved significantly to address these challenges:

- **Graphene-Enhanced Composites:** Utilizing the exceptional in-plane thermal conductivity of graphene (2000-5000 W/m·K) in polymer matrices, these TIMs achieve bulk thermal conductivities of 10-50 W/m·K while maintaining electrical insulation and mechanical compliance. The anisotropic nature of graphene alignment enables preferential heat transfer in the thickness direction while minimizing in-plane electrical conduction risks.
- **Liquid Metal Alloys:** Typically based on gallium-indium-tin compositions, these materials offer exceptional thermal conductivities of 15-40 W/m·K in their liquid state, effectively conforming to surface irregularities. Implementation challenges include containment to prevent electrical shorting, interfacial wetting characteristics, and long-term stability against oxidation and intermetallic formation.
- **Reflowable Solder TIMs:** Utilizing low-melting-point solders (typically indium-based with melting points of 100-160°C), these materials provide ultra-low thermal resistances of 0.01-0.05 K·cm^2/W through the formation of metallurgical bonds. Applications are limited to compatible surfaces and must accommodate coefficient of thermal expansion mismatches and rework considerations.

Optimal TIM selection requires balancing multiple competing requirements: thermal performance (0.05-0.5 K·cm^2/W interface resistance), mechanical compliance (modulus typically 0.1-10 MPa), electrical isolation (breakdown strength ¿ 1 kV/mm), manufacturability (cure requirements, dispensability, rework capability), and long-term reliability under thermal cycling conditions (-40°C to 150°C for automotive applications). Advanced characterization techniques including laser flash analysis, infrared thermography, and acoustic microscopy are employed to validate performance and predict lifetime behavior under the harsh operating conditions encountered in in-wheel motor applications.

2.5 MATHEMATICAL MODELING OF IN-WHEEL SUSPENSION SYSTEM

Comprehensive mathematical modeling is essential for understanding, predicting, and optimizing the complex dynamic behavior of IWSS [9]. These integrated systems represent a tightly coupled electromechanical environment where traditional suspension dynamics interact strongly with electric motor operation, power electronics, and thermal management systems [3, 15]. We develop increasingly sophisticated multi-physics models that capture the essential physical phenomena while maintaining computational tractability for design optimization and real-time control applications [23]. The modeling hierarchy progresses from simplified lumped-parameter representations suitable for initial sizing and control design to high-fidelity finite element models capable of predicting local stress concentrations and thermal hotspots. Each modeling approach serves distinct purposes throughout the development cycle, with varying trade-offs between computational efficiency, predictive accuracy, and physical insight [7, 19].

2.5.1 VEHICLE DYNAMICS MODELING

2.5.1.1 Quarter-Car Model

The quarter-car model provides fundamental insights into vertical dynamics while maintaining analytical simplicity, serving as the cornerstone for initial suspension tuning and ride comfort analysis [9, 14, 15]. This two-degree-of-freedom system captures the essential dynamics between the sprung mass (vehicle body) and unsprung mass (wheel assembly), with particular relevance to in-wheel systems where the unsprung mass increases significantly due to motor integration. The system equations governing the motion are derived from Newton's second law:

$$m_s \ddot{z}_s = -k_s(z_s - z_u) - c_s(\dot{z}_s - \dot{z}_u) + F_a - m_s g \tag{2.56}$$

$$m_u \ddot{z}_u = k_s(z_s - z_u) + c_s(\dot{z}_s - \dot{z}_u) - k_t(z_u - z_r) - F_a - m_u g - F_{motor} \tag{2.57}$$

where m_s and m_u represent the sprung and unsprung masses respectively, k_s and c_s are the suspension spring stiffness and damping coefficient, k_t is the tire vertical stiffness, z_s, z_u, and z_r denote the vertical displacements of the sprung mass, unsprung mass, and road profile, F_a represents any active suspension force, and F_{motor} accounts for additional forces generated by the in-wheel motor assembly. The state-space representation provides a convenient formulation for control design and frequency domain analysis:

$$\dot{\mathbf{x}} = \mathbf{A}\mathbf{x} + \mathbf{B}\mathbf{u} + \mathbf{E}\mathbf{w} \tag{2.58}$$

where $\mathbf{x} = [z_s - z_u, \dot{z}_s, z_u - z_r, \dot{z}_u]^T$ is the state vector, $\mathbf{u} = [F_a]$ is the control input vector, and $\mathbf{w} = [\dot{z}_r]$ is the disturbance input. The system matrices are defined as:

$$\mathbf{A} = \begin{bmatrix} 0 & 1 & 0 & -1 \\ -\frac{k_s}{m_s} & -\frac{c_s}{m_s} & 0 & \frac{c_s}{m_s} \\ 0 & 0 & 0 & 1 \\ \frac{k_s}{m_u} & \frac{c_s}{m_u} & -\frac{k_t}{m_u} & -\frac{c_s}{m_u} \end{bmatrix}, \quad \mathbf{B} = \begin{bmatrix} 0 \\ \frac{1}{m_s} \\ 0 \\ -\frac{1}{m_u} \end{bmatrix}, \quad \mathbf{E} = \begin{bmatrix} 0 \\ 0 \\ -1 \\ 0 \end{bmatrix}$$

This formulation enables the calculation of key performance metrics including ride comfort (characterized by sprung mass acceleration PSD), road holding (quantified by dynamic tire load variations), and suspension working space (limited by mechanical stops), providing crucial insights during preliminary design stages [14].

2.5.1.2 Half-Car Model

The half-car model extends the quarter-car formulation to include pitch dynamics and axle coupling effects, offering significantly improved representation of actual vehicle behavior during acceleration, braking, and road input scenarios [9, 14]. This four-degree-of-freedom system captures the essential coupling between front and rear suspensions through the vehicle body, with equations of motion:

$$m\ddot{z} = -F_{sf} - F_{sr} - mg \tag{2.59}$$

$$I_{yy}\ddot{\theta} = aF_{sf} - bF_{sr} + M_{motor} \tag{2.60}$$

$$m_{uf}\ddot{z}_{uf} = F_{sf} - k_{tf}(z_{uf} - z_{rf}) - m_{uf}g - F_{motor,f} \tag{2.61}$$

$$m_{ur}\ddot{z}_{ur} = F_{sr} - k_{tr}(z_{ur} - z_{rr}) - m_{ur}g - F_{motor,r} \tag{2.62}$$

where m is the total sprung mass, I_{yy} is the pitch moment of inertia, z is the heave displacement of the center of mass, θ is the pitch angle, a and b represent the distances from the center of mass to the front and rear axles, F_{sf} and F_{sr} are the front and rear suspension forces (incorporating spring, damper, and active force components), M_{motor} accounts for pitch moments generated by unequal front/rear motor torques, and $F_{motor,f}$ and $F_{motor,r}$ represent vertical force components from front and rear in-wheel motors. The half-car model enables analysis of weight transfer during longitudinal maneuvers, anti-dive and anti-squat characteristics, and the coupling between pitch motion and vertical vibration modes, making it particularly valuable for assessing the impact of increased unsprung mass on vehicle dynamics.

2.5.1.3 Full-Car Model

The full-car model provides a complete seven-degree-of-freedom representation of vehicle dynamics, capturing all three rigid body modes of the sprung mass (heave, pitch, roll) along with four independent unsprung mass vertical motions [9, 15]. This comprehensive formulation is essential for accurate prediction of real-world vehicle behavior, particularly during combined maneuvers involving coupled lateral and longitudinal dynamics, and becomes critically important for in-wheel motor applications where the significant increase in unsprung mass dramatically alters the vehicle's dynamic characteristics.

The generalized coordinate vector for the 7-DOF model is defined as:

$$\mathbf{q} = [z_s, \theta, \phi, z_{ufl}, z_{ufr}, z_{url}, z_{urr}]^T \tag{2.63}$$

where z_s represents the sprung mass heave displacement at the center of gravity, θ denotes the pitch angle (positive nose-up), ϕ signifies the roll angle (positive right-side-down), and z_{ufl}, z_{ufr}, z_{url}, z_{urr} represent the vertical displacements of the front-left, front-right, rear-left, and rear-right unsprung masses respectively.

The equations of motion are derived using Lagrangian mechanics, resulting in the following system:

$$\mathbf{M}\ddot{\mathbf{q}} + \mathbf{C}\dot{\mathbf{q}} + \mathbf{K}\mathbf{q} = \mathbf{F}_{ext} + \mathbf{F}_{motor} + \mathbf{F}_{active} + \mathbf{F}_{gravity} \tag{2.64}$$

The mass matrix $\mathbf{M} \in \mathbb{R}^{7\times 7}$ is given by:

$$\mathbf{M} = \begin{bmatrix} m_s & 0 & 0 & 0 & 0 & 0 & 0 \\ 0 & I_{yy} & -I_{xy} & 0 & 0 & 0 & 0 \\ 0 & -I_{xy} & I_{xx} & 0 & 0 & 0 & 0 \\ 0 & 0 & 0 & m_{ufl} & 0 & 0 & 0 \\ 0 & 0 & 0 & 0 & m_{ufr} & 0 & 0 \\ 0 & 0 & 0 & 0 & 0 & m_{url} & 0 \\ 0 & 0 & 0 & 0 & 0 & 0 & m_{urr} \end{bmatrix} \tag{2.65}$$

where m_s is the total sprung mass, I_{xx} and I_{yy} are the roll and pitch moments of inertia about the center of gravity, I_{xy} represents the product of inertia, and m_{uij} are the unsprung masses at each corner.

The stiffness matrix $\mathbf{K} \in \mathbb{R}^{7\times 7}$ incorporates suspension springs, anti-roll bars, and tire vertical stiffnesses [9]:

$$\mathbf{K} = \begin{bmatrix} k_{11} & k_{12} & k_{13} & -k_{sfl} & -k_{sfr} & -k_{srl} & -k_{srr} \\ k_{21} & k_{22} & k_{23} & ak_{sfl} & ak_{sfr} & -bk_{srl} & -bk_{srr} \\ k_{31} & k_{32} & k_{33} & -\frac{t_f}{2}k_{sfl} & \frac{t_f}{2}k_{sfr} & -\frac{t_r}{2}k_{srl} & \frac{t_r}{2}k_{srr} \\ -k_{sfl} & ak_{sfl} & -\frac{t_f}{2}k_{sfl} & k_{sfl}+k_{tfl} & 0 & 0 & 0 \\ -k_{sfr} & ak_{sfr} & \frac{t_f}{2}k_{sfr} & 0 & k_{sfr}+k_{tfr} & 0 & 0 \\ -k_{srl} & -bk_{srl} & -\frac{t_r}{2}k_{srl} & 0 & 0 & k_{srl}+k_{trl} & 0 \\ -k_{srr} & -bk_{srr} & \frac{t_r}{2}k_{srr} & 0 & 0 & 0 & k_{srr}+k_{trr} \end{bmatrix} \tag{2.66}$$

where k_{sij} are the suspension spring rates, k_{tij} are the tire vertical stiffnesses, a and b represent the longitudinal distances from the center of gravity to the front and rear axles, t_f and t_r are the front and rear track widths, and the coupling terms k_{mn} are given by:

$$k_{11} = k_{sfl} + k_{sfr} + k_{srl} + k_{srr} \tag{2.67}$$

$$k_{12} = k_{21} = a(k_{sfl} + k_{sfr}) - b(k_{srl} + k_{srr}) \tag{2.68}$$

$$k_{13} = k_{31} = -\frac{t_f}{2}k_{sfl} + \frac{t_f}{2}k_{sfr} - \frac{t_r}{2}k_{srl} + \frac{t_r}{2}k_{srr} \tag{2.69}$$

$$k_{22} = a^2(k_{sfl} + k_{sfr}) + b^2(k_{srl} + k_{srr}) + k_{arb,f} + k_{arb,r} \tag{2.70}$$

$$k_{23} = k_{32} = -\frac{at_f}{2}k_{sfl} + \frac{at_f}{2}k_{sfr} + \frac{bt_r}{2}k_{srl} - \frac{bt_r}{2}k_{srr} \tag{2.71}$$

$$k_{33} = \frac{t_f^2}{4}(k_{sfl} + k_{sfr}) + \frac{t_r^2}{4}(k_{srl} + k_{srr}) + \frac{k_{arb,f}t_f^2}{2} + \frac{k_{arb,r}t_r^2}{2} \tag{2.72}$$

with $k_{arb,f}$ and $k_{arb,r}$ representing the front and rear anti-roll bar stiffnesses.

The damping matrix $\mathbf{C} \in \mathbb{R}^{7\times7}$ follows a similar structure to the stiffness matrix:

$$\mathbf{C} = \begin{bmatrix} c_{11} & c_{12} & c_{13} & -c_{sfl} & -c_{sfr} & -c_{srl} & -c_{srr} \\ c_{21} & c_{22} & c_{23} & ac_{sfl} & ac_{sfr} & -bc_{srl} & -bc_{srr} \\ c_{31} & c_{32} & c_{33} & -\frac{t_f}{2}c_{sfl} & \frac{t_f}{2}c_{sfr} & -\frac{t_r}{2}c_{srl} & \frac{t_r}{2}c_{srr} \\ -c_{sfl} & ac_{sfl} & -\frac{t_f}{2}c_{sfl} & c_{sfl} & 0 & 0 & 0 \\ -c_{sfr} & ac_{sfr} & \frac{t_f}{2}c_{sfr} & 0 & c_{sfr} & 0 & 0 \\ -c_{srl} & -bc_{srl} & -\frac{t_r}{2}c_{srl} & 0 & 0 & c_{srl} & 0 \\ -c_{srr} & -bc_{srr} & \frac{t_r}{2}c_{srr} & 0 & 0 & 0 & c_{srr} \end{bmatrix} \tag{2.73}$$

where c_{sij} are the suspension damping coefficients, and the terms c_{mn} are analogous to their stiffness counterparts k_{mn}.

The external force vector $\mathbf{F}_{ext}$ accounts for road inputs and aerodynamic forces:

$$\mathbf{F}_{ext} = [0,0,0,k_{tfl}z_{rfl},k_{tfr}z_{rfr},k_{trl}z_{rrl},k_{trr}z_{rrr}]^T + \mathbf{F}_{aero} \tag{2.74}$$

where z_{rij} represent the road elevation inputs at each wheel and $\mathbf{F}_{aero}$ captures aerodynamic downforce and moments.

The motor force vector $\mathbf{F}_{motor}$ specifically addresses the unique characteristics of in-wheel motor systems [15]:

$$\mathbf{F}_{motor} = [0,M_{pitch},M_{roll},F_{mfl},F_{mfr},F_{mrl},F_{mrr}]^T \tag{2.75}$$

where $M_{pitch} = (F_{mfl}+F_{mfr})a-(F_{mrl}+F_{mrr})b$ and $M_{roll} = (F_{mfr}+F_{mrr})\frac{t}{2} - (F_{mfl}+F_{mrl})\frac{t}{2}$ represent the pitch and roll moments generated by vertical force components from motor operation, and F_{mij} are the vertical force components at each wheel due to motor torque reaction, gyroscopic effects, and unsprung mass inertial forces.

The active suspension force vector $\mathbf{F}_{active}$ enables the implementation of advanced control strategies:

$$\mathbf{F}_{active} = [0,0,0,F_{afl},F_{afr},F_{arl},F_{arr}]^T \tag{2.76}$$

where F_{aij} represent the controllable forces generated by active suspension actuators at each corner.

The gravity force vector completes the system:

$$\mathbf{F}_{gravity} = [-m_s g,0,0,-m_{ufl}g,-m_{ufr}g,-m_{url}g,-m_{urr}g]^T \tag{2.77}$$

This comprehensive 7-DOF model enables detailed analysis of roll center behavior, dynamic wheel load variations during cornering and braking, the influence of unsprung mass distribution on handling characteristics, and the complex interactions between multiple vibration modes [14, 15]. For in-wheel motor applications, the model is particularly valuable for assessing the impact of significant unsprung mass increases on vehicle stability, wheel hop phenomena, and the coupling between vertical dynamics and road holding capability, especially during transient maneuvers on irregular road surfaces.

2.5.2 ELECTROMECHANICAL COUPLING

The integration of suspension and motor dynamics creates important coupling effects that must be captured in comprehensive models to accurately predict system behavior and ensure optimal performance.

2.5.2.1 Wheel Dynamics

The rotational dynamics of each wheel incorporates the complex interactions between propulsion, braking, and vertical forces, with particular significance for in-wheel systems where the motor is directly subject to suspension motions [20]:

$$I_w\dot{\omega} = T_m - T_b - F_x R - T_{loss} - T_{inertial} - T_{gyroscopic} \tag{2.78}$$

where I_w is the total rotational inertia of the wheel assembly (including tire, rim, brake rotor, and motor rotor), ω is the angular velocity, T_m is the motor electromagnetic torque, T_b is the brake torque, F_x is the longitudinal tire-road contact force, R is the effective rolling radius, T_{loss} accounts for various loss torques (bearing friction, windage, motor iron losses), $T_{inertial}$ represents torque components due to angular acceleration of the unsprung mass in vertical direction, and $T_{gyroscopic}$ captures gyroscopic moments that become significant at high rotational speeds. The motor torque T_m is itself a complex function of electrical inputs, rotor position, temperature, and magnetic saturation [10, 16]:

$$T_m = \frac{3}{2}\frac{P}{2}\left[\lambda_{pm} i_q + (L_d - L_q) i_d i_q\right] \eta_{torque}(T, \omega, i) \tag{2.79}$$

where P is the number of poles, λ_{pm} is the permanent magnet flux linkage, L_d and L_q are the d-axis and q-axis inductances, i_d and i_q are the d-axis and q-axis currents, and η_{torque} represents the torque efficiency factor accounting for temperature effects, magnetic saturation, and harmonic losses.

2.5.2.2 Tire Modeling

The tire forces represent the critical interface between the vehicle and road, with modeling accuracy profoundly influencing the predictive capability of the overall vehicle model. Semi-empirical approaches such as the Pacejka Magic Formula provide an effective balance between physical insight and empirical accuracy [9, 17]:

$$F_x = D_x \sin[C_x \arctan(B_x\kappa - E_x(B_x\kappa - \arctan(B_x\kappa)))] + S_{vx} \tag{2.80}$$

$$F_y = D_y \sin[C_y \arctan(B_y\alpha - E_y(B_y\alpha - \arctan(B_y\alpha)))] + S_{vy} \tag{2.81}$$

where κ is the longitudinal slip ratio, α is the slip angle, B, C, D, and E are parameters characterizing the shape of the force-slip curves, and S_v represents vertical shifts. For in-wheel motor applications, several important extensions to the basic Magic Formula are necessary: (1) inclusion of load sensitivity effects where peak friction coefficients decrease with increasing normal load, (2) incorporation of combined slip behavior where longitudinal and lateral forces interact nonlinearly, (3) modeling of transient tire dynamics using relaxation length concepts to capture the delay between slip inputs and force generation, and (4) accounting for the significant inertia of the in-wheel motor assembly which affects high-frequency force transmission. Additionally, the vertical force F_z exhibits dynamic variations due to suspension motions [14, 15]:

$$F_z = k_t(z_u - z_r) + c_t(\dot{z}_u - \dot{z}_r) + F_{z,static} + \Delta F_{z,inertia} + \Delta F_{z,motor} \tag{2.82}$$

where k_t and c_t are the tire vertical stiffness and damping, $F_{z,static}$ is the static load, $\Delta F_{z,inertia}$ represents load transfer due to longitudinal and lateral acceleration, and $\Delta F_{z,motor}$ accounts for vertical force components generated by the in-wheel motor.

2.5.3 MULTIPHYSICS MODELING FRAMEWORK

A complete multiphysics model integrates multiple physical domains into a unified simulation framework, capturing the complex interactions that characterize IWSS behavior [4, 15].

2.5.3.1 Electro-Thermal Coupling

The coupling between electrical and thermal domains occurs through temperature-dependent parameters that significantly influence motor performance and efficiency [18]. This bidirectional coupling requires simultaneous solution of electrical and thermal equations:

$$R(T) = R_0[1 + \alpha_{cu}(T - T_0) + \beta_{cu}(T - T_0)^2] \tag{2.83}$$

$$\lambda_{pm}(T) = \lambda_{pm0}[1 + \alpha_{pm}(T - T_0) + \beta_{pm}(T - T_0)^2] \tag{2.84}$$

$$B_{sat}(T) = B_{sat0}[1 + \alpha_{fe}(T - T_0)] \tag{2.85}$$

where R_0, λ_{pm0}, and B_{sat0} are reference values at temperature T_0, α_{cu} and β_{cu} are the linear and quadratic temperature coefficients for copper resistivity, α_{pm} and β_{pm} characterize the reversible and irreversible temperature dependence of permanent magnet flux, and α_{fe} represents the temperature dependence of iron saturation flux density. The thermal dynamics themselves are governed by the heat equation with distributed losses as source terms:

$$\rho c_p \frac{\partial T}{\partial t} = \nabla \cdot (k \nabla T) + q_{gen}(I, \omega, T) \tag{2.86}$$

where ρ is the material density, c_p is the specific heat capacity, k is the thermal conductivity, and q_{gen} represents the volumetric heat generation rate from copper losses, iron losses, magnet losses, and mechanical losses, all of which exhibit complex dependence on current I, speed ω, and temperature T [4].

2.5.3.2 Thermo-Mechanical Coupling

Thermal expansion affects clearances, preloads, and stress distributions throughout the in-wheel assembly, with particular significance for air gap maintenance in the motor and bearing longevity [4]:

$$\delta L = L_0 \left[\int_{T_0}^{T} \alpha_{th}(T) dT + \epsilon_{plastic} + \epsilon_{creep} \right] \tag{2.87}$$

where L_0 is the initial dimension at reference temperature T_0, $\alpha_{th}(T)$ is the temperature-dependent coefficient of thermal expansion, $\epsilon_{plastic}$ accounts for permanent deformation from thermal cycling, and ϵ_{creep} represents time-dependent deformation under sustained thermal loads. Thermal stresses develop due to constrained expansion and temperature gradients:

$$\sigma_{th} = E\alpha_{th}\Delta T + \frac{E}{1-\nu}\left[\alpha_{th}\Delta T - \frac{1}{V}\int_V \alpha_{th}\Delta T\, dV\right] + \sigma_{th,gradient} \tag{2.88}$$

where E is the Young's modulus, ν is Poisson's ratio, the first term represents stress from uniformly constrained expansion, the second term accounts for stress due to differential expansion in statically indeterminate structures, and $\sigma_{th,gradient}$ captures additional stresses arising from spatial temperature variations [7]. These thermomechanical effects are particularly critical at the wheel bearing interface, where thermal expansion can alter preload conditions and affect bearing life, and at the motor air gap, where differential expansion between stator and rotor can potentially cause rubbing contact under extreme thermal conditions.

2.5.3.3 Integrated System Model

The complete IWSS can be represented as a coupled set of nonlinear differential equations encompassing mechanical, electrical, thermal, and magnetic domains [4, 15]:

$$\begin{aligned}
\frac{d\mathbf{x}_{mech}}{dt} &= \mathbf{f}_{mech}(\mathbf{x}_{mech}, \mathbf{x}_{elec}, \mathbf{x}_{thermal}, \mathbf{x}_{mag}, \mathbf{u}, t)\\
\frac{d\mathbf{x}_{elec}}{dt} &= \mathbf{f}_{elec}(\mathbf{x}_{elec}, \mathbf{x}_{mech}, \mathbf{x}_{thermal}, \mathbf{x}_{mag}, \mathbf{u}, t)\\
\frac{d\mathbf{x}_{thermal}}{dt} &= \mathbf{f}_{thermal}(\mathbf{x}_{thermal}, \mathbf{x}_{mech}, \mathbf{x}_{elec}, \mathbf{x}_{mag}, \mathbf{u}, t)\\
\frac{d\mathbf{x}_{mag}}{dt} &= \mathbf{f}_{mag}(\mathbf{x}_{mag}, \mathbf{x}_{mech}, \mathbf{x}_{elec}, \mathbf{x}_{thermal}, \mathbf{u}, t)
\end{aligned} \tag{2.89}$$

where $\mathbf{x}_{mech}$ encompasses mechanical states (positions, velocities, accelerations), $\mathbf{x}_{elec}$ includes electrical states (currents, voltages, flux linkages), $\mathbf{x}_{thermal}$ contains thermal states (temperatures, heat fluxes), $\mathbf{x}_{mag}$ represents magnetic states (local flux densities, field intensities), and $\mathbf{u}$ comprises control inputs (inverter switching, active suspension forces, braking commands). The coupling functions $\mathbf{f}_{mech}$, $\mathbf{f}_{elec}$, $\mathbf{f}_{thermal}$, and $\mathbf{f}_{mag}$ capture the complex interdisciplinary interactions, including the influence of vibration on motor performance, the effect of temperature on material properties, the impact of magnetic forces on mechanical stresses, and the generation of heat from various loss mechanisms. This integrated modeling approach enables comprehensive system optimization, identification of potential failure modes, development of robust control strategies, and accurate prediction of real-world performance across the entire operating envelope [3, 23].

2.6 ADVANCED MODELING TECHNIQUES FOR IN-WHEEL SUSPENSION SYSTEM

Sophisticated modeling techniques are indispensable for accurately predicting the complex multiphysics behavior of IWSS and supporting comprehensive optimization efforts across multiple performance domains. These advanced computational approaches transcend traditional analytical methods by capturing nonlinearities, complex boundary conditions, and interdisciplinary coupling effects that are essential for realistic performance prediction [9]. The integration of finite element analysis, multibody dynamics, and control-oriented modeling frameworks enables a holistic understanding of system behavior from component-level stress concentrations to full-vehicle dynamic response, providing the foundation for robust design validation and performance optimization throughout the development cycle.

2.6.1 FINITE ELEMENT ANALYSIS

Finite element analysis provides detailed insights into structural, thermal, and electromagnetic behavior at high spatial resolution, enabling the prediction of local phenomena that cannot be captured by lumped-parameter approaches. This computational technique discretizes complex geometries into manageable subdomains, transforming continuous partial differential equations into systems of algebraic equations that can be solved numerically [7, 24].

2.6.1.1 Structural FEA

The structural finite element formulation for dynamic analysis follows the semi-discrete governing equations:

$$\mathbf{M}\ddot{\mathbf{u}} + \mathbf{C}\dot{\mathbf{u}} + \mathbf{K}\mathbf{u} = \mathbf{F}_{ext} + \mathbf{F}_{body} + \mathbf{F}_{contact} \tag{2.90}$$

where $\mathbf{M} \in \mathbb{R}^{n \times n}$ is the consistent mass matrix obtained through $\mathbf{M} = \int_V \rho \mathbf{N}^T \mathbf{N} dV$ with $\mathbf{N}$ representing the shape functions, $\mathbf{C} \in \mathbb{R}^{n \times n}$ is the proportional damping matrix typically formulated as $\mathbf{C} = \alpha \mathbf{M} + \beta \mathbf{K}$ according to Rayleigh damping assumptions, $\mathbf{K} \in \mathbb{R}^{n \times n}$ is the stiffness matrix computed via $\mathbf{K} = \int_V \mathbf{B}^T \mathbf{D} \mathbf{B} dV$ with $\mathbf{B}$ being the strain-displacement matrix and $\mathbf{D}$ the constitutive matrix, $\mathbf{u} \in \mathbb{R}^n$ is the nodal displacement vector, $\mathbf{F}_{ext}$ represents externally applied forces, $\mathbf{F}_{body}$ accounts for body forces including gravity and centrifugal effects, and $\mathbf{F}_{contact}$ handles complex contact interactions between components. For in-wheel suspension applications, structural FEA is particularly valuable for predicting stress concentrations in the hub carrier under combined loading conditions, analyzing fatigue life of suspension linkages, optimizing lightweight structures through topology optimization, and validating component designs against structural integrity requirements under extreme loading scenarios including pothole impacts and curb strikes [1].

2.6.1.2 Electromagnetic FEA

The electromagnetic analysis solves the full set of Maxwell's equations in their weak form, with the magnetic vector potential formulation providing an efficient approach for motor and actuator design [24]:

$$\nabla \times \left(\frac{1}{\mu(\mathbf{B})} \nabla \times \mathbf{A} \right) = \mathbf{J}_{source} - \sigma \frac{\partial \mathbf{A}}{\partial t} + \nabla \times \mathbf{M}_{pm} \tag{2.91}$$

where $\mathbf{A}$ is the magnetic vector potential satisfying $\mathbf{B} = \nabla \times \mathbf{A}$, $\mu(\mathbf{B})$ is the nonlinear magnetic permeability dependent on flux density $\mathbf{B}$, $\mathbf{J}_{source}$ represents the source current density in windings, σ is the electrical conductivity, and $\mathbf{M}_{pm}$ is the magnetization vector of permanent magnets.

The resulting discretized formulation becomes:

$$\mathbf{K}_A(\mathbf{A})\mathbf{A} + \mathbf{C}_A\dot{\mathbf{A}} = \mathbf{F}_A \tag{2.92}$$

where $\mathbf{K}_A$ is the nonlinear stiffness matrix, $\mathbf{C}_A$ is the damping matrix accounting for eddy currents, and $\mathbf{F}_A$ contains source terms. Electromagnetic FEA enables precise prediction of motor performance parameters including torque production, cogging torque, back-EMF waveform quality, core losses, and magnet eddy current losses. For in-wheel applications, this analysis is crucial for optimizing motor geometry to achieve target torque-speed characteristics while minimizing radial forces that could excite structural vibrations and acoustic noise, particularly important given the direct mounting of the motor within the wheel assembly [15].

2.6.1.3 Thermal FEA

The thermal analysis solves the transient heat conduction equation with appropriate boundary conditions to predict temperature distributions throughout the system [4]:

$$\rho c_p \frac{\partial T}{\partial t} = \nabla \cdot (k \nabla T) + \dot{q}_{gen} + \dot{q}_{contact} + \dot{q}_{phase} \tag{2.93}$$

where ρ is the material density, c_p is the temperature-dependent specific heat capacity, k is the anisotropic thermal conductivity tensor, $\dot{q}_{gen}$ represents volumetric heat generation from electromagnetic losses, mechanical friction, and hysteresis, $\dot{q}_{contact}$ accounts for heat generation at contacting interfaces, and $\dot{q}_{phase}$ captures latent heat effects in phase change materials.

The finite element discretization yields:

$$\mathbf{C}_T\dot{\mathbf{T}} + \mathbf{K}_T\mathbf{T} = \mathbf{Q}_{gen} + \mathbf{Q}_{boundary} \tag{2.94}$$

where $\mathbf{C}_T$ is the thermal capacitance matrix, $\mathbf{K}_T$ is the thermal conductance matrix, $\mathbf{T}$ is the nodal temperature vector, $\mathbf{Q}_{gen}$ contains heat generation terms, and $\mathbf{Q}_{boundary}$ represents boundary heat fluxes including convection $\mathbf{Q}_{conv} = \mathbf{H}_{conv}(\mathbf{T} - T_\infty)$ and radiation $\mathbf{Q}_{rad} = \mathbf{H}_{rad}(\mathbf{T}^4 - T_{sur}^4)$. Thermal FEA is essential for predicting hot spot temperatures in motor windings and power electronics, optimizing cooling system designs, analyzing thermal stress distributions, and ensuring component reliability under worst-case thermal loading scenarios, particularly critical in the thermally challenging environment of an in-wheel motor where heat rejection pathways are severely constrained [4].

2.6.2 MULTIBODY DYNAMICS

Multibody dynamics simulations capture the complex kinematic and dynamic behavior of suspension systems through a rigorous mathematical framework that combines rigid body dynamics with kinematic constraints [9, 19]. The general formulation for a constrained multibody system is given by the differential-algebraic equations:

$$\mathbf{M}(\mathbf{q})\ddot{\mathbf{q}} + \mathbf{C}(\mathbf{q}, \dot{\mathbf{q}}) + \mathbf{K}(\mathbf{q}) = \mathbf{Q}_{ext} + \mathbf{Q}_{gyro} + \mathbf{Q}_{coriolis} + \mathbf{J}^T \lambda \tag{2.95}$$

subject to the holonomic and nonholonomic constraint equations:

$$\mathbf{\Phi}(\mathbf{q}) = \mathbf{0}, \quad \mathbf{\Psi}(\mathbf{q}, \dot{\mathbf{q}}) = \mathbf{0} \tag{2.96}$$

where $\mathbf{M}(\mathbf{q}) \in \mathbb{R}^{n \times n}$ is the configuration-dependent mass matrix, $\mathbf{q} \in \mathbb{R}^n$ is the generalized coordinate vector, $\mathbf{C}(\mathbf{q}, \dot{\mathbf{q}})$ and $\mathbf{K}(\mathbf{q})$ represent velocity-dependent and position-dependent forces, $\mathbf{Q}_{ext}$ contains externally applied forces and torques, $\mathbf{Q}_{gyro}$ and $\mathbf{Q}_{coriolis}$ account for gyroscopic and Coriolis effects in rotating components, $\mathbf{J} = \partial\mathbf{\Phi}/\partial\mathbf{q}$ is the constraint Jacobian matrix, and λ represents the Lagrange multipliers corresponding to constraint forces. For suspension system analysis, the constraint equations $\mathbf{\Phi}(\mathbf{q}) = \mathbf{0}$ precisely define the kinematic joints (spherical, universal, revolute) and mechanical linkages that govern the suspension's motion characteristics. Multibody dynamics enables comprehensive analysis of suspension kinematics including camber and toe variation throughout wheel travel, anti-dive and anti-squat characteristics, roll center migration, and the interaction between suspension compliance and vehicle handling dynamics. For in-wheel motor applications, this approach is particularly valuable for assessing the dynamic loading on motor mounts, predicting the effect of increased unsprung mass on suspension behavior, and optimizing suspension geometry to compensate for the unique inertial properties introduced by the integrated motor assembly [15].

2.6.3 CONTROL-ORIENTED MODELING

Control-oriented models strategically balance physical accuracy with computational efficiency to support the design and validation of real-time control algorithms for active suspension, torque vectoring, and vehicle stability systems. These models abstract the essential dynamics while preserving the key characteristics necessary for effective controller synthesis and performance evaluation [23].

2.6.3.1 Linear Parameter Varying (LPV) Models

LPV models capture system nonlinearities and time-varying characteristics through explicit parameter dependence, providing a powerful framework for gain-scheduled controller design [23]:

$$\dot{\mathbf{x}} = \mathbf{A}(\theta(t))\mathbf{x} + \mathbf{B}(\theta(t))\mathbf{u} + \mathbf{B}_d(\theta(t))\mathbf{d} \tag{2.97}$$

$$\mathbf{y} = \mathbf{C}(\theta(t))\mathbf{x} + \mathbf{D}(\theta(t))\mathbf{u} \tag{2.98}$$

where $\mathbf{x} \in \mathbb{R}^n$ is the state vector, $\mathbf{u} \in \mathbb{R}^m$ is the control input vector, $\mathbf{y} \in \mathbb{R}^p$ is the output vector, $\mathbf{d} \in \mathbb{R}^q$ represents disturbance inputs, and $\theta(t) \in \mathbb{R}^r$ is the time-varying parameter vector that captures operating condition dependencies such as vehicle speed, payload mass, road surface conditions, and tire characteristics. The parameter dependence is typically affine:

$$\mathbf{A}(\theta) = \mathbf{A}_0 + \theta_1\mathbf{A}_1 + \cdots + \theta_r\mathbf{A}_r \tag{2.99}$$

with similar expansions for the other system matrices. For in-wheel suspension control, LPV models enable the design of controllers that automatically adapt to changing operating conditions, such as adjusting active suspension damping based on vehicle speed and road roughness, or modifying torque vectoring strategies according to tire-road friction conditions and vehicle loading state.

2.6.3.2 Reduced-Order Models

Model order reduction techniques create compact representations of high-dimensional systems while preserving input-output behavior for specific frequency ranges or operating conditions [7]:

$$\mathbf{x} \approx \mathbf{V}\mathbf{z} \quad \text{where} \quad \dim(\mathbf{z}) \ll \dim(\mathbf{x}) \tag{2.100}$$

The reduced-order system becomes [23]:

$$\dot{\mathbf{z}} = \mathbf{V}^T\mathbf{A}\mathbf{V}\mathbf{z} + \mathbf{V}^T\mathbf{B}\mathbf{u}, \quad \mathbf{y}_r = \mathbf{C}\mathbf{V}\mathbf{z} \tag{2.101}$$

where $\mathbf{V} \in \mathbb{R}^{n\times r}$ is the projection matrix obtained through methods such as Proper Orthogonal Decomposition (POD), Balanced Truncation, or Krylov subspace techniques. For large-scale finite element models of in-wheel systems, model reduction enables real-time simulation for hardware-in-the-loop testing and efficient controller implementation while maintaining predictive accuracy in the frequency ranges relevant to vehicle dynamics and control system bandwidth [3, 23]. Specific applications include creating efficient real-time models of the coupled electro-thermal dynamics for model predictive control of motor temperature, developing compact representations of flexible body dynamics for active vibration control, and generating fast-running models for parameter estimation and fault detection algorithms in vehicle stability systems.

2.7 CONCLUSION

The successful development and implementation of IWSS represents a paradigm shift in automotive engineering, demanding an integrated, multidisciplinary design methodology that simultaneously addresses the complex interplay of mechanical, electromagnetic, thermal, and control challenges. This chapter has systematically presented a comprehensive framework for understanding, analyzing, and optimizing these sophisticated systems, establishing an architectural blueprint that balances competing performance objectives while respecting the severe constraints imposed by the wheel envelope. The journey through the various subsystems—from suspension kinematics and structural design to electromagnetic integration and thermal management—has illuminated the intricate web of dependencies that characterize these integrated mechatronic assemblies.

The architectural design of in-wheel suspensions involves fundamental and often competing trade-offs between kinematic performance, packaging efficiency, and mass optimization that must be carefully balanced for each specific application. The comparative analysis of suspension topologies—from the kinematic precision of double wishbone designs to the packaging efficiency of MacPherson struts and the tunable compliance of multi-link systems—reveals that no single architecture dominates all performance metrics. The selection process must therefore be driven by application-specific priorities, whether they favor uncompromised handling precision, maximized space for motor and brake components, or optimal vibration isolation. Furthermore, the material selection for critical structural components like the hub carrier presents another layer of optimization, where the high strength-to-weight ratio of forged aluminum or composite materials must be weighed against the cost-effectiveness of cast iron or the extreme performance of titanium alloys, all while ensuring structural integrity under complex multiaxial loading conditions.

Electromagnetic integration requires sophisticated multi-objective optimization to achieve the high torque density demanded by traction applications while simultaneously managing substantial thermal loads and stringent structural requirements. The analysis of motor topologies, particularly the comparative benefits of radial flux versus axial flux configurations, demonstrates that the optimal electromagnetic design is profoundly influenced by the available packaging space and the specific torque-speed requirements of the application. The selection of permanent magnet materials, especially the high-energy-density NdFeB magnets, necessitates careful consideration of their temperature sensitivity and the implementation of coercivity-enhancing strategies to prevent demagnetization under extreme operating conditions. The winding design must balance electromagnetic performance against manufacturability constraints, while the comprehensive loss modeling must accurately capture the complex interdependencies between electrical, magnetic, and thermal phenomena to enable predictive performance simulation across the entire operating envelope.

Thermal management emerges not merely as an ancillary consideration but as a critical enabling technology, with overall system performance often limited by heat rejection capability rather than fundamental electromagnetic design. The concentrated heat generation within the thermally challenging environment of the wheel assembly demands innovative cooling strategies, ranging from advanced liquid cooling jackets with optimized flow passages to phase-change cooling systems that leverage latent heat effects for handling transient thermal loads. The thermal interfaces between components frequently constitute the dominant thermal resistance in the system, necessitating the application of advanced thermal interface materials with enhanced conductivity and reliable performance under thermal cycling conditions. The successful thermal management strategy must therefore adopt a system-level perspective, integrating heat generation, conduction, convection, and rejection into a cohesive framework that ensures reliable operation under the most demanding conditions.

Comprehensive mathematical modeling provides essential insights for system optimization and control design, but must carefully balance physical accuracy with computational tractability to remain practical for engineering development. The hierarchical modeling approach—progressing from simplified quarter-car models for initial control design to sophisticated full-vehicle models for handling analysis and high-fidelity finite element models for component stress prediction—enables targeted analysis at appropriate levels of detail throughout the development process. The multiphysics modeling framework that captures the electro-thermal-mechanical coupling effects is particularly crucial for in-wheel systems, where the traditional boundaries between engineering domains become blurred. Advanced modeling techniques, including linear parameter-varying formulations for control design and model order reduction for real-time implementation, provide the mathematical foundation for developing sophisticated control algorithms that can fully exploit the unique capabilities of individual wheel control.

The inherently multiphysics nature of in-wheel systems necessitates integrated design approaches that consider interactions between traditionally separate engineering domains from the earliest conceptual stages. The historical practice of sequential design—where mechanical, electrical, and software components are developed independently and integrated later—proves inadequate for these tightly coupled systems. Instead, a concurrent engineering methodology that embraces the fundamental interdependencies between suspension dynamics, electromagnetic performance, thermal behavior, and control system functionality is essential for achieving optimal system-level performance. This integrated perspective reveals synergistic opportunities, such as using the motor as a dynamic vibration absorber or leveraging suspension kinematics to enhance cooling airflow, that would remain hidden within traditional disciplinary silos.

Looking forward, the future development of in-wheel suspension technology will likely focus on several key research and innovation vectors. Further miniaturization through advanced materials and additive manufacturing techniques will continue to push the boundaries of power and torque density within the constrained wheel volume. Improved thermal management strategies, potentially incorporating micro-channel cooling, advanced phase-change materials, and hybrid cooling architectures, will enable sustained operation at higher power levels. Enhanced integration through mechatronic co-design methodologies will further blur the lines between mechanical and electrical domains, creating truly optimized systems rather than merely connected components. Sophisticated control algorithms, potentially incorporating artificial intelligence and machine learning techniques, will fully exploit the capabilities of individual wheel control to achieve unprecedented levels of vehicle stability, safety, and performance. As these technologies mature and production volumes increase, resulting in improved economies of scale and manufacturing efficiency, IWSS are poised to play an increasingly important role in enabling next-generation vehicle architectures with superior performance, energy efficiency, and design flexibility. The comprehensive architectural blueprint and analytical framework presented in this chapter provide the essential foundation for these future developments, establishing both the fundamental principles and practical methodologies necessary to navigate the complex trade-offs and interdisciplinary challenges that define this exciting frontier in automotive engineering.

With the architectural and multiphysics model now established, the discussion advances toward the analytical formulation of system dynamics and control. Chapter 3 develops the theoretical underpinnings that connect mechanical, electromagnetic, and control domains into a unified mathematical structure.

PROBLEMS

Problem 1: A new electric vehicle platform is being designed with in-wheel motors. Compare and contrast the double wishbone and MacPherson strut topologies for this application.

(a) Identify two key advantages of the double wishbone configuration for in-wheel implementation.
(b) Identify two key advantages of the MacPherson strut configuration for in-wheel implementation.
(c) For a high-performance sports EV, which topology would you recommend, and why?

Problem 2: For a double wishbone in-wheel suspension with the following parameters:

- Upper wishbone length: 0.25 m
- Lower wishbone length: 0.30 m
- Wheel travel: ±0.08 m
- Initial camber angle: -0.5°

(a) Derive the expression for camber change as a function of wheel travel.
(b) Calculate the camber angle at maximum compression and maximum rebound.
(c) Explain how this camber gain affects the tire contact patch during cornering.

Problem 3: An IWSS has the following mass components:

- Conventional unsprung mass: 42 kg
- Additional motor mass: 18 kg
- Additional suspension mechanism mass: 8 kg
- Mass reduction from removed components: 12 kg

(a) Calculate the net change in unsprung mass.
(b) The tire stiffness is 220,000 N/m. Calculate the change in unsprung mass natural frequency.
(c) Propose two strategies to mitigate the negative effects of increased unsprung mass.

Problem 4: A permanent magnet synchronous hub motor has the following parameters:

- Number of poles: 8
- Permanent magnet flux linkage: 0.12 Wb
- d-axis inductance: 0.8 mH
- q-axis inductance: 1.2 mH
- Maximum phase current: 250 A

(a) Calculate the maximum electromagnetic torque using the PMSM torque equation.
(b) If the motor operates at 1500 RPM, calculate the mechanical power output.
(c) Explain why outer rotor configurations are preferred for in-wheel applications.

Problem 5: A hub motor generates 3.5 kW of losses during continuous operation. The cooling system uses a liquid coolant with:

- Flow rate: 2.5 L/min
- Specific heat capacity: 3500 J/kg·K
- Density: 1050 kg/m^3
- Inlet temperature: 65°C

(a) Calculate the temperature rise of the coolant.
(b) If the maximum allowable coolant temperature is 85°C, what is the maximum allowable loss generation?
(c) Propose two methods to improve the heat transfer efficiency.

Problem 6: Consider the quarter-car model with the following parameters:

- Sprung mass (m_s): 320 kg
- Unsprung mass (m_u): 45 kg
- Spring stiffness (k_s): 25,000 N/m
- Damper coefficient (c_s): 1800 N·s/m
- Tire stiffness (k_t): 210,000 N/m

(a) Derive the state-space representation matrices A, B, and E.
(b) Calculate the natural frequencies of the sprung and unsprung masses.
(c) Explain the physical significance of each state variable in your model.

Problem 7: Compare three candidate materials for a hub carrier application:

- Cast Aluminum: $E = 72$ GPa, $\rho = 2700$ kg/m^3, $\sigma_y = 280$ MPa, $Cost factor = 2.0$

- Forged Aluminum: $E = 72$ GPa, $\rho = 2700$ kg/m^3, $\sigma_y = 380$ MPa, Cost factor = 3.0
- Titanium Alloy: $E = 115$ GPa, $\rho = 4500$ kg/m^3, $\sigma_y = 950$ MPa, Cost factor = 12.0

(a) Calculate the specific stiffness (E/ρ) for each material.
(b) Calculate the specific strength (σ_y/ρ) for each material.
(c) Justify which material you would select for a high-performance application and why.

Problem 8: A hub motor operates with the following loss components:

- Copper losses: 1200 W
- Iron losses: 650 W
- Magnet losses: 180 W
- Stray losses: 120 W
- Windage losses: 80 W

The motor has an active volume of 0.008 m^3 and operates for 30 minutes at peak load.

(a) Calculate the total power loss and heat generation density.
(b) If the motor efficiency is 92% at this operating point, what is the output power?
(c) Explain how temperature affects copper losses and the implications for thermal design.

Problem 9: A semi-active in-wheel suspension uses magnetorheological dampers. The damper force can be modeled as:

$$F_d = c_1 v + c_2 v^2 + F_{MR}$$

where F_{MR} is the controllable magnetorheological force component.

(a) Derive the equations of motion for the quarter-car model including this damper model.
(b) Design a skyhook control strategy for this system and write the control law.
(c) Explain how individual wheel control can improve vehicle stability during cornering.

Problem 10: A hub carrier must withstand the following extreme loads:

- Vertical load (bump): 5.5 g
- Lateral load (cornering): 1.8 g
- Braking load: 1.2 g
- Vehicle mass per corner: 450 kg

(a) Calculate the magnitude of each force component.
(b) Using superposition, estimate the worst-case combined loading scenario.
(c) Describe the finite element analysis process you would use to validate the hub carrier design.

Problem 11: A hub motor requires cooling for 4.2 kW of losses. The available space allows for a cooling jacket with:

- Heat transfer area: 0.15 m^2
- Coolant convection coefficient: 850 $W/m^2 \cdot K$
- Maximum allowable temperature difference: 35°C

(a) Calculate the maximum heat transfer capability of this cooling system.
(b) Is the cooling system adequate? If not, propose two design modifications.
(c) Calculate the required coolant flow rate if the allowable temperature rise is 15°C ($c_p = 3500$ J/kg·K, $\rho = 1050$ kg/m^3).

Problem 12: In an in-wheel system, several coupling effects occur between physical domains:

(a) Explain the electro-thermal coupling in the motor and how it creates a positive feedback loop.
(b) Describe the thermo-mechanical coupling in the bearing system and its effect on clearance.
(c) Discuss how road-induced vibrations affect motor performance and lifespan.

Problem 13: An in-wheel system has the following dynamic properties:

- Sprung mass natural frequency: 1.4 Hz
- Unsprung mass natural frequency: 12.5 Hz
- Motor rotational speed range: 0–2000 RPM

(a) Convert the motor speed range to excitation frequency in Hz.
(b) Identify any potential resonance conditions and their implications.
(c) Propose two methods to avoid resonance problems in the system design.

Problem 14: Compare two hub motor designs for the same torque requirement:

- Design A: Outer rotor, 18 kg, Torque = 420 N·m, Volume = 0.012 m^3
- Design B: Inner rotor, 15 kg, Torque = 380 N·m, Volume = 0.010 m^3

(a) Calculate the torque density (N·m/kg) for each design.
(b) Calculate the power density (kW/L) assuming both operate at 1500 RPM.
(c) Considering the overall system integration, which design is more suitable for in-wheel application, and why?

Problem 15: Designing an in-wheel suspension requires balancing multiple competing requirements:

(a) Create a requirements trade-off matrix considering unsprung mass, packaging volume, thermal performance, cost, and reliability.
(b) Propose an integrated design methodology that addresses mechanical, electromagnetic, and thermal domains simultaneously.
(c) How would you validate your design using the modeling techniques discussed in this chapter?

REFERENCES

1. Park, Hoonmin, Reza Langari, and Hak Yi. "Design and testing of double-wishbone suspension for enhanced outdoor maneuver stability of a six-wheeled mobile robot." Mechatronics 103 (2024): 103237.
2. Upadhyay, Pranav, et al. "Design and analysis of double wishbone suspension system." IOP Conference Series: Materials Science and Engineering. Vol. 748. No. 1. IOP Publishing, 2020.
3. Haseen, S. Fahira, P. Lakshmi, and S. Madhupreetha. "Efficient Vibration Control: Model Predictive Controller Design for In-Wheel Motor-Driven Electric Vehicle Suspension Systems." JOURNAL OF VIBRATION ENGINEERING & TECHNOLOGIES 13.7 (2025).
4. Fakhfakh, Mohamed Amine, et al. "Thermal analysis of a permanent magnet synchronous motor for electric vehicles." Journal of Asian Electric Vehicles 6.2 (2008): 1145-1151.
5. Liu, Chunhua, Christopher HT Lee, and Mu Chen. "Comparison of outer-rotor permanent magnet machines for in-wheel drives." 2013 IEEE International Symposium on Industrial Electronics. IEEE, 2013.
6. Termini, Luca. Modeling and simulation of an hydraulic interconnected suspension: analysis of vertical and roll characteristic. Diss. Politecnico di Torino, 2022.
7. Vasco, Joel C. "Additive manufacturing for the automotive industry." Additive manufacturing. Elsevier, 2021. 505-530.
8. Shao, Lingyun, et al. "Design and construction of axial-flux permanent magnet motors for electric propulsion applications—A review." Ieee Access 9 (2021): 158998-159017.
9. Pacejka, Hans B. "Tire and vehicle dynamics." Elsevier, 2005.
10. Zhou, Chenxi, et al. "Design consideration of fractional slot concentrated winding interior permanent magnet synchronous motor for EV and HEV applications." IEEE Access 9 (2021): 64116-64126.
11. "Front SLA ($2pcLCA$)." MotionView Documentation, Altair Engineering, Inc., https://help.altair.com/hwdesktop/hwx/topics/motionview/front_sla_2pc_lca_r.htm. Accessed 24 Sep. 2025.
12. Bucchi, Francesco, and Basilio Lenzo. "Analytical derivation and analysis of vertical and lateral installation ratios for swing axle, McPherson and double wishbone suspension architectures." Actuators. Vol. 11. No. 8. MDPI, 2022.
13. Tokgoz, Furkan, et al. "Feasibility Analysis of Multiphase Machines for Electric Vehicle Applications." IEEE Transactions on Industry Applications (2025).
14. Gillespie, Thomas D. "Fundamentals of vehicle dynamics." Society of Automotive Engineers, 1992.
15. Hu, Yiming, et al. "Analysis and suppression of in-wheel motor electromagnetic excitation of IWM-EV." Proceedings of the Institution of Mechanical Engineers, Part D: Journal of Automobile Engineering 235.6 (2021): 1552-1572.
16. Elgowainy, Amgad, ed. Electric, hybrid, and fuel cell vehicles. New York, NY, USA: Springer, 2021.
17. Wong, J. Y. "Theory of ground vehicles." John Wiley & Sons, 2008.
18. Raja, Selvakumar, et al. Thermal analysis of an electric Motor in an electric vehicle. No. 2023-01-0532. SAE Technical Paper, 2023.
19. Milliken, William F., and Douglas L. Milliken. "Race car vehicle dynamics." Society of Automotive Engineers, 1995.

20. Hasanien, Hany M. "Torque ripple minimization of permanent magnet synchronous motor using digital observer controller." Energy Conversion and Management 51.1 (2010): 98-104.
21. Bastow, Donald, Geoffrey Howard, and John P. Whitehead. Car suspension and handling. Warrendale: SAE international, 2004.
22. Côté, Olivier, Ahmed Chebak, and Jean-François Méthot. "Design and Optimization of a high torque in-wheel surface-mounted PM synchronous motor using concentrated winding." 2013 International Electric Machines & Drives Conference. IEEE, 2013.
23. Rajamani, Rajesh. Vehicle dynamics and control. Boston, MA: Springer US, 2006.
24. Hendershot, James R., and Timothy John Eastham Miller. Design of brushless permanent-magnet motors. Oxford university press, 1995.
25. Reimpell, Jornsen, Helmut Stoll, and Jurgen Betzler. The automotive chassis: engineering principles. Elsevier, 2001.
26. "Suspension Kinematics and Compliance (K&C) Tester." *Universal Mechanism Plugins Catalogue.* Universal Mechanism, https://www. universalmechanism.com/en/plugins/catalogue/index.php?id=2. Accessed 24 Dec. 2025.
27. Jahns, Thomas M., and Wen L. Soong. "Pulsating torque minimization techniques for permanent magnet AC motor drives-a review." IEEE transactions on industrial electronics 43.2 (2002): 321-330.
28. Gaeid, Khalaf S., Takialddin Al Smadi, and Ukashatu Abubakar. "Double control strategy of PMSM rotor speed-based traction drive using resolver." Results in Control and Optimization 13 (2023): 100301.
29. Chen, Rutong, Jinhui Jiang, and Ke Wang. "A semi-active vibration isolation system with adaptive anti-resonance control for enhanced performance." Journal of Vibration Engineering & Technologies 13.3 (2025): 1-15.

3 Theoretical Foundations for the Multiphysics Design and Control of Active In-Wheel Suspension System

3.1 INTRODUCTION: THE MATHEMATICAL IMPERATIVE FOR SYSTEM SYNTHESIS

The IWSS constitutes a tightly integrated mechatronic module where performance is dictated by the complex, nonlinear coupling of multiple physical domains. Its design and control cannot be approached through compartmentalized engineering disciplines or iterative prototyping; it demands a first-principles, mathematically rigorous framework from the outset [1, 13]. This chapter establishes such a framework, presenting a comprehensive theoretical treatment that spans from the fundamental laws of electromagnetism and continuum mechanics to the pinnacle of modern multivariable control theory [14, 16].

The core challenge is one of *system synthesis under conflicting constraints*: achieving high torque density and bandwidth while minimizing unsprung mass and volume, and simultaneously ensuring ride comfort, road holding, and vehicle stability. We address this challenge through a dual-pronged theoretical exploration [4]:

1. **Multiphysics System Design:** Developing a coupled-field formulation for the hub motor, integrating electromagnetic, thermal, and structural mechanics via the Finite Element Method (FEM), and employing mathematical optimization to derive Pareto-optimal geometries [12, 23].
2. **Hierarchical Control Synthesis:** Formulating the suspension control problem within a state-space framework and deriving optimal, robust, and adaptive control laws that explicitly manage the performance trade-offs inherent in vehicle dynamics [7, 14].

This chapter will not merely present equations but will derive them from first principles, illustrating the logical progression from physical laws to implementable computational models and control algorithms [2, 22]. This depth is necessary to provide engineers with the foundational knowledge required to innovate, rather than just apply, in this complex design space.

DOI: 10.1201/9781003780533-3

3.2 FIRST-PRINCIPLES AND COMPUTATIONAL PHYSICS FOR HUB MOTOR DESIGN

The hub motor is a quintessential multiphysics device. Its performance is governed by the simultaneous interaction of electromagnetic fields, heat transfer, and structural stresses. A superficial treatment is insufficient; we must begin with the most general governing equations and systematically derive the simplified models used in practice [1, 3].

3.2.1 FOUNDATIONAL ELECTROMAGNETIC FIELD THEORY

The behavior of electromagnetic fields is universally described by Maxwell's equations. In their differential form, they are [1]:

$$\begin{aligned}
&(\text{G1}) \quad \nabla \cdot \mathbf{D} = \rho_v \\
&(\text{G2}) \quad \nabla \cdot \mathbf{B} = 0 \\
&(\text{G3}) \quad \nabla \times \mathbf{E} = -\frac{\partial \mathbf{B}}{\partial t} \\
&(\text{G4}) \quad \nabla \times \mathbf{H} = \mathbf{J} + \frac{\partial \mathbf{D}}{\partial t}
\end{aligned} \tag{3.1}$$

where:

- $\mathbf{D}$ is the electric flux density (C/m^2)
- ρ_v is the free electric charge density (C/m^3)
- $\mathbf{B}$ is the magnetic flux density (T or Wb/m^2)
- $\mathbf{E}$ is the electric field intensity (V/m)
- $\mathbf{H}$ is the magnetic field intensity (A/m)
- $\mathbf{J}$ is the electric current density (A/m^2)

For low-frequency applications like motor design (typically ¡ 1 kHz), the displacement current term $\frac{\partial \mathbf{D}}{\partial t}$ in (G4) is negligible compared to the conduction current $\mathbf{J}$. This is the *magnetoquasistatic* approximation [3]. We also introduce the constitutive relations for linear, isotropic materials:

$$\mathbf{D} = \epsilon \mathbf{E}, \quad \mathbf{B} = \mu \mathbf{H}, \quad \mathbf{J} = \sigma \mathbf{E}$$

where ϵ is permittivity, μ is permeability, and σ is conductivity.

From (G2), we know that $\mathbf{B}$ is solenoidal (divergence-free). This allows us to define the *magnetic vector potential* $\mathbf{A}$ such that:

$$\mathbf{B} = \nabla \times \mathbf{A}$$

Substituting this into (G3) gives:

$$\nabla \times \mathbf{E} = -\frac{\partial}{\partial t}(\nabla \times \mathbf{A}) \Rightarrow \nabla \times \left(\mathbf{E} + \frac{\partial \mathbf{A}}{\partial t}\right) = 0$$

A vector whose curl is zero can be expressed as the gradient of a scalar. Thus, we define the *electric scalar potential* V:

$$\mathbf{E}+\frac{\partial \mathbf{A}}{\partial t}=-\nabla V \Rightarrow \mathbf{E}=-\nabla V-\frac{\partial \mathbf{A}}{\partial t}$$

Now, substituting $\mathbf{B}=\nabla\times\mathbf{A}$ and $\mathbf{H}=\frac{1}{\mu}\mathbf{B}=\frac{1}{\mu}\nabla\times\mathbf{A}$ into the simplified Ampère's Law (G4) (neglecting $\frac{\partial \mathbf{D}}{\partial t}$), and using $\mathbf{J}=\sigma\mathbf{E}=\sigma(-\nabla V-\frac{\partial \mathbf{A}}{\partial t})$, we arrive at the fundamental PDE for the A-V formulation [1, 5]:

$$\nabla\times\left(\frac{1}{\mu}\nabla\times\mathbf{A}\right)+\sigma\frac{\partial \mathbf{A}}{\partial t}+\sigma\nabla V=0 \tag{3.2}$$

In regions with prescribed source current density $\mathbf{J}_s$ (e.g., stator windings), the current is often known a priori, so the equation becomes:

$$\nabla\times\left(\frac{1}{\mu}\nabla\times\mathbf{A}\right)=\mathbf{J}_s$$

For time-harmonic problems, assuming $e^{j\omega t}$ time dependence, Equation 3.2 simplifies to [3]:

$$\nabla\times\left(\frac{1}{\mu}\nabla\times\mathbf{A}\right)+j\omega\sigma\mathbf{A}=\mathbf{J}_s \tag{3.3}$$

This is the cornerstone equation for frequency-domain FEM analysis of electric machines [1].

3.2.2 THE FINITE ELEMENT METHOD FOR ELECTROMAGNETICS: A DETAILED DERIVATION

The FEM solves Equation 3.3 numerically by discretizing the domain Ω and approximating the solution [2, 13]. We use the *weighted residuals* method, specifically the Galerkin approach. The *residual* $\mathbf{R}$ of Equation 3.3 is:

$$\mathbf{R}=\nabla\times\left(\frac{1}{\mu}\nabla\times\mathbf{A}\right)+j\omega\sigma\mathbf{A}-\mathbf{J}_s \tag{3.4}$$

We require the weighted average of this residual over the domain to be zero, using vector weighting functions $\mathbf{W}_i$ (which are often the same as the shape functions $\mathbf{N}_i$):

$$\int_\Omega \mathbf{W}_i\cdot\mathbf{R}\,d\Omega=0 \quad \text{for all } i \tag{3.5}$$

$$\Rightarrow \int_\Omega \mathbf{W}_i\cdot\left[\nabla\times\left(\frac{1}{\mu}\nabla\times\mathbf{A}\right)\right]d\Omega+\int_\Omega \mathbf{W}_i\cdot(j\omega\sigma\mathbf{A})d\Omega=\int_\Omega \mathbf{W}_i\cdot\mathbf{J}_s d\Omega$$

The first term contains a second-order derivative. We can reduce the continuity requirement on $\mathbf{A}$ by applying a vector identity (Green's first identity for the curl operator) [5]:

$$\int_{\Omega} \mathbf{W}_i \cdot \left[\nabla\times\left(\frac{1}{\mu}\nabla\times\mathbf{A}\right)\right] d\Omega = \int_{\Omega} [\nabla\times\mathbf{W}_i]\cdot\left[\frac{1}{\mu}\nabla\times\mathbf{A}\right] d\Omega - \oint_{\Gamma} \mathbf{W}_i \cdot \left[\mathbf{n}\times\left(\frac{1}{\mu}\nabla\times\mathbf{A}\right)\right] d\Gamma \tag{3.6}$$

The surface integral term vanishes for properly chosen boundary conditions (e.g., magnetic symmetry, Dirichlet $\mathbf{n}\times\mathbf{A} = 0$). This gives the *weak form* [1, 13]:

$$\int_{\Omega} \left(\frac{1}{\mu}(\nabla\times\mathbf{W}_i)\cdot(\nabla\times\mathbf{A}) + j\omega\sigma\mathbf{W}_i\cdot\mathbf{A}\right) d\Omega = \int_{\Omega} \mathbf{W}_i \cdot \mathbf{J}_s d\Omega \tag{3.7}$$

We now discretize the domain into finite elements (e.g., tetrahedra) and approximate the solution within each element using vector shape functions $\mathbf{N}_j$ which have tangential continuity (edge elements) [5]:

$$\mathbf{A} = \sum_{j=1}^{n} \mathbf{N}_j A_j \tag{3.8}$$

Using the Galerkin method ($\mathbf{W}_i = \mathbf{N}_i$), and substituting the discretization into the weak form, we get for each weighting function i:

$$\sum_{j=1}^{n} \left[\int_{\Omega} \left(\frac{1}{\mu}(\nabla\times\mathbf{N}_i)\cdot(\nabla\times\mathbf{N}_j) + j\omega\sigma\mathbf{N}_i\cdot\mathbf{N}_j\right) d\Omega\right] A_j = \int_{\Omega} \mathbf{N}_i \cdot \mathbf{J}_s d\Omega \tag{3.9}$$

This is a system of linear equations [2]:

$$(\mathbf{K}_\nu + j\omega\mathbf{K}_\sigma)\mathbf{a} = \mathbf{f} \tag{3.10}$$

where: - $\mathbf{K}_\nu[i,j] = \int_{\Omega} \frac{1}{\mu}(\nabla\times\mathbf{N}_i)\cdot(\nabla\times\mathbf{N}_j)d\Omega$ is the *reluctivity stiffness matrix*. - $\mathbf{K}_\sigma[i,j] = \int_{\Omega} \sigma\mathbf{N}_i\cdot\mathbf{N}_j d\Omega$ is the *conductivity (mass) matrix*. - $\mathbf{f}[i] = \int_{\Omega} \mathbf{N}_i\cdot\mathbf{J}_s d\Omega$ is the *source vector*. - $\mathbf{a}$ is the vector of unknown nodal (edge) potentials A_j.

Solving this complex-valued system yields the magnetic vector potential $\mathbf{A}$ throughout the domain, from which all other quantities ($\mathbf{B}, \mathbf{H}$, forces, losses) are derived [1].

3.2.3 POST-PROCESSING: FORCE, TORQUE, AND LOSS CALCULATIONS

The magnetic flux density is calculated as $\mathbf{B} = \nabla\times\mathbf{A}$. The Maxwell Stress Tensor (MST) method is a robust technique for calculating force and torque [3]. The magnetic stress tensor $\overleftrightarrow{\mathbf{T}}$ is defined as:

$$\overleftrightarrow{\mathbf{T}} = \frac{1}{\mu_0}\begin{bmatrix} B_x^2 - \frac{1}{2}B^2 & B_x B_y & B_x B_z \\ B_y B_x & B_y^2 - \frac{1}{2}B^2 & B_y B_z \\ B_z B_x & B_z B_y & B_z^2 - \frac{1}{2}B^2 \end{bmatrix}$$

The force on a body enclosed by surface S is given by:

$$\mathbf{F} = \oint_S \overleftrightarrow{\mathbf{T}} \cdot d\mathbf{S} \tag{3.11}$$

And the torque is:

$$\mathbf{T} = \oint_S \mathbf{r} \times (\overleftrightarrow{\mathbf{T}} \cdot d\mathbf{S}) \tag{3.12}$$

where **r** is the position vector from the axis of rotation. In practice, this surface integral is evaluated over an air-gap cylinder [1].

For core losses, the improved Generalized Steinmetz Equation (iGSE) offers accuracy for non-sinusoidal flux waveforms [3]:

$$P_{v,fe} = \frac{1}{T}\int_0^T \left(k_h \left|\frac{dB}{dt}\right| |B(t)|^{\alpha-1} + k_c \left(\frac{dB}{dt}\right)^2 \right) dt \tag{3.13}$$

where $P_{v,fe}$ is the time-average power loss per unit volume, T is the period of the waveform, and k_h, k_c, α are material parameters determined from measured loss data.

3.2.4 THERMAL FIELD THEORY AND FEM FORMULATION

The governing equation for heat transfer is the energy balance equation. Assuming no internal mass transfer, it is given by [13]:

$$\rho c_p \frac{\partial T}{\partial t} = \nabla \cdot (k \nabla T) + \dot{q}_{gen} \tag{3.14}$$

where ρ is density, c_p is specific heat capacity, k is the thermal conductivity tensor, T is temperature, and $\dot{q}_{gen}$ is the volumetric heat generation rate (from Ohmic and core losses: $\dot{q}_{gen} = \frac{|J|^2}{\sigma} + P_{v,fe}$).

Boundary conditions are crucial. The most common is convection:

$$-k\frac{\partial T}{\partial n} = h(T - T_\infty) \tag{3.15}$$

where h is the convective heat transfer coefficient and T_∞ is the coolant temperature. Radiation can also be significant:

$$-k\frac{\partial T}{\partial n} = \epsilon \sigma_{SB}(T^4 - T_{sur}^4)$$

where ϵ is emissivity and σ_{SB} is the Stefan-Boltzmann constant.

The FEM formulation follows a similar path to the electromagnetic case [13, 22]. Using scalar shape functions N_i and the Galerkin method, the weak form is derived by multiplying Equation 3.14 by a weighting function W_i and integrating:

$$\int_\Omega W_i \left(\rho c_p \frac{\partial T}{\partial t} \right) d\Omega = \int_\Omega W_i \left(\nabla \cdot (k \nabla T) \right) d\Omega + \int_\Omega W_i \dot{q}_{gen} d\Omega \tag{3.16}$$

Applying Green's theorem (integration by parts) to the conduction term [2]:

$$\int_\Omega W_i \left(\nabla \cdot (k \nabla T) \right) d\Omega = -\int_\Omega \nabla W_i \cdot (k \nabla T) d\Omega + \oint_\Gamma W_i (k \nabla T \cdot \mathbf{n}) d\Gamma \tag{3.17}$$

The surface integral incorporates the boundary conditions. For a convection boundary, $k\nabla T \cdot \mathbf{n} = -h(T - T_\infty)$. Substituting and discretizing $T = \sum N_j T_j$ and $W_i = N_i$ yields the matrix system [13]:

$$\mathbf{C}\dot{\mathbf{T}} + (\mathbf{K}_t + \mathbf{K}_h)\mathbf{T} = \mathbf{Q}_{gen} + \mathbf{Q}_h \tag{3.18}$$

where:

- $\mathbf{C}[i,j] = \int_\Omega \rho c_p N_i N_j d\Omega$ is the thermal capacitance matrix.
- $\mathbf{K}_t[i,j] = \int_\Omega (\nabla N_i) \cdot k(\nabla N_j) d\Omega$ is the thermal conductivity matrix.
- $\mathbf{K}_h[i,j] = \oint_{\Gamma_{conv}} h N_i N_j d\Gamma$ is the convection surface matrix.
- $\mathbf{Q}_{gen}[i] = \int_\Omega N_i \dot{q}_{gen} d\Omega$ is the heat generation load vector.
- $\mathbf{Q}_h[i] = \oint_{\Gamma_{conv}} h T_\infty N_i d\Gamma$ is the convection load vector.

For coolant flow, h is not a constant but a function of flow dynamics. It is correlated using the Nusselt number Nu. For turbulent flow in smooth tubes, the Gnielinski correlation is accurate [13]:

$$Nu = \frac{(f/8)(Re_D - 1000)Pr}{1 + 12.7(f/8)^{1/2}(Pr^{2/3} - 1)} \tag{3.19}$$

where $f = (0.79\ln(Re_D) - 1.64)^{-2}$ is the Darcy friction factor, valid for $3000 < Re_D < 5 \times 10^6$ and $0.5 < Pr < 2000$.

3.2.5 STRUCTURAL MECHANICS AND ROTORDYNAMICS

The governing equation for linear elastodynamics is [13, 26]:

$$\nabla \cdot \boldsymbol{\sigma} + \mathbf{f}_b = \rho \ddot{\mathbf{u}} \tag{3.20}$$

where $\boldsymbol{\sigma}$ is the Cauchy stress tensor, $\mathbf{f}_b$ is the body force vector (e.g., gravity, centrifugal force), and $\mathbf{u}$ is the displacement vector.

The relationship between stress and strain is given by the constitutive law. For linear elastic materials, this is Hooke's Law, generalized in tensor form:

$$\boldsymbol{\sigma} = \mathbf{C} : \boldsymbol{\epsilon}$$

where $\mathbf{C}$ is the fourth-order stiffness tensor, and $\boldsymbol{\epsilon}$ is the infinitesimal strain tensor, defined as:

$$\boldsymbol{\epsilon} = \frac{1}{2}\left(\nabla\mathbf{u} + (\nabla\mathbf{u})^T\right)$$

For isotropic materials, the constitutive law simplifies, and can be written in matrix form as $\{\boldsymbol{\sigma}\} = [\mathbf{D}]\{\boldsymbol{\epsilon}\}$, where $[\mathbf{D}]$ is the material matrix [26].

The centrifugal body force for a rotor spinning at angular velocity ω is [21]:

$$\mathbf{f}_b^{cent} = \rho\omega \times (\omega \times \mathbf{r}) \tag{3.21}$$

where $\mathbf{r}$ is the position vector from the axis of rotation. The magnitude is $\rho\omega^2 r$ radially outward.

The FEM formulation uses vector shape functions $\mathbf{N}_i^u$ for displacement. The weak form is derived via the principle of virtual work, leading to [13, 22]:

$$\mathbf{M}\ddot{\mathbf{U}} + \mathbf{C}_d\dot{\mathbf{U}} + \mathbf{K}_s\mathbf{U} = \mathbf{F}_{ext} \tag{3.22}$$

where:

- $\mathbf{M}[i,j] = \int_\Omega \rho(\mathbf{N}_i^u)^T \mathbf{N}_j^u d\Omega$ is the mass matrix.
- $\mathbf{K}_s[i,j] = \int_\Omega (\mathbf{B}_i^u)^T \mathbf{D}(\mathbf{B}_j^u) d\Omega$ is the structural stiffness matrix ($\mathbf{B}^u$ is the strain-displacement matrix).
- $\mathbf{C}_d$ is the damping matrix, often modeled as Rayleigh damping: $\mathbf{C}_d = \alpha_R\mathbf{M} + \beta_R\mathbf{K}_s$.
- $\mathbf{F}_{ext}$ is the external force vector, including body forces and surface tractions.

Modal analysis solves the undamped, free-vibration eigenvalue problem:

$$(\mathbf{K}_s - \omega_n^2\mathbf{M})\boldsymbol{\phi}_n = 0 \tag{3.23}$$

This yields natural frequencies ω_n and mode shapes $\boldsymbol{\phi}_n$, which are critical for avoiding resonance.

3.2.6 STRONGLY COUPLED MULTIPHYSICS FORMULATION

The physics are coupled [4]. Electromagnetic losses ($\dot{q}_{gen}$) depend on temperature (T) through electrical resistivity $\rho_e(T) = \rho_{e,0}[1 + \alpha(T - T_0)]$. Thermal expansion induces thermal strains $\boldsymbol{\epsilon}_{th} = \alpha_{th}(T - T_{ref})\mathbf{I}$, which add to the mechanical strain: $\boldsymbol{\epsilon}_{total} = \boldsymbol{\epsilon}_{mech} + \boldsymbol{\epsilon}_{th}$. Centrifugal force depends on rotational speed ω, which is controlled by the drive cycle.

A monolithic (fully coupled) approach solves all fields simultaneously. The residual of the coupled system can be written as [4]:

$$\mathbf{R}(\mathbf{a},\mathbf{T},\mathbf{U}) = \begin{bmatrix} \mathbf{R}_{EM}(\mathbf{a},\mathbf{T}) \\ \mathbf{R}_{TH}(\mathbf{T},\mathbf{a},\dot{\mathbf{a}}) \\ \mathbf{R}_{STR}(\mathbf{U},\mathbf{T}) \end{bmatrix} = 0$$

This is a highly nonlinear system solved using the Newton-Raphson method [23]:

$$\begin{bmatrix} \frac{\partial \mathbf{R}_{EM}}{\partial \mathbf{a}} & \frac{\partial \mathbf{R}_{EM}}{\partial \mathbf{T}} & 0 \\ \frac{\partial \mathbf{R}_{TH}}{\partial \mathbf{a}} & \frac{\partial \mathbf{R}_{TH}}{\partial \mathbf{T}} & 0 \\ 0 & \frac{\partial \mathbf{R}_{STR}}{\partial \mathbf{T}} & \frac{\partial \mathbf{R}_{STR}}{\partial \mathbf{U}} \end{bmatrix} \begin{bmatrix} \Delta\mathbf{a} \\ \Delta\mathbf{T} \\ \Delta\mathbf{U} \end{bmatrix} = - \begin{bmatrix} \mathbf{R}_{EM} \\ \mathbf{R}_{TH} \\ \mathbf{R}_{STR} \end{bmatrix}$$

The off-diagonal terms $\frac{\partial \mathbf{R}_{EM}}{\partial \mathbf{T}}$ (dependence of EM system on temperature) and $\frac{\partial \mathbf{R}_{STR}}{\partial \mathbf{T}}$ (thermal stress) are the coupling terms. Solving this monolithic system is computationally expensive but most accurate [4].

3.2.7 THEORETICAL BASIS OF TOPOLOGY OPTIMIZATION

Topology optimization seeks to find the optimal material distribution $\rho(\mathbf{x})$ within a design domain Ω that minimizes an objective function f subject to constraints $g_j \leq 0$. Using the SIMP method, a pseudo-density $\tilde{\rho}_e$ is assigned to each element, which modulates the material properties. For structural problems [12]:

$$E_e(\tilde{\rho}_e) = E_{min} + \tilde{\rho}_e^p (E_0 - E_{min})$$

where p is the penalization power (typically $p = 3$) to drive the solution to 0 (void) or 1 (solid). E_{min} is a small value to avoid numerical singularity.

The optimization problem for minimum compliance (maximize global stiffness) with a volume constraint is:

$$\begin{aligned} \min_{\tilde{\rho}} \quad & f(\tilde{\rho}) = \mathbf{F}^T \mathbf{U}(\tilde{\rho}) \\ \text{subject to} \quad & \mathbf{K}_s(\tilde{\rho})\mathbf{U} = \mathbf{F} \\ & g(\tilde{\rho}) = \frac{\sum_{e=1}^{N} v_e \tilde{\rho}_e}{V_0} - f_r \leq 0 \\ & 0 \leq \tilde{\rho}_e \leq 1, \quad e = 1, \ldots, N \end{aligned} \tag{3.24}$$

where v_e is element volume, V_0 is total domain volume, and f_r is the volume fraction constraint.

The sensitivity of the objective function with respect to the design variable is found using the adjoint method [12, 23]:

$$\frac{\partial f}{\partial \tilde{\rho}_e} = -\mathbf{U}^T \frac{\partial \mathbf{K}_s}{\partial \tilde{\rho}_e} \mathbf{U} \tag{3.25}$$

Since $\mathbf{K}_s = \sum_{e=1}^{N} \mathbf{K}_e^0 E_e(\tilde{\rho}_e)$, then $\frac{\partial \mathbf{K}_s}{\partial \tilde{\rho}_e} = \mathbf{K}_e^0 \frac{\partial E_e}{\partial \tilde{\rho}_e} = \mathbf{K}_e^0 p \tilde{\rho}_e^{p-1}(E_0 - E_{min})$. Thus:

$$\frac{\partial f}{\partial \tilde{\rho}_e} = -p\tilde{\rho}_e^{p-1}(E_0 - E_{min})\mathbf{u}_e^T \mathbf{K}_e^0 \mathbf{u}_e$$

where $\mathbf{u}_e$ is the element displacement vector. These sensitivities are used in gradient-based optimization algorithms like the Method of Moving Asymptotes (MMA) to update the design variables [23].

3.3 THEORETICAL FOUNDATIONS OF ADVANCED CONTROL SYSTEMS FOR ACTIVE IN-WHEEL SUSPENSION

The control of an active IWSS represents a sophisticated multi-objective optimization problem that must reconcile conflicting performance criteria under time-varying operating conditions and substantial parametric uncertainties [14, 16]. This section develops a comprehensive theoretical framework for suspension control, progressing systematically from fundamental state-space representations to advanced robust and nonlinear control strategies [7, 25].

3.3.1 MATHEMATICAL MODELING OF VEHICLE DYNAMICS

This section presents three fundamental vehicle suspension models with increasing complexity: the quarter-car, half-car, and full-car models. Each model serves specific purposes in vehicle dynamics analysis and control system design, with trade-offs between computational complexity and modeling accuracy.

3.3.1.1 Quarter-Car Model

In vehicular suspension analysis, the quarter-car model in Figure 3.1 constitutes a classical two-degree-of-freedom mechanical system that isolates the vertical dynamics of one wheel station. The model represents the vehicle body as a sprung mass and the wheel assembly as an unsprung mass, interconnected through linear or nonlinear spring and damper elements. This simplified representation enables analytical and numerical investigation of fundamental suspension performance metrics, including natural frequencies, damping ratios, transmissibility, and dynamic tire load variations.

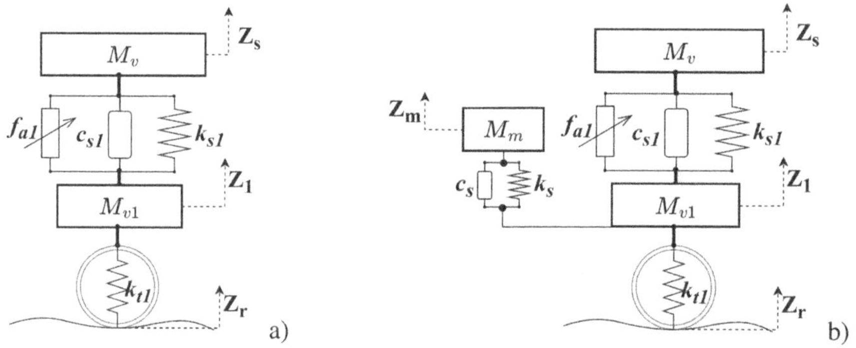

Figure 3.1: Quarter-car suspension models: (a) 2 degrees of freedom; (b) 3 degrees of freedom

To serve as a foundational benchmark for evaluating suspension concepts, control laws, and parameter sensitivity before proceeding to higher-fidelity models. Preliminary assessment of ride comfort using ISO-weighted filters, design of skyhook and groundhook controllers, and study of resonance phenomena in the sprung and unsprung mass subsystems. The formulation assumes decoupled corners, ignores chassis pitch and roll, and typically employs linearized tire and suspension characteristics. Consequently, while invaluable for conceptual design and educational purposes, the quarter-car model must be superseded by half-car or full-car representations when analyzing handling, roll stability, or coupled vertical–longitudinal dynamics.

3.3.1.2 Half-Car Model

The half-car model, depicted in Figure 3.2, advances suspension analysis by integrating pitch and/or roll degree-of-freedom with vertical dynamics, thereby capturing the interdependent behavior of front-rear or left-right suspensions. This four-degree-of-freedom system—comprising sprung mass bounce and pitch/roll, plus vertical displacements of two unsprung masses—provides a necessary framework for studying phenomena such as pitch oscillations during transient maneuvers, anti-dive/anti-squat geometry effects, and front-rear or left-right dampers coordination strategies.

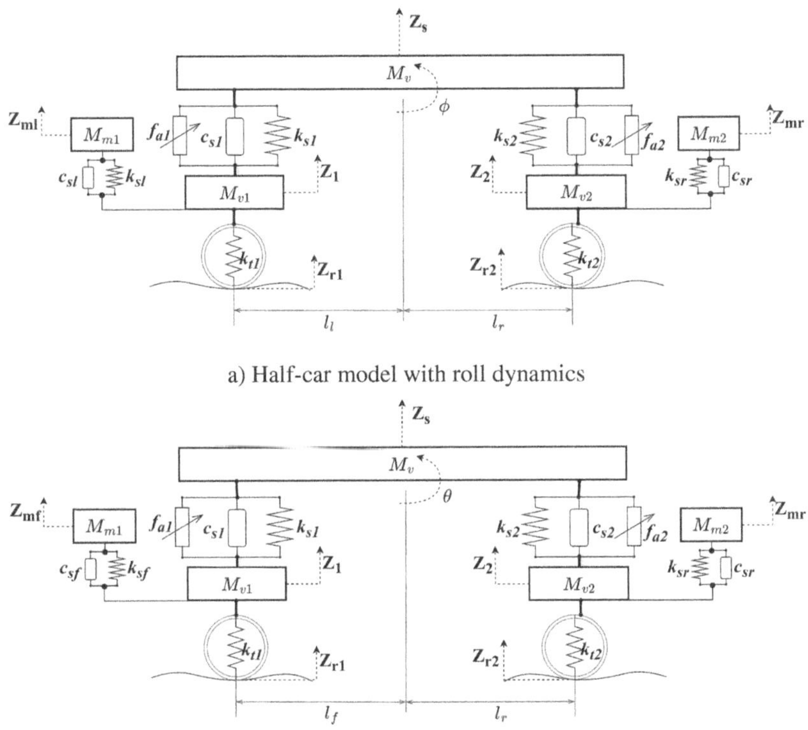

a) Half-car model with roll dynamics

b) Half-car model with pitch dynamics

Figure 3.2: Half-car model with 4 degrees of freedom

Compared to the quarter-car model, the half-car formulation offers substantially improved realism for longitudinal dynamics applications, albeit at the cost of increased model order and parametric complexity. Its principal limitation remains the omission of roll motion and lateral load transfer, restricting its utility to planar longitudinal-vertical analyses. Nevertheless, it serves as an indispensable intermediate model for suspension development, control system design, and vehicle pitch stability assessment.

3.3.1.3 Full-Car Model

Representing the highest-fidelity lumped-parameter approach in vertical vehicle dynamics, the full-car model illustrated in Figure 3.3 integrates seven degrees of freedom: three for the sprung mass (heave, pitch, and roll) and four for the unsprung masses (vertical displacement at each wheel). This formulation enables a complete characterization of the dynamic interactions between all suspension corners, facilitating accurate analysis of whole-vehicle response to road inputs and driver commands.

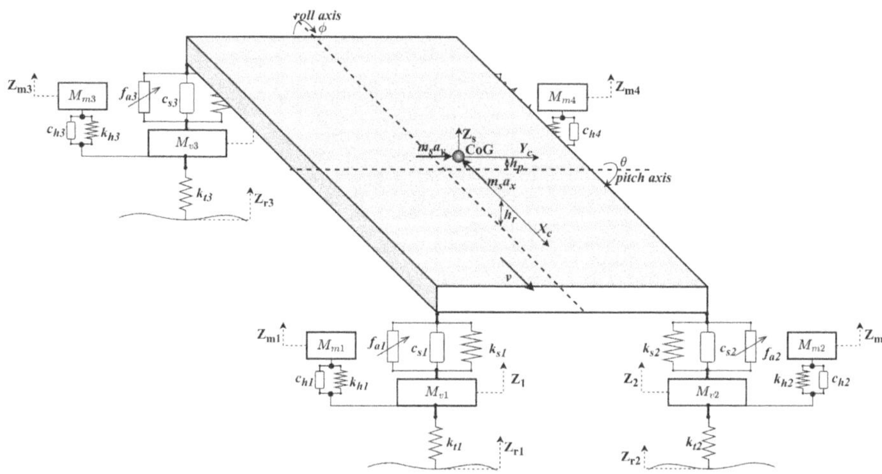

Figure 3.3: Full-car model with heave, pitch, and roll dynamics—7 degrees of freedom

In contrast to reduced-order models, the full-car representation accounts for critical coupling mechanisms absent in both quarter-car and half-car models—namely, roll dynamics and lateral load transfer. These capabilities make it particularly suitable for applications such as rollover prevention system design, advanced active suspension development, and ride–handling compromise optimization. Its comprehensive nature also supports the study of cross-coupling effects during combined longitudinal and lateral maneuvers, which are essential for modern integrated vehicle dynamics control.

While the full-car model's 14-state structure increases computational burden compared to simpler models, this complexity is warranted for applications requiring high predictive accuracy. The model thus serves as a vital tool in the latter stages of suspension system development, controller calibration, and performance verification, bridging the gap between theoretical analysis and real vehicle implementation.

3.3.1.4 Comparative Analysis and Model Selection Methodology

The strategic selection of vehicle dynamics models represents a critical engineering decision that fundamentally influences the efficacy of suspension system analysis and control synthesis. This comparative assessment delineates the operational domains and technical considerations governing model deployment across the fidelity spectrum.

Quarter-Car Model: Analytical Foundation and Computational Primitive

- **Optimal Deployment Domain:** Serves as the fundamental analytical construct for preliminary suspension characterization, providing essential insights into vertical vibration modes and basic ride comfort metrics. Its computational parsimony renders it indispensable for extensive parametric optimization studies, real-time estimation algorithms, and educational demonstrations of core suspension principles.
- **Theoretical Limitations:** Exhibits significant phenomenological deficiencies by neglecting critical inertial coupling mechanisms. The model's inability to capture pitch-roll interconnections and cross-axle dynamic interactions fundamentally constrains its predictive accuracy for whole-vehicle response analysis. Consequently, it proves inadequate for handling characterization, rollover propensity assessment, or comprehensive stability analyses.

Half-Car Model: Longitudinal Dynamics Specialization

- **Optimal Deployment Domain:** Provides enhanced fidelity for investigating longitudinal weight transfer phenomena during acceleration and braking maneuvers. The incorporation of pitch dynamics enables accurate prediction of anti-dive and anti-squat suspension characteristics, making it particularly valuable for powertrain integration studies and brake system interaction analyses.
- **Dimensional Constraints:** While capturing front-rear dynamic coupling, the model's planar formulation inherently disregards critical lateral dynamics. This omission proves particularly limiting for cornering analysis, roll stability assessment, and any scenario involving asymmetric road excitation or lateral load transfer mechanisms.

Full-Car Model (7-DOF): Holistic Vehicle Dynamics Representation

- **Comprehensive Capabilities:** Embodies the most complete representation of vehicle dynamic behavior within the linear modeling paradigm. The simultaneous incorporation of heave, pitch, and roll degrees of freedom enables accurate prediction of complex inertial coupling effects, making it indispensable for advanced suspension control synthesis and whole-vehicle dynamic response prediction.
- **Implementation Considerations:** The model's increased state dimensionality (14 states) necessitates sophisticated computational infrastructure for controller implementation and real-time execution. While modern processing capabilities have mitigated these constraints, careful consideration must be given to sampling rate requirements, memory allocation, and computational throughput for embedded system deployment.

Systematic Selection Framework:

The model selection process should follow a systematic methodology based on specific application requirements:

1. **Performance Metric Requirements:** Quarter-car models suffice for basic ride comfort optimization, while full-car representations become necessary when handling metrics or roll stability indices must be considered.
2. **Excitation Characteristics:** Symmetric road inputs permit lower-fidelity models, whereas spatially-correlated or asymmetric excitations demand full-vehicle representations to capture cross-coupling effects.
3. **Control Architecture Complexity:** Decentralized control strategies may utilize simplified models, while coordinated multi-objective control necessitates higher-fidelity representations to account for dynamic interactions.
4. **Implementation Constraints:** Computational resources, sampling rate requirements, and memory limitations impose practical constraints on model complexity for real-time applications.

For advanced suspension control applications requiring comprehensive vehicle dynamics characterization, the 7-DOF full-car model establishes the benchmark for fidelity-complexity trade-offs. Its capacity to capture essential dynamic couplings while maintaining computational tractability renders it the predominant choice for sophisticated active and semi-active suspension system development in contemporary automotive engineering practice.

3.3.2 CLASSICAL CONTROL: PID THEORY AND LIMITATIONS

The Proportional-Integral-Derivative (PID) controller represents the classical approach to feedback control [16]. The continuous-time control law is:

$$u(t) = K_p e(t) + K_i \int_0^t e(\tau)d\tau + K_d \frac{de(t)}{dt} \tag{3.26}$$

In the Laplace domain [16], this corresponds to:

$$C(s) = \frac{U(s)}{E(s)} = K_p + \frac{K_i}{s} + K_d s \tag{3.27}$$

For suspension control, typical error signals include:

- Suspension deflection: $e(t) = z_s - z_u$
- Sprung mass acceleration: $e(t) = \ddot{z}_s$
- Mixed error criteria

The fundamental limitation of PID control for suspension systems lies in its inability to optimally manage the inherent conflict between ride comfort (minimizing $\ddot{z}_s$) and road holding (minimizing $z_u - z_r$). This can be analyzed through the root locus of the closed-loop system.

Consider the sensitivity function:

$$S(s) = \frac{1}{1 + P(s)C(s)}$$

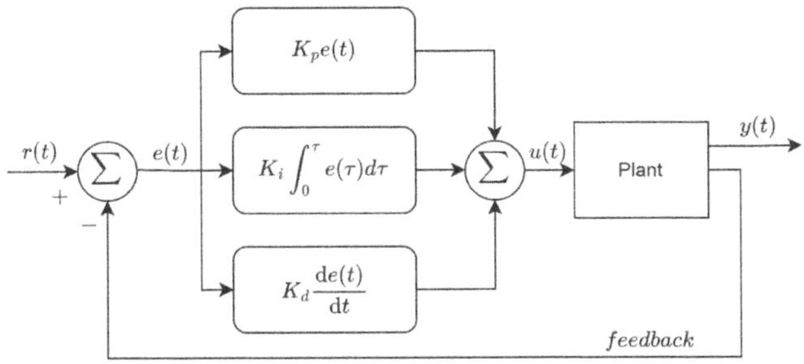

Figure 3.4: PID control theory structure

and complementary sensitivity:

$$T(s) = \frac{P(s)C(s)}{1+P(s)C(s)}$$

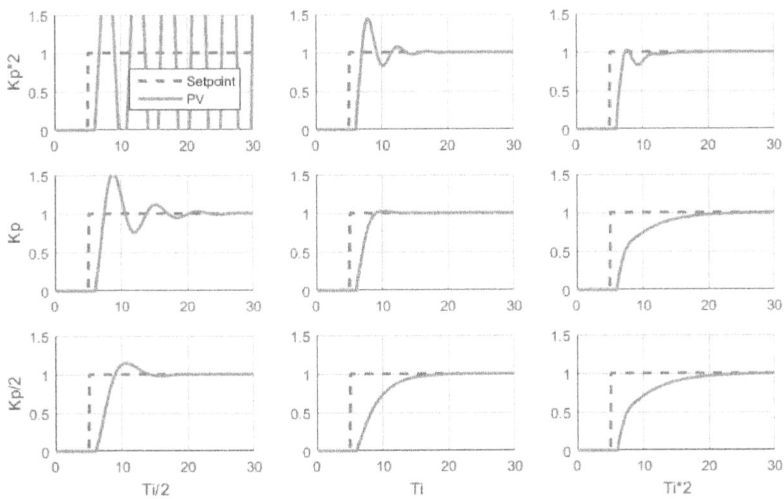

Figure 3.5: The effect of P, I and D in a PID controller [27]

The performance limitations are captured by the Bode Integral Theorem:

$$\int_0^\infty \ln|S(j\omega)|d\omega = \pi \sum \text{Re}(p_k) \tag{3.28}$$

where p_k are the unstable open-loop poles. This fundamental limitation shows that reducing sensitivity at one frequency range necessarily increases it at another, explaining why fixed-gain PID controllers cannot provide consistent performance across varying road conditions and operating points.

3.3.3 MODERN OPTIMAL CONTROL: LQR FORMULATION

The Linear Quadratic Regulator (LQR) provides a systematic framework for handling multi-objective control problems [14, 24]. The control law $u = -\mathbf{K}\mathbf{x}$ minimizes the quadratic cost function:

$$J = \int_0^{\infty} \left(\mathbf{x}^T(t)\mathbf{Q}\mathbf{x}(t) + u^T(t)Ru(t) + 2\mathbf{x}^T(t)\mathbf{N}u(t)\right) dt \tag{3.29}$$

For suspension systems, the weighting matrix $\mathbf{Q}$ is typically diagonal:

$$\mathbf{Q} = \mathrm{diag}(q_1, q_2, q_3, q_4) = \mathrm{diag}(q_{susp}, q_{\ddot{z}_s}, q_{tire}, q_{\dot{z}_u})$$

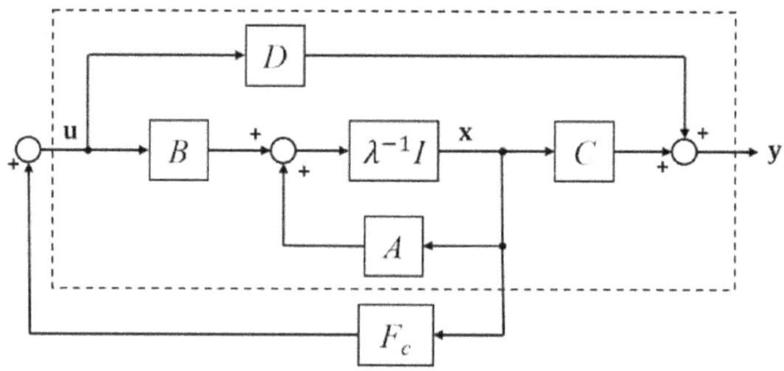

Figure 3.6: LQR Full-state feedback control structure

The relationship between states and performance outputs is:

$$\ddot{z}_s = [-\frac{k_s}{m_s} \;\; -\frac{c_s}{m_s} \;\; 0 \;\; \frac{c_s}{m_s}]\mathbf{x} + \frac{1}{m_s}u$$

Thus, weighting q_2 effectively penalizes sprung mass acceleration. The optimal gain matrix $\mathbf{K}$ is obtained by solving the Algebraic Riccati Equation (ARE) [14]:

$$\mathbf{A}^T\mathbf{P} + \mathbf{P}\mathbf{A} - (\mathbf{P}\mathbf{B} + \mathbf{N})R^{-1}(\mathbf{B}^T\mathbf{P} + \mathbf{N}^T) + \mathbf{Q} = 0 \tag{3.30}$$

The optimal feedback gain is then:

$$\mathbf{K} = R^{-1}(\mathbf{B}^T\mathbf{P} + \mathbf{N}^T)$$

The closed-loop system:

$$\dot{\mathbf{x}} = (\mathbf{A} - \mathbf{B}\mathbf{K})\mathbf{x}$$

is guaranteed to be asymptotically stable, and the cost J is minimized.

3.3.4 STOCHASTIC ESTIMATION AND CONTROL: LQG THEORY

The Linear Quadratic Gaussian (LQG) controller extends LQR to handle systems with process and measurement noise [6, 14]. The system model becomes:

$$\begin{aligned} \dot{\mathbf{x}} &= \mathbf{A}\mathbf{x} + \mathbf{B}u + \mathbf{L}\mathbf{w} \\ \mathbf{y} &= \mathbf{C}\mathbf{x} + \mathbf{D}u + \mathbf{v} \end{aligned} \tag{3.31}$$

where $\mathbf{w} \sim \mathcal{N}(0, \mathbf{W})$ and $\mathbf{v} \sim \mathcal{N}(0, \mathbf{V})$ are uncorrelated white Gaussian noise processes.

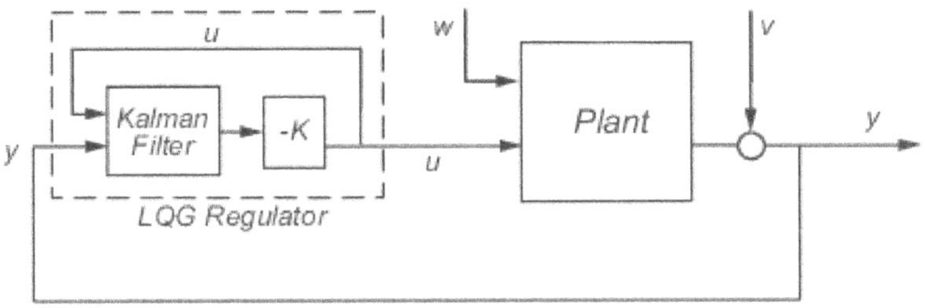

Figure 3.7: Linear Quadratic Gaussian control configuration

The LQG controller consists of two components:

1. **Kalman Filter for State Estimation:**

$$\dot{\hat{\mathbf{x}}} = \mathbf{A}\hat{\mathbf{x}} + \mathbf{B}u + \mathbf{K}_f(\mathbf{y} - \mathbf{C}\hat{\mathbf{x}} - \mathbf{D}u)$$

The Kalman gain $\mathbf{K}_f$ is computed by solving the Filter Algebraic Riccati Equation:

$$\mathbf{A}\mathbf{P}_f + \mathbf{P}_f\mathbf{A}^T - \mathbf{P}_f\mathbf{C}^T\mathbf{V}^{-1}\mathbf{C}\mathbf{P}_f + \mathbf{L}\mathbf{W}\mathbf{L}^T = 0 \tag{3.32}$$

Then:

$$\mathbf{K}_f = \mathbf{P}_f\mathbf{C}^T\mathbf{V}^{-1}$$

2. **LQR State Feedback:**

$$u = -\mathbf{K}\hat{\mathbf{x}}$$

By the Separation Principle, the optimal estimator and controller can be designed independently, and the combined system remains stable.

3.3.5 ROBUST CONTROL: H_∞ SYNTHESIS

H_∞ control addresses performance robustness under model uncertainties and disturbance rejection [7, 18]. The standard control configuration is:

$$\begin{aligned} \dot{\mathbf{x}} &= \mathbf{A}\mathbf{x} + \mathbf{B}_1\mathbf{w} + \mathbf{B}_2u \\ \mathbf{z} &= \mathbf{C}_1\mathbf{x} + \mathbf{D}_{11}\mathbf{w} + \mathbf{D}_{12}u \\ \mathbf{y} &= \mathbf{C}_2\mathbf{x} + \mathbf{D}_{21}\mathbf{w} + \mathbf{D}_{22}u \end{aligned} \tag{3.33}$$

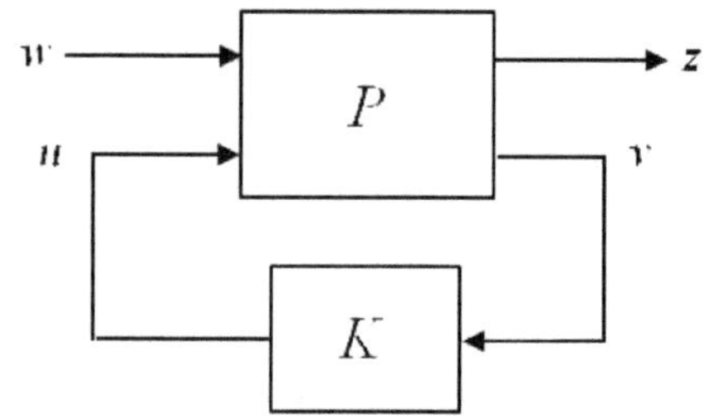

Figure 3.8: Generalize H_∞ control configuration

where **w** includes disturbances and noise, **z** represents performance outputs, and **y** contains measured outputs.

The H_∞ control problem is to find a stabilizing controller $K(s)$ that minimizes:

$$||T_{zw}(s)||_\infty = \sup_\omega \bar{\sigma}(T_{zw}(j\omega))$$

where $T_{zw}(s)$ is the closed-loop transfer function from **w** to **z**.

The solution involves solving two Hamiltonians [7]. For a given performance level $\gamma > 0$, a stabilizing controller exists if:

1. There exists $\mathbf{X}_\infty \geq 0$ satisfying:

$$\mathbf{A}^T\mathbf{X}_\infty + \mathbf{X}_\infty\mathbf{A} + \mathbf{C}_1^T\mathbf{C}_1 - \mathbf{X}_\infty(\mathbf{B}_2\mathbf{B}_2^T - \frac{1}{\gamma^2}\mathbf{B}_1\mathbf{B}_1^T)\mathbf{X}_\infty = 0$$

2. There exists $\mathbf{Y}_\infty \geq 0$ satisfying:

$$\mathbf{A}\mathbf{Y}_\infty + \mathbf{Y}_\infty\mathbf{A}^T + \mathbf{B}_1\mathbf{B}_1^T - \mathbf{Y}_\infty(\mathbf{C}_2^T\mathbf{C}_2 - \frac{1}{\gamma^2}\mathbf{C}_1^T\mathbf{C}_1)\mathbf{Y}_\infty = 0$$

3. The spectral radius condition: $\rho(\mathbf{X}_\infty\mathbf{Y}_\infty) < \gamma^2$

The central H_∞ controller is then given by:

$$K(s) = \left[\begin{array}{c|c} \mathbf{A} - \mathbf{B}_2\mathbf{K}_c - \mathbf{K}_f\mathbf{C}_2 & \mathbf{K}_f \\ \hline -\mathbf{K}_c & 0 \end{array}\right]$$

where:

$$\mathbf{K}_c = \mathbf{B}_2^T\mathbf{X}_\infty, \quad \mathbf{K}_f = \mathbf{Y}_\infty\mathbf{C}_2^T$$

3.3.6 ROBUST CONTROL: H_2 SYNTHESIS

H_2 control focuses on optimal performance in terms of minimizing the variance of output signals under stochastic disturbances and measurement noise [19, 20]. The generalized plant structure remains identical to the H_∞ formulation:

$$\begin{aligned} \dot{\mathbf{x}} &= \mathbf{A}\mathbf{x} + \mathbf{B}_1\mathbf{w} + \mathbf{B}_2u \\ \mathbf{z} &= \mathbf{C}_1\mathbf{x} + \mathbf{D}_{11}\mathbf{w} + \mathbf{D}_{12}u \\ \mathbf{y} &= \mathbf{C}_2\mathbf{x} + \mathbf{D}_{21}\mathbf{w} + \mathbf{D}_{22}u \end{aligned} \tag{3.34}$$

where **w** represents white noise disturbances with unit intensity, **z** denotes performance outputs, and **y** contains measured outputs.

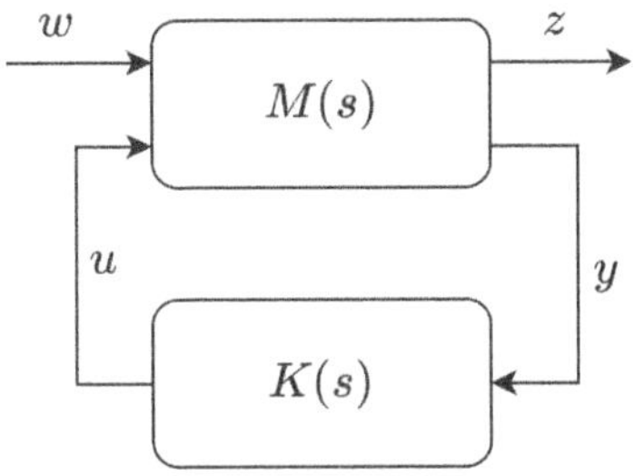

Figure 3.9: H_2 Control configuration

The H_2 control problem aims to find a stabilizing controller $K(s)$ that minimizes the H_2-norm:

$$\|T_{zw}(s)\|_2 = \left(\frac{1}{2\pi} \int_{-\infty}^{\infty} \text{trace}[T_{zw}(j\omega)^* T_{zw}(j\omega)] d\omega \right)^{1/2}$$

where $T_{zw}(s)$ is the closed-loop transfer function from **w** to **z**. This norm represents the root-mean-square (RMS) value of the output **z** when the input **w** is a white noise process.

For the solution to exist, the following standard assumptions are required [19]:

1. $(\mathbf{A}, \mathbf{B}_2)$ is stabilizable and $(\mathbf{C}_2, \mathbf{A})$ is detectable
2. $\mathbf{D}_{12}^T [\mathbf{C}_1 \ \mathbf{D}_{12}] = [0 \ \mathbf{I}]$
3. $\mathbf{D}_{21} [\mathbf{B}_1^T \ \mathbf{D}_{21}^T] = [0 \ \mathbf{I}]$

The optimal H_2 controller is obtained by solving two algebraic Riccati equations:

1. The control Riccati equation:

$$\mathbf{A}^T \mathbf{X}_2 + \mathbf{X}_2 \mathbf{A} + \mathbf{C}_1^T \mathbf{C}_1 - \mathbf{X}_2 \mathbf{B}_2 \mathbf{B}_2^T \mathbf{X}_2 = 0$$

2. The estimation Riccati equation:

$$\mathbf{A} \mathbf{Y}_2 + \mathbf{Y}_2 \mathbf{A}^T + \mathbf{B}_1 \mathbf{B}_1^T - \mathbf{Y}_2 \mathbf{C}_2^T \mathbf{C}_2 \mathbf{Y}_2 = 0$$

The optimal H_2 controller is then given by the separation principle:

$$K(s) = \left[\begin{array}{c|c} \mathbf{A} - \mathbf{B}_2 \mathbf{K}_c - \mathbf{K}_f \mathbf{C}_2 & \mathbf{K}_f \\ \hline -\mathbf{K}_c & 0 \end{array} \right]$$

where:

$$\mathbf{K}_c = \mathbf{B}_2^T \mathbf{X}_2, \quad \mathbf{K}_f = \mathbf{Y}_2 \mathbf{C}_2^T$$

The H_2 controller provides optimal disturbance rejection for Gaussian white noise inputs and minimizes the RMS values of the performance outputs, making it particularly suitable for applications where variance minimization is critical [20, 21].

3.3.7 NONLINEAR AND INTELLIGENT CONTROL STRATEGIES

3.3.7.1 Fuzzy Logic Control Theoretical Foundation

Fuzzy Logic Control (FLC) is based on fuzzy set theory. A fuzzy set A in universe X is characterized by membership function $\mu_A : X \to [0,1]$.

The FLC consists of [8, 19]:

1. **Fuzzification:** Maps crisp inputs to fuzzy sets
2. **Fuzzy Inference:** Applies IF-THEN rules
3. **Defuzzification:** Converts fuzzy output to crisp value

For suspension control, typical rules are:

$$\text{IF } e_1 \text{ is } A_1^j \text{ AND } e_2 \text{ is } A_2^j \text{ THEN } u \text{ is } B^j$$

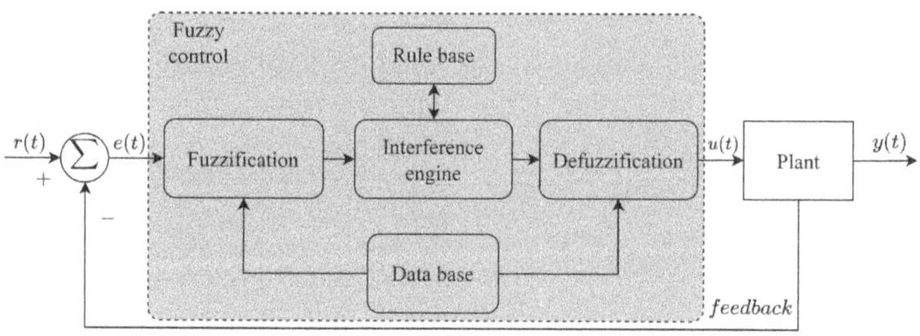

Figure 3.10: Fuzzy logic control configuration

Using product inference and center-average defuzzification [19]:

$$u^* = \frac{\sum_{j=1}^{M} \omega_j c_j}{\sum_{j=1}^{M} \omega_j} \tag{3.35}$$

where $\omega_j = \prod_{i=1}^{n} \mu_{A_i^j}(e_i)$ is the firing strength and c_j is the centroid of output fuzzy set B^j.

3.3.7.2 Sliding Mode Control Formulation

Sliding Mode Control (SMC) provides robustness to parameter variations and disturbances [9, 20]. Define a sliding surface:

$$s = \left(\frac{d}{dt} + \lambda\right)^{n-1} e \tag{3.36}$$

For a second-order system, this becomes [9]:

$$s = \dot{e} + \lambda e$$

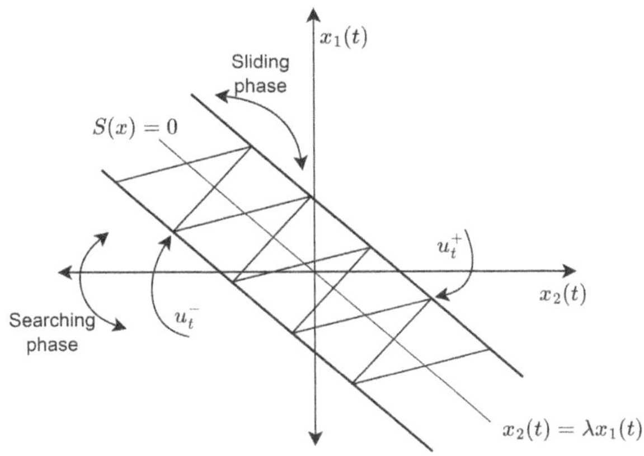

Figure 3.11: Basic structure of sliding mode control

The control law has two components:

$$u = u_{eq} + u_{sw} \tag{3.37}$$

where u_{eq} is the equivalent control that maintains $s = 0$, and u_{sw} is the switching control that drives the system to the sliding surface:

$$u_{sw} = -K \cdot \text{sat}(s/\Phi)$$

The stability is guaranteed by the Lyapunov function $V = \frac{1}{2}s^2$, requiring [9, 25]:

$$\dot{V} = s\dot{s} \leq -\eta|s|$$

3.3.8 ADAPTIVE CONTROL FORMULATIONS

3.3.8.1 Model Reference Adaptive Control

Model Reference Adaptive Control (MRAC) represents a sophisticated approach for handling system uncertainties and parameter variations in active suspension systems. This method employs a reference model that defines the desired dynamic behavior, with the adaptive mechanism continuously adjusting controller parameters to minimize the discrepancy between the actual system response and the reference model output. This approach is particularly valuable for IWASS applications where vehicle parameters may vary significantly due to changing load conditions and operating environments.

For Model Reference Adaptive Control (MRAC), we define a reference model [10, 11, 15]:

$$\dot{\mathbf{x}}_m = \mathbf{A}_m \mathbf{x}_m + \mathbf{B}_m r \tag{3.38}$$

The controller structure is:

$$u = \mathbf{K}_x^T \mathbf{x} + \mathbf{K}_r^T r \tag{3.39}$$

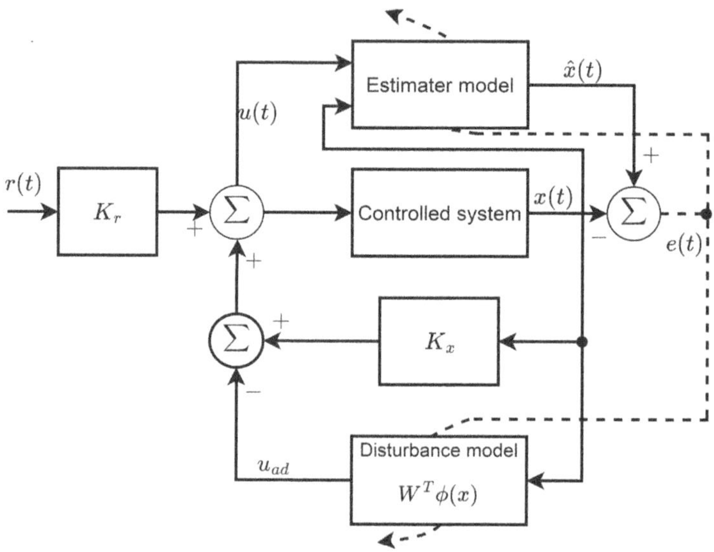

Figure 3.12: Model reference adaptive control configuration

The adaptation laws using the MIT rule [10]:

$$\begin{aligned} \dot{\mathbf{K}}_x &= -\gamma \mathbf{x}\mathbf{e}^T \mathbf{P}\mathbf{B} \\ \dot{\mathbf{K}}_r &= -\gamma r \mathbf{e}^T \mathbf{P}\mathbf{B} \end{aligned} \tag{3.40}$$

where $\mathbf{e} = \mathbf{x} - \mathbf{x}_m$ is the tracking error, and $\mathbf{P} > 0$ satisfies the Lyapunov equation:

$$\mathbf{A}_m^T \mathbf{P} + \mathbf{P}\mathbf{A}_m = -\mathbf{Q}$$

The adaptation gain γ plays a crucial role in determining the convergence rate and stability of the adaptive process, requiring careful selection to balance responsiveness against potential oscillations in the control parameters.

3.3.9 LINEAR PARAMETER VARYING CONTROL

For advanced suspension systems, operational parameters such as vehicle load, ride height, and driving speed are not constant but vary significantly during normal use. These variations can drastically alter the system's dynamics, potentially degrading the performance and stability of a controller designed for a single, fixed operating

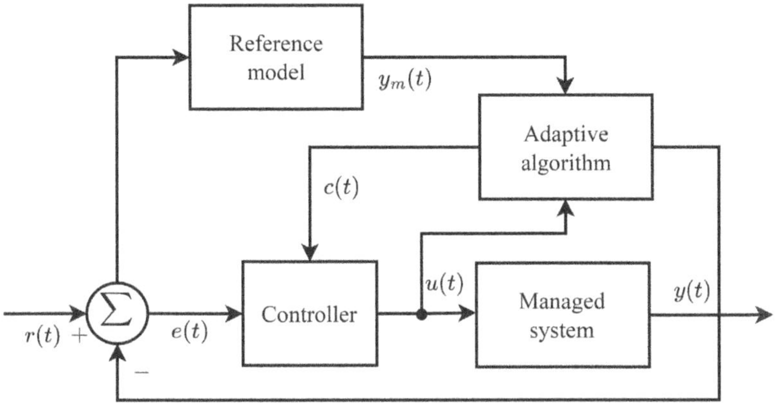

Figure 3.13: Basic structure of Model reference adaptive control

condition. To address this challenge, Linear Parameter Varying (LPV) control provides a powerful and systematic framework. The core idea of LPV control is to treat the varying parameters not as uncertainties, but as measurable, time-varying signals that can be used to schedule the controller in real-time, thereby maintaining optimal performance across the entire operating envelope.

Within the LPV framework, the suspension system is modeled as a linear system whose state-space matrices depend explicitly on a vector of time-varying parameters, denoted as $\theta(t)$. This parameter vector can include, for instance, the sprung mass (representing passenger and cargo load) and the vehicle speed. The general LPV system representation is given by the following equations:

$$\begin{aligned} \dot{\mathbf{x}}(t) &= \mathbf{A}(\theta(t))\mathbf{x}(t) + \mathbf{B}(\theta(t))u(t) \\ \mathbf{y}(t) &= \mathbf{C}(\theta(t))\mathbf{x}(t) + \mathbf{D}(\theta(t))u(t) \end{aligned} \tag{3.41}$$

Here, $\mathbf{x}(t)$ is the state vector, $u(t)$ is the control input, and $\mathbf{y}(t)$ is the output vector. The key feature of this model is that the system matrices **A**, **B**, **C**, and **D** are functions of the time-varying parameter $\theta(t)$, which is assumed to be measurable or estimable in real-time.

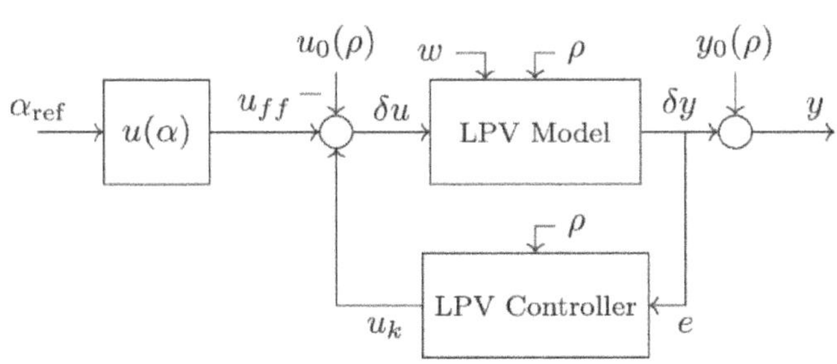

Figure 3.14: Structure of LPV model and LPV controller with inputs and outputs

The synthesis of an LPV controller involves designing a control law that itself is scheduled by the same parameter $\theta(t)$. This is mathematically expressed as:

$$u(t) = \mathbf{K}(\theta(t))\mathbf{x}(t) \tag{3.42}$$

The primary goal of the synthesis procedure is to find this parameter-dependent controller gain $\mathbf{K}(\theta)$ such that closed-loop stability and performance are guaranteed for all possible trajectories of the parameter $\theta(t)$ within a predefined, bounded set. This robust guarantee is typically achieved through the use of parameter-dependent Lyapunov functions. Unlike classical control methods that use a single, fixed Lyapunov function, LPV synthesis searches for a function $V(\mathbf{x},\theta) = \mathbf{x}^T\mathbf{P}(\theta)\mathbf{x}$ that proves stability for every possible value of θ. This approach is less conservative than robust $\mathcal{H}_\infty$ control, as it leverages the knowledge of the varying parameters to provide a less pessimistic and often higher-performing control solution, ensuring ride comfort and handling remain consistent regardless of changes in vehicle load or speed.

3.3.10 MODEL PREDICTIVE CONTROL

For advanced suspension systems operating under real-world conditions, the control strategy must not only handle system constraints but also anticipate future system behavior to proactively counteract road disturbances. While traditional controllers react to past errors, this predictive capability is essential for optimizing the conflicting objectives of ride comfort, road holding, and actuator usage. Model Predictive Control (MPC) provides a comprehensive framework that explicitly addresses these challenges through online optimization of future control actions while systematically respecting all system constraints.

The fundamental principle of MPC involves solving a finite-horizon optimal control problem at each sampling instant based on the current system state. Using an internal model of the plant dynamics, the controller predicts the future behavior of the system over a prediction horizon and computes a sequence of control moves that minimizes a specified cost function. The first element of this optimized sequence is then applied to the system, and the entire process is repeated at the next time step in a receding horizon fashion.

The MPC formulation for the suspension system begins with a discrete-time state-space representation:

$$\begin{aligned}\mathbf{x}(k+1) &= \mathbf{A}\mathbf{x}(k) + \mathbf{B}u(k) + \mathbf{E}w(k)\\ \mathbf{y}(k) &= \mathbf{C}\mathbf{x}(k) + \mathbf{D}u(k)\end{aligned} \tag{3.43}$$

where $\mathbf{x}(k)$ represents the state vector, $u(k)$ is the control input, $w(k)$ denotes road disturbances, and $\mathbf{y}(k)$ contains the performance outputs. The core MPC optimization problem at each time step k is formulated as:

$$\begin{aligned}&\min_{\mathbf{U}} \sum_{i=1}^{N_p} \|\mathbf{y}(k+i) - \mathbf{y}_{ref}(k+i)\|_{\mathbf{Q}}^2 + \sum_{i=0}^{N_c-1} \|u(k+i)\|_{\mathbf{R}}^2\\ &\text{subject to: } \mathbf{x}(k+i+1) = \mathbf{A}\mathbf{x}(k+i) + \mathbf{B}u(k+i) + \mathbf{E}w(k+i)\\ &u_{min} \le u(k+i) \le u_{max}\\ &|z_s(k+i) - z_u(k+i)| \le \delta_{max}\\ &|z_u(k+i) - z_r(k+i)| \le \Delta_{max}\end{aligned} \tag{3.44}$$

Here, N_p denotes the prediction horizon, N_c represents the control horizon, $\mathbf{Q}$ and $\mathbf{R}$ are weighting matrices that balance performance objectives, while the constraints ensure that actuator limits, suspension travel, and tire deflection bounds are respected.

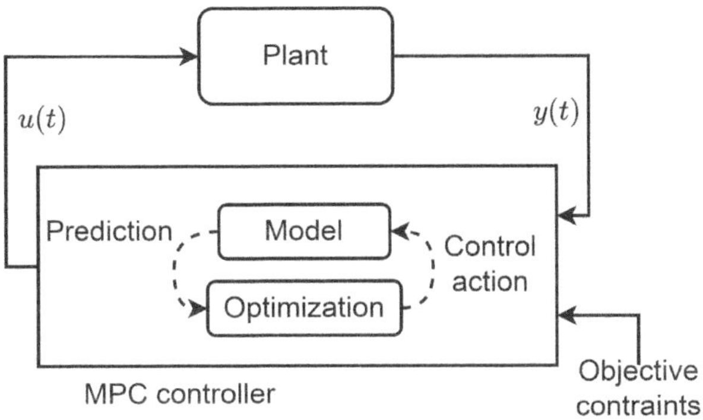

Figure 3.15: Structure of MPC principle

The distinctive feature of MPC is its explicit constraint handling capability, which ensures that the control system operates within safe physical limits under all conditions. Unlike traditional control approaches that treat constraints as afterthoughts, MPC incorporates them directly into the optimization problem, providing guaranteed constraint satisfaction. The receding horizon implementation provides inherent feedback and robustness to model uncertainties, as the optimization is continually updated with new measurements. This approach is particularly advantageous for suspension systems, as it enables proactive control actions that anticipate and mitigate the effects of road disturbances before they significantly impact vehicle dynamics, thereby maintaining consistent performance across varying operating conditions.

3.4 INTEGRATED CHASSIS CONTROL: A MULTIVARIABLE FRAMEWORK

While individual subsystem controllers offer significant improvements, the ultimate theoretical framework for vehicle dynamics management lies in Integrated Chassis Control (ICC). This holistic approach seeks to overcome the limitations of isolated control systems by coordinating the actions of the suspension, steering, and powertrain subsystems. The central premise is that these subsystems are intrinsically coupled through the vehicle's dynamics; for instance, a steering maneuver induces roll and pitch that affect tire loads, while powertrain torque can influence yaw and ride balance. By designing a unified controller that accounts for these interactions, ICC aims to resolve conflicts and exploit synergies between subsystems, thereby achieving a global optimum in vehicle performance, safety, and comfort that is unattainable with decentralized strategies [7, 14, 16].

The foundation of the ICC framework is the creation of a comprehensive vehicle model that amalgamates the dynamics of each chassis subsystem. This is achieved by constructing an augmented state vector, $\mathbf{x}_{total}$, which concatenates the states from the suspension, steering, and powertrain systems. Mathematically, this is represented as:

$$\mathbf{x}_{total} = [\mathbf{x}_{susp}^T, \mathbf{x}_{steer}^T, \mathbf{x}_{powertrain}^T]^T$$

This combined state vector provides a complete description of the vehicle's dynamic posture at any given time. Similarly, the control inputs from each subsystem are aggregated into a unified control input vector. The integrated control law is then formulated as a full-state feedback problem for this multivariable system. The control objective is to compute an optimal gain matrix, $\mathbf{K}_{int}$, that maps the entire state vector to all available actuators simultaneously. The control law is given by:

$$\begin{bmatrix} \mathbf{u}_{susp} \\ \mathbf{u}_{steer} \\ \mathbf{u}_{torque} \end{bmatrix} = -\mathbf{K}_{int}\mathbf{x}_{total} \tag{3.45}$$

The synthesis of the high-dimensional gain matrix $\mathbf{K}_{int}$ is a complex optimization task, typically addressed by solving a large-scale LQR or H_∞ problem for the combined system. In the LQR framework, the control engineer defines weighting matrices $\mathbf{Q}$ and $\mathbf{R}$ that penalize state deviations and control efforts, respectively, across all integrated subsystems. The controller $\mathbf{K}_{int}$ is then the solution to the associated algebraic Riccati equation, inherently balancing the performance trade-offs between ride comfort, handling agility, and actuator energy. For robustness against model uncertainties and external disturbances, an $\mathcal{H}_\infty$ synthesis can be employed, which minimizes the worst-case amplification of disturbance inputs to performance outputs, ensuring stable and predictable vehicle behavior across a wide range of operating conditions.

The primary challenge in implementing this elegant theoretical framework is its computational and practical complexity. The high order of the combined system model leads to a large-scale optimization problem, and the resulting controller may be complex to implement in real-time electronic control units (ECU). Furthermore, this centralized approach requires a sophisticated sensor suite and a reliable, high-bandwidth communication network to share data between subsystems. Despite these challenges, the multivariable ICC framework represents the pinnacle of vehicle dynamics control, offering a mathematically rigorous path towards achieving synergistic performance where the vehicle's chassis acts as a truly unified system rather than a collection of competing components.

3.5 CONCLUSION: UNIFIED THEORETICAL FRAMEWORK

This chapter has established a comprehensive theoretical foundation for the enhancement of IWASS through the integration of advanced multiphysics design and sophisticated control strategies. By progressing systematically from the first-principles derivation of Maxwell's equations and the principles of continuum mechanics to the sophisticated formulations of modern optimal and robust control theory, the presented

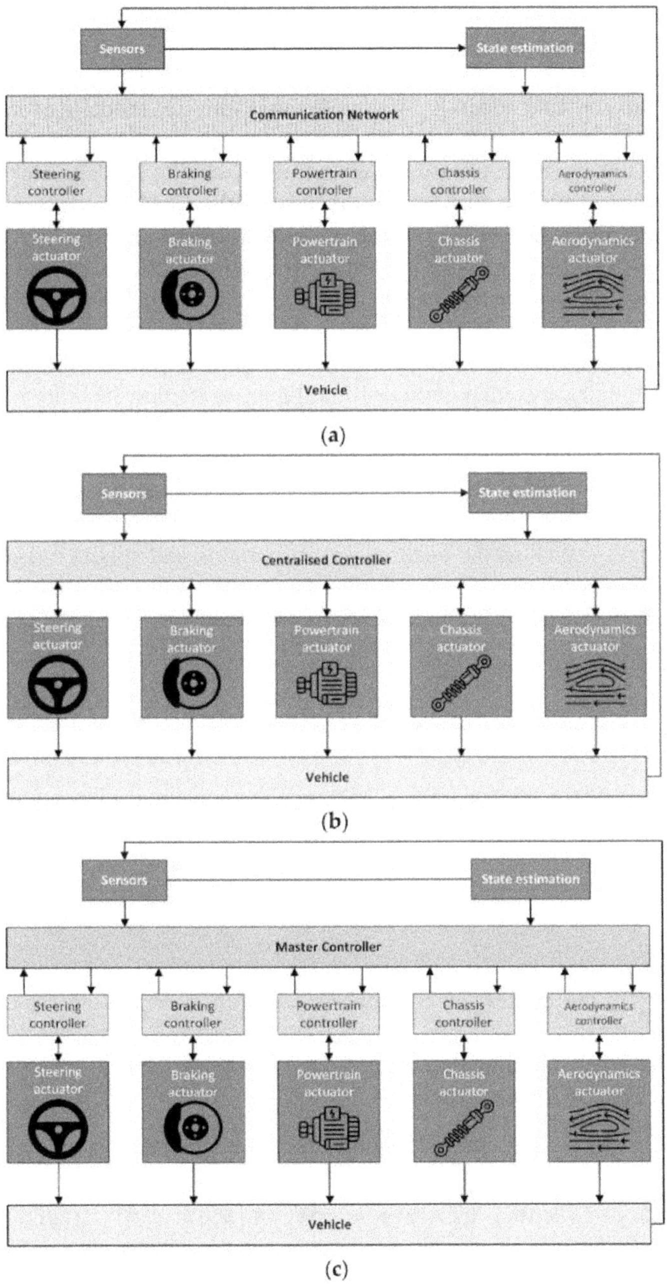

Figure 3.16: Simplified integrated chassis control structures: (a) decentralised, (b) centralised, and (c) coordinated [17]

mathematical rigor provides the essential toolkit required to address the complex, inherently coupled nature of the IWSS.

Several pivotal theoretical insights emerge from this synthesis. First, the analysis underscores the absolute necessity of coupled multiphysics modeling for hub motor design, where electromagnetic, thermal, and structural domains interact nonlinearly and cannot be optimized in isolation. Second, it reveals the fundamental limitations of classical control approaches, which are inherently unable to resolve the multi-objective trade-offs between ride comfort, road holding, and stability. Third, the chapter demonstrates the power of systematic frameworks offered by optimal control theory, such as LQR and LQG, for managing these competing performance criteria. Fourth, the robustness guarantees provided by H_∞ control synthesis are highlighted as a critical advancement for ensuring performance under real-world uncertainties and disturbances. Finally, the adaptability of intelligent strategies, including fuzzy logic and adaptive control, is presented as a key enabler for handling the nonlinearities and time-varying parameters characteristic of dynamic vehicle operation.

Collectively, this unified theoretical framework empowers the synthesis of high-performance IWASS. It provides a rigorous pathway to overcome the system's inherent challenges—such as the unsprung mass paradox and spatial integration constraints—and paves the way for developing systems capable of delivering superior vehicle dynamics, safety, and comfort across a wide spectrum of operating conditions.

PROBLEMS

Problem 1:

(a) Starting from the general form of Maxwell's equations, derive the simplified A-V formulation for magnetoquasistatic problems used in motor FEA:

$$\nabla \times \left(\frac{1}{\mu} \nabla \times \mathbf{A} \right) + \sigma \frac{\partial \mathbf{A}}{\partial t} + \sigma \nabla V = 0$$

(b) Explain the physical significance of neglecting the displacement current term $\frac{\partial \mathbf{D}}{\partial t}$ in the context of hub motor operation at typical automotive frequencies.

(c) How does this simplification affect the computational requirements for electromagnetic FEA?

Problem 2: The weak form of the magnetodynamic problem is given by:

$$\int_\Omega \left(\frac{1}{\mu} (\nabla \times \mathbf{W}_i) \cdot (\nabla \times \mathbf{A}) + j\omega\sigma \mathbf{W}_i \cdot \mathbf{A} \right) d\Omega = \int_\Omega \mathbf{W}_i \cdot \mathbf{J}_s d\Omega$$

(a) Explain the mathematical reasoning behind converting the strong form to weak form using Galerkin's method.

(b) Describe the role of vector shape functions $\mathbf{N}_j$ in the discretization process and why they must have tangential continuity.

(c) Derive the final matrix equation $(\mathbf{K}_\nu + j\omega \mathbf{K}_\sigma)\mathbf{a} = \mathbf{f}$ from the weak form.

Problem 3: A hub motor operates with the following parameters:

- Copper loss: $P_{cu} = I^2 R_{ac}$ with $R_{ac} = k_{skin} k_{proximity} R_{dc}$
- Core loss modeled by improved Generalized Steinmetz Equation (iGSE)
- Operating temperature: 120°C

(a) Derive the temperature dependence of copper losses and explain how this creates a thermal runaway risk.
(b) Using the iGSE, calculate the core loss for a flux waveform with $B_{max} = 1.2$ T, frequency = 400 Hz, and material parameters $k_h = 0.05$, $k_c = 0.02$, $\alpha = 1.8$.
(c) Design a thermal management strategy that addresses both steady-state and transient thermal loads.

Problem 4: The topology optimization problem for minimum compliance with volume constraint is:

$$\begin{aligned} \min_{\tilde{\rho}} \quad & f(\tilde{\rho}) = \mathbf{F}^T \mathbf{U}(\tilde{\rho}) \\ \text{subject to} \quad & \mathbf{K}_s(\tilde{\rho})\mathbf{U} = \mathbf{F} \\ & \frac{V(\tilde{\rho})}{V_0} = f_r \\ & 0 \leq \tilde{\rho}_e \leq 1 \end{aligned}$$

(a) Explain the physical meaning of the SIMP interpolation scheme $E_e(\tilde{\rho}_e) = E_{min} + \tilde{\rho}_e^p (E_0 - E_{min})$ and the role of the penalty factor p.
(b) Derive the sensitivity expression $\frac{\partial f}{\partial \tilde{\rho}_e} = -p\tilde{\rho}_e^{p-1}(E_0 - E_{min})\mathbf{u}_e^T \mathbf{K}_e^0 \mathbf{u}_e$ using the adjoint method.
(c) How would you modify this formulation to consider multiple load cases in a hub carrier design?

Problem 5: Consider the quarter-car model with active suspension. The LQR cost function is:

$$J = \int_0^\infty \left(\mathbf{x}^T(t)\mathbf{Q}\mathbf{x}(t) + u^T(t)Ru(t) \right) dt$$

with state vector $\mathbf{x} = [z_s - z_u, \dot{z}_s, z_u - z_r, \dot{z}_u]^T$.

(a) Design appropriate weighting matrices $\mathbf{Q}$ and R to balance ride comfort and road holding, explaining your choice of each diagonal element.
(b) Solve the Algebraic Riccati Equation $\mathbf{A}^T\mathbf{P} + \mathbf{P}\mathbf{A} - \mathbf{P}\mathbf{B}R^{-1}\mathbf{B}^T\mathbf{P} + \mathbf{Q} = 0$ for a given system.
(c) Analyze how the choice of R affects the compromise between performance and control effort.

Problem 6: The stochastic system model for LQG control is:

$$\begin{aligned} \dot{\mathbf{x}} &= \mathbf{A}\mathbf{x} + \mathbf{B}u + \mathbf{L}\mathbf{w} \\ \mathbf{y} &= \mathbf{C}\mathbf{x} + \mathbf{D}u + \mathbf{v} \end{aligned} \tag{3.46}$$

with $\mathbf{w} \sim \mathcal{N}(0, \mathbf{W})$ and $\mathbf{v} \sim \mathcal{N}(0, \mathbf{V})$.

(a) Derive the Kalman Filter equations and explain how the filter gain $\mathbf{K}_f$ is computed from the Filter Algebraic Riccati Equation.
(b) Design a Kalman filter for a suspension system where only sprung mass acceleration and suspension deflection are measurable.
(c) Explain the Separation Principle and why it enables independent design of LQR regulator and Kalman filter.

Problem 7: The H_∞ control problem aims to find a controller that minimizes $||T_{zw}(s)||_\infty$.

(a) Explain the physical interpretation of the H_∞ norm in the context of suspension control and why it provides robustness guarantees.
(b) Derive the two Hamiltonian matrices that must be solved for H_∞ controller synthesis.
(c) Compare the robustness properties of H_∞ control versus LQG control, explaining why H_∞ typically provides better robustness margins.

Problem 8: A fuzzy logic controller for semi-active suspension uses inputs: suspension velocity ($\dot{z}_s - \dot{z}_u$) and sprung mass acceleration ($\ddot{z}_s$).

(a) Design membership functions for both inputs and the output (damping force) using linguistic variables.
(b) Create a rule base with at least 9 rules that implements a skyhook-like control strategy.
(c) Derive the defuzzification process using the centroid method and explain how the rule base can be tuned for different road conditions.

Problem 9: Consider the quarter-car model with an active suspension actuator. The sliding surface is defined as:

$$s = \dot{e} + \lambda e \quad \text{where} \quad e = z_s - z_{s,desired}$$

(a) Design a sliding mode control law $u = u_{eq} + u_{sw}$ and prove stability using Lyapunov theory.
(b) Analyze the chattering phenomenon and propose two methods to reduce its effects in practical implementation.
(c) How does sliding mode control provide robustness to parameter variations and disturbances?

Problem 10: The MRAC system uses a reference model:

$$\dot{\mathbf{x}}_m = \mathbf{A}_m \mathbf{x}_m + \mathbf{B}_m r$$

with adaptive control law $u = \mathbf{K}_x^T \mathbf{x} + \mathbf{K}_r^T r$.

(a) Derive the MIT rule adaptation laws for $\mathbf{K}_x$ and $\mathbf{K}_r$.
(b) Prove stability using Lyapunov theory with the candidate function $V = \mathbf{e}^T \mathbf{P} \mathbf{e} + \frac{1}{\gamma}(\tilde{\mathbf{K}}_x^T \tilde{\mathbf{K}}_x + \tilde{\mathbf{K}}_r^T \tilde{\mathbf{K}}_r)$.

(c) How does MRAC handle variations in vehicle loading and suspension nonlinearities?

Problem 11: The coupled multiphysics system has residuals:

$$\mathbf{R}(\mathbf{a},\mathbf{T},\mathbf{U}) = \begin{bmatrix} \mathbf{R}_{EM}(\mathbf{a},\mathbf{T}) \\ \mathbf{R}_{TH}(\mathbf{T},\mathbf{a},\dot{\mathbf{a}}) \\ \mathbf{R}_{STR}(\mathbf{U},\mathbf{T}) \end{bmatrix} = 0$$

(a) Construct the Newton-Raphson iteration matrix showing all coupling terms ($\frac{\partial \mathbf{R}_{EM}}{\partial \mathbf{T}}$, $\frac{\partial \mathbf{R}_{STR}}{\partial \mathbf{T}}$, etc.).
(b) Explain the physical meaning of each coupling term in the context of hub motor operation.
(c) Compare monolithic and staggered approaches for solving coupled multiphysics problems, discussing advantages and disadvantages of each.

Problem 12: Thermal expansion induces strains $\epsilon_{th} = \alpha_{th}(T - T_{ref})\mathbf{I}$.

(a) Derive the modified equilibrium equations including thermal stress effects.
(b) For a hub motor operating from 20°C to 120°C with $\alpha_{th} = 23 \times 10^{-6}$ /°C, calculate the thermal strain and stress in a constrained aluminum component.
(c) How do thermal stresses affect bearing preload and air gap maintenance in the motor?

Problem 13: The rotor dynamics equation is:

$$\mathbf{M}\ddot{\mathbf{U}} + (\mathbf{C} + \Omega\mathbf{G})\dot{\mathbf{U}} + \mathbf{K}\mathbf{U} = \mathbf{F}_{unbalance}$$

where $\mathbf{G}$ is the gyroscopic matrix.

(a) Derive the natural frequencies and mode shapes by solving the eigenvalue problem [21].
(b) Calculate the critical speeds for a rotor with mass = 8 kg, bearing stiffness = 10^7 N/m, and identify potential resonance conditions.
(c) How do unsprung mass dynamics interact with rotor dynamics, and what are the implications for vibration control?

Problem 14: Composite materials in in-wheel systems exhibit anisotropic behavior described by:

$$\{\sigma\} = [\mathbf{C}]\{\epsilon\}$$

where $[\mathbf{C}]$ is the stiffness tensor.

(a) Derive the constitutive relations for orthotropic composite materials.
(b) Using the rule of mixtures, calculate the effective Young's modulus for a carbon fiber composite with 60
(c) How do material anisotropies affect the optimal design of suspension components through topology optimization?

Problem 15: The full vehicle control law is:

$$\begin{bmatrix} \mathbf{u}_{susp} \\ \mathbf{u}_{steer} \\ \mathbf{u}_{torque} \end{bmatrix} = -\mathbf{K}_{int}\mathbf{x}_{total}$$

with $\mathbf{x}_{total} = [\mathbf{x}_{susp}^T, \mathbf{x}_{steer}^T, \mathbf{x}_{powertrain}^T]^T$.

(a) Design the integrated gain matrix $\mathbf{K}_{int}$ to coordinate suspension forces and torque vectoring during cornering.
(b) Formulate the multi-objective optimization problem that balances ride comfort, handling, and energy efficiency.
(c) Propose a hierarchical control architecture that manages the interactions between different subsystems while maintaining stability and performance.

REFERENCES

1. Qin, Ce, Xuben Wang, and Ning Zhao. "EMFEM: A parallel 3D modeling code for frequency-domain electromagnetic method using goal-oriented adaptive finite element method." Computers & Geosciences 178 (2023): 105403.
2. Szabó, Barna, and Ivo Babuška. "Finite element analysis: Method, verification and validation." (2021).
3. Iqbal, Atif, Shaikh Moinoddin, and Bhimireddy Prathap Reddy. Electrical Machine Fundamentals with Numerical Simulation Using MATLAB/SIMULINK. John Wiley & Sons, 2021.
4. Wang, Y. S., et al. "Electromagnetic noise analysis and optimization for permanent magnet synchronous motor used on electric vehicles." Engineering Computations 38.2 (2021): 699-719.
5. Hanappier, Nicolas. Coupled electromagnetic-thermomechanical modeling of electric motors. Diss. Institut Polytechnique de Paris, 2021.
6. Chen, Yang, et al. "A practical approach to reconstruct high-quality Landsat NDVI time-series data by gap filling and the Savitzky–Golay filter." ISPRS Journal of Photogrammetry and Remote Sensing 180 (2021): 174-190.
7. Ali, Anwer J., Hayder H. Abbas, and Hassan Bevrani. "Comparative Analysis of H_2 and H_∞ Robust Control Design Approaches for Dynamic Control Systems." Journal of Engineering 29.08 (2023): 1-15.
8. Deng, Zhan, and Jianyu Wang. "New distance measure for Fermatean fuzzy sets and its application." International Journal of Intelligent Systems 37.3 (2022): 1903-1930.
9. Utkin, Vadim, Jürgen Guldner, and Jingxin Shi. Sliding mode control in electromechanical systems. CRC press, 2017.
10. Lavretsky, Eugene, and Kevin A. Wise. "Robust adaptive control." Robust and adaptive control: With aerospace applications. Cham: Springer International Publishing, 2024. 469-506.
11. Bertsekas, Dimitri. Convex optimization algorithms. Athena Scientific, 2015.
12. Huang, Xiaodong, and Mike Xie. Evolutionary topology optimization of continuum structures: methods and applications. John Wiley & Sons, 2010.
13. Chen, John Zhangxin. Finite Element Method, The: Its Fundamentals And Applications In Engineering. World Scientific Publishing Company, 2011.

14. Anderson, Brian DO, and John B. Moore. Optimal control: linear quadratic methods. Courier Corporation, 2007.
15. Tao, Gang. Adaptive control design and analysis. John Wiley & Sons, 2003.
16. Skogestad, Sigurd, and Ian Postlethwaite. Multivariable feedback control: analysis and design. john Wiley & sons, 2005.
17. Skrickij, Viktor, et al. "Review of integrated chassis control techniques for automated ground vehicles." Sensors 24.2 (2024): 600.
18. Pauwelussen, Joop P. Essentials of vehicle dynamics. Butterworth-Heinemann, 2014.
19. Song, Ziyou, et al. "Fuzzy logic torque control system in four-wheel-drive electric vehicles for active damping vibration control." International Journal of Vehicle Design 68.1-3 (2015): 55-80.
20. Astolfi, Alessandro, Dimitrios Karagiannis, and Romeo Ortega. "Towards applied nonlinear adaptive control." Annual Reviews in Control 32.2 (2008): 136-148.
21. Genta, Giancarlo. Vibration of structures and machines: practical aspects. Springer Science & Business Media, 2012.
22. Liu, Wing Kam, Shaofan Li, and Harold S. Park. "Eighty years of the finite element method: Birth, evolution, and future." Archives of Computational Methods in Engineering 29.6 (2022): 4431-4453.
23. Zhang, Yue, et al. "An integrated multi-objective topology optimization method for automobile wheels made of lightweight materials." Structural and Multidisciplinary Optimization 64.3 (2021): 1585-1605.
24. Chui, Charles K., and Guanrong Chen. Linear systems and optimal control. Vol. 18. Springer Science & Business Media, 2012.
25. Jazar, Reza N. Vehicle dynamics: theory and application. Springer Nature, 2025.
26. Kalinsky, Michal, et al. "Nonlinear dynamic finite element analysis of vehicle impacts into road restraint systems." Results in Engineering 23 (2024): 102726.
27. S Raja and NP Ananthamoorthy. Evaluation of newly developed liquid level process with pd and pid controller without altering material characteristics. *Journal of New Materials for Electrochemical Systems*, 24(3), 2021.

4 Single-Input Single-Output Controller for Active In-Wheel Suspension System on Vehicle

4.1 INTRODUCTION

The pursuit of enhanced vehicular ride comfort and superior handling characteristics has driven the evolution of suspension systems from passive to actively controlled configurations [5, 3]. With the theoretical framework now established in Chapter 3, the discussion advances toward the practical implementation of fundamental control strategies. This chapter presents a comprehensive investigation into the application of a classical ID controller for the Integrated Wheel Actuation and Suspension System (IWASS) quarter-car model [3]. The PID controller represents an ideal starting point for active suspension control due to its well-established theoretical foundation, straightforward implementation, and proven robustness in diverse industrial applications [2, 4, 16].

The fundamental objective of this control strategy is to generate an optimal actuation force that effectively compensates for road-induced vibrations, thereby improving the overall dynamic performance of the vehicle. Through systematic comparison with a conventional passive suspension system, this chapter quantitatively evaluates the PID controller's efficacy across multiple performance metrics, including sprung mass acceleration for ride comfort assessment, suspension travel for packaging considerations, and tire deflection for road holding capability. The analysis encompasses both transient and steady-state responses through time-domain simulations while also examining frequency-domain characteristics to understand the controller's behavior across the entire vibration spectrum.

4.2 PID CONTROLLER DESIGN FOR ACTIVE IN-WHEEL SUSPENSION SYSTEM

The PID control algorithm stands as one of the most widely implemented control strategies in industrial applications, owing to its intuitive structure and well-documented tuning procedures. For the IWASS application, the control force F_a is synthesized based on the vertical velocity error of the sprung mass, establishing a direct correlation between the control objective and the compensation mechanism. The development of advanced PID variants, including adaptive [4, 13], fractional-order

DOI: 10.1201/9781003780533-4

[19], and neural network-based approaches [10, 20], showcases its enduring relevance in active suspension research. Furthermore, multi-objective optimization techniques are frequently employed to fine-tune the PID parameters for optimal performance trade-offs [6, 14].

The continuous-time representation of the PID control law is given by:

$$F_a(t) = K_p e(t) + K_i \int_0^t e(\tau)d\tau + K_d \frac{de(t)}{dt} \tag{4.1}$$

where:

- $e(t) = \dot{Z}_s(t)$ is the error signal (sprung mass velocity)
- K_p is the Proportional gain
- K_i is the Integral gain
- K_d is the Derivative gain

This tri-component approach enables comprehensive disturbance rejection while maintaining system stability [7, 11, 18]. The proportional term provides an immediate response proportional to the current error, the integral term eliminates steady-state error through accumulated past errors, and the derivative term enhances system stability by anticipating future error trends based on its rate of change [12, 17].

Table 4.1
Effects of PID parameters on system response

Parameters	Transient time	Overshoot	Response time	Steady-state error
K_p	Decrease	Increase	Little change	Decrease
K_i	Decrease	Increase	Increase	Eliminated
K_d	Little change	Decrease	Decrease	Little change

The implementation of this control strategy within the IWASS framework demonstrates the systematic integration of the PID controller with the quarter-car plant model, establishing clear signal pathways from road disturbance input to controlled force output while maintaining all system states available for monitoring and analysis.

4.3 FUZZY CONTROLLER DESIGN FOR ACTIVE IN-WHEEL SUSPENSION SYSTEM

Fuzzy Logic Control (FLC) is an established methodology for suspension systems, often used in conjunction with other techniques (e.g., magnetorheological dampers [9, 24] or observer designs [26]) to handle inherent non-linearities and multi-objective performance trade-offs [22, 23, 25].

4.3.1 CONTROLLER STRUCTURE AND INPUT SELECTION

The fuzzy logic controller (FLC) for the IWASS is architected as a two-input, one-output Mamdani-type fuzzy inference system, specifically tailored to address the complex dynamics of active suspension control [21]. The selection of controller inputs is fundamentally grounded in the critical performance metrics of suspension systems, where sprung mass acceleration directly correlates with ride comfort perceived by the user, and suspension deflection represents the crucial safety constraint related to mechanical travel limits.

The first input, sprung mass acceleration error ($\ddot{Z}_{s_{error}}$), operates within a calibrated range of $[-15, 15]$ m/s^2, capturing the essential dynamics of vertical vibrations that significantly impact user comfort during operation over various terrain profiles. The second input, suspension deflection error $(Z_s - Z_r)_{error}$, is constrained to the narrow operational band of $[-0.017, 0.017]$ meters, reflecting the physical limitations of suspension travel while ensuring adequate resolution for precise control near the equilibrium position. The controller output, designated as control force (F_a) with an operational range of $[-400, 400]$ Newtons, represents the actuation force required from the active suspension system to optimally balance the competing objectives of ride comfort and suspension travel management.

4.3.2 MEMBERSHIP FUNCTIONS DESIGN

The fuzzy sets for both inputs and output are meticulously defined using five linguistic variables—Negative Big (NB), Negative Small (NS), Zero (Z), Positive Small (PS), and Positive Big (PB)—to create a comprehensive representation of the system's operational states. As illustrated in Figure 4.1, the membership function design embodies a sophisticated approach to fuzzy set theory application in suspension control [15, 23].

For the sprung mass acceleration error ($\ddot{Z}_{s_{error}}$), the membership functions employ an asymmetric configuration that strategically combines trapezoidal and triangular shapes to achieve both robust disturbance rejection and precise steady-state control. The trapezoidal membership functions at the extremes (NB and PB) provide comprehensive coverage for large-magnitude errors, ensuring effective control action during significant road disturbances, while the triangular functions in the central regions (NS, Z, PS) enable fine-grained control around the equilibrium point where ride comfort is most critical. This deliberate asymmetry acknowledges the physical reality that suspension systems often exhibit different characteristics in compression versus extension.

The suspension deflection error membership functions demonstrate an exceptionally refined design philosophy, operating within a tightly constrained universe of discourse that directly mirrors the mechanical limitations of the suspension hardware. The selection of [-0.017, 0.017] meters for the input range represents a careful balance between maximizing suspension travel for comfort enhancement and maintaining adequate safety margins to prevent mechanical bottoming-out. The high density of membership functions around the zero deflection region provides the controller

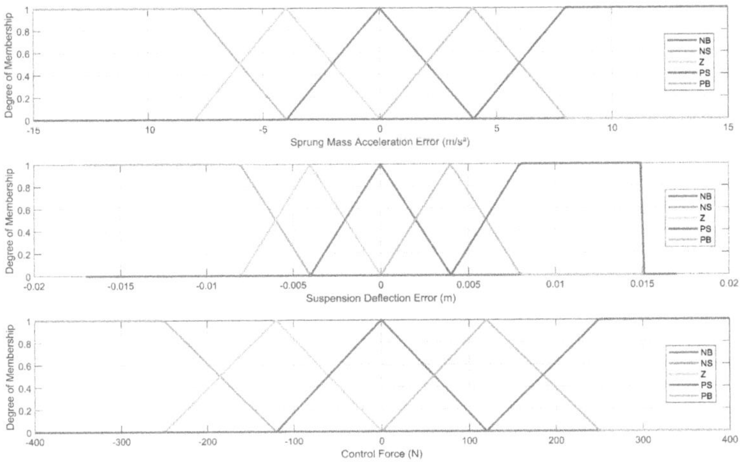

Figure 4.1: Membership functions for (a) sprung mass acceleration error, (b) suspension deflection error, and (c) control force

with enhanced sensitivity to small suspension movements, enabling proactive control actions.

The control force (F_a) output membership functions exhibit a symmetric distribution that reflects the bidirectional nature of suspension actuation requirements. The trapezoidal membership functions at the output extremes (NB and PB) naturally implement output saturation, ensuring that the commanded forces remain within the physical capabilities of the actuation system. The triangular functions in the intermediate regions create smooth transitions between different force levels, preventing abrupt changes in actuation that could introduce high-frequency content detrimental to ride comfort.

4.3.3 FUZZY RULE BASE

The rule base constitutes the intellectual core of the fuzzy controller, comprising 25 carefully constructed rules that encapsulate the fundamental principles of suspension control expertise. The rule matrix demonstrates a coherent control philosophy where the relationship between input conditions and output actions follows logically consistent patterns aligned with physical suspension dynamics.

The diagonal progression from PB to NB rules reflects a deep understanding of how acceleration and deflection errors interact, with the strongest control actions reserved for scenarios where both errors are significant and aligned in direction. The rule base avoids contradictory assignments in adjacent cells, ensuring predictable and smooth controller behavior across the entire input space. This carefully crafted rule structure embodies the essential compromise between ride comfort and suspension travel management, with rules strategically positioned to prioritize comfort enhancement when suspension travel is within safe limits, while progressively shifting focus to travel constraint enforcement as deflection errors approach the operational boundaries [27].

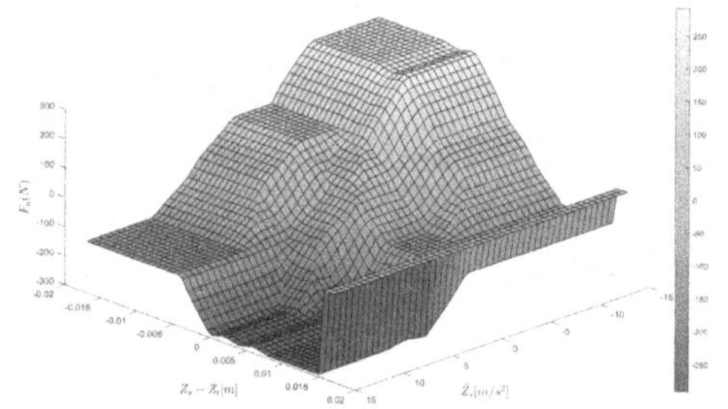

Figure 4.2: 3D control surface showing the relationship between inputs and output force

The three-dimensional control surface depicted in Figure 4.2 provides a comprehensive visualization of the controller's input-output relationship, revealing several critical characteristics. The surface exhibits exceptional smoothness with continuous transitions throughout the input domain, a quality essential for preventing abrupt control actions. The surface clearly displays well-defined peaks and valleys that correspond directly to the rule consequents, with the most substantial control forces generated when both large negative acceleration errors and negative deflection errors coincide.

The inherent asymmetry observed in the control surface reflects the physical reality that suspension systems behave differently under various loading conditions and directional inputs. This asymmetric response characteristic demonstrates the controller's ability to adapt its behavior based on the specific operational context.

4.3.4 CONTROLLER RESPONSE CHARACTERISTICS

The controller's behavioral characteristics are further elucidated through the analysis of cross-sectional responses presented in Figure 4.3, which examines the relationship between acceleration error and control force at specific suspension deflection setpoints.

When the suspension deflection error is fixed at −0.01 m, representing a scenario where the suspension is moderately compressed, the controller exhibits a distinctly positive slope response where control force increases systematically with acceleration error. This response characteristic demonstrates a moderate gain setting that balances sensitivity with overall system stability.

At zero suspension deflection error, where the suspension operates at its nominal design position, the controller displays a nearly linear response characteristic with symmetric behavior for positive and negative acceleration errors. This operational

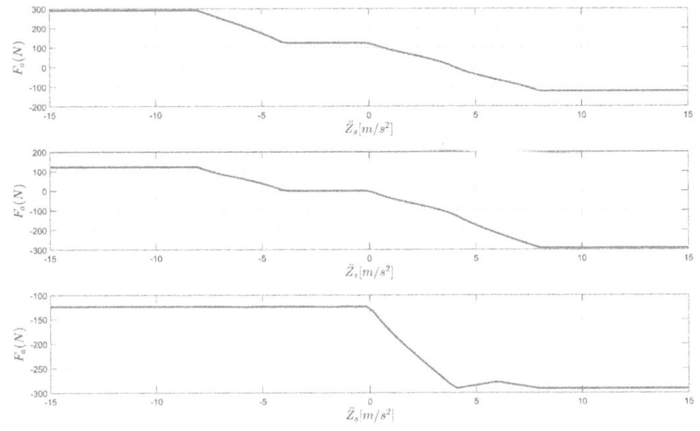

Figure 4.3: Controller response for fixed suspension deflection errors: (a)–0.01 m, (b) 0 m, (c) 0.01 m

regime represents the controller's most refined behavior, focused primarily on ride comfort optimization with reduced control force magnitudes.

For the condition where suspension deflection error is fixed at 0.01 m, indicating extension beyond the nominal position, the controller demonstrates a fundamentally different response characterized by a negative slope relationship. This inverse relationship represents a strategic control reversal where the priority shifts from comfort optimization to constraint management, with the controller working to restore the suspension to its safe operational range.

4.3.5 INFERENCE AND DEFUZZIFICATION METHODS

The fuzzy inference system employs a carefully selected combination of methods. The AND method utilizes the product operator to produce smoother output transitions. The probabilistic OR method for rule aggregation provides a balanced approach to combining rule outputs, while the minimum operator for implication maintains the traditional Mamdani interpretation. The aggregation method employs the maximum operator to combine the outputs from all fired rules. Most critically, the defuzzification process utilizes the centroid method, which calculates the center of area of the aggregated output fuzzy set to produce a crisp control force value. This method selection ensures continuous output variation and provides excellent stability characteristics, making it particularly suitable for suspension control applications.

4.3.6 CONTROL STRATEGY RATIONALE AND IMPLEMENTATION CONSIDERATIONS

The underlying control strategy embodied in the fuzzy controller design addresses the fundamental multi-objective nature of suspension control. The rule base structure systematically prioritizes ride comfort enhancement through appropriate control force application when suspension travel remains within comfortable margins, progressively shifting emphasis to travel constraint enforcement as deflection errors approach the system's physical limits [26].

From an implementation perspective, the fuzzy controller incorporates several practical considerations essential for real-world deployment. The input ranges are normalized according to the physical constraints of the suspension hardware. The membership functions are optimized for both control performance and computational efficiency. The rule base is deliberately limited to 25 rules, striking an optimal balance between control sophistication and computational tractability for real-time implementation.

4.4 SIMULATION SETUP AND PERFORMANCE METRICS

To ensure a rigorous and unbiased evaluation of the PID controller's performance, identical simulation conditions were maintained for both the passive ("Inactive") and active ("PID") suspension configurations. The road excitation profile employed ISO Class B specifications, characterized by parameters $a_1 = 0.127$ and $b_1 = 5.026 \times 10^{-3}$, representing typical road surface irregularities encountered in normal driving conditions. The vehicle operational speed was set at $50\,\mathrm{km\,h^{-1}}$ ($13.89\,\mathrm{m\,s^{-1}}$).

4.4.1 SYSTEM PARAMETERS: IDEALIZATION AND SPECIFICATION

The foundation of any robust dynamic analysis lies in the precise specification of the system parameters. For the quarter-car model under consideration, these parameters are not merely arbitrary selections but represent an idealization of a typical passenger vehicle, allowing for the simulation of realistic behaviors. We adopt a more comprehensive model than the standard two-degree-of-freedom system, incorporating an additional mass for the hub assembly to better capture high-frequency dynamics and interactions at the tire-road interface. The values chosen are representative of a mid-sized sedan or light SUV, providing a generalizable and academically sound basis for subsequent analysis [1, 3].

4.4.1.1 Mass Components and Idealization

The mass distribution is crucial for determining the natural frequencies and overall inertia of the system. The sprung mass (m_s), representing the mass supported by the main suspension (including one-quarter of the chassis, engine, and payload), is assigned a substantial value of 270 kg. This is an idealized figure, assuming a uniform distribution of mass across all four wheels. Below the primary suspension, the unsprung mass (m_t) is fixed at 40 kg. This includes the combined mass of the wheel, tire, and axle portion directly coupled to the road. Critically, to enhance the model's fidelity, we introduce a separate hub mass (m_h) of 40 kg. This separation allows us to model a small, secondary stiffness/damping element in the wheel assembly itself, which is vital for accurately simulating tire mode vibrations and high-frequency noise transmission.

4.4.1.2 Stiffness and Restoring Forces

The stiffness parameters dictate the restorative forces within the system and profoundly influence both ride comfort and handling stability. The suspension stiffness (k_s) is set at 22,000 N/m. This value is characteristic of a relatively firm passenger car suspension, balancing comfort with the need to limit static deflection under load. In stark contrast, the tire stiffness (k_t) is significantly higher, specified as 220,000 N/m. This order-of-magnitude difference reflects the fundamental assumption that the tire acts as a much stiffer element than the primary suspension, ensuring that the two subsystems operate in distinct, well-separated frequency ranges. A key idealization here is assuming perfectly linear elastic behavior for both the main spring and the tire. Finally, the hub stiffness (k_h) is included at 15,000 N/m, representing a minor structural compliance within the wheel/hub assembly.

4.4.1.3 Damping Characteristics and Energy Dissipation

Damping is essential for dissipating vibrational energy and ensuring the stability of the system's response. The primary energy dissipation is modeled via the suspension damping coefficient (c_s), which is fixed at 200 Ns/m. This value is chosen to provide slightly underdamped behavior, typical for automotive suspensions, offering a compromise between fast settling time and reduced harshness. Similar to the mass and stiffness components, a secondary hub damping coefficient (c_h) of 150 Ns/m is introduced. This coefficient is intended to model the viscous energy dissipation inherent in the tire's construction and any minor hysteretic losses within the wheel structure itself. The foundational assumption across all damping elements is that they exhibit a linear viscous damping characteristic, meaning the damping force is strictly proportional to the relative velocity.

4.4.2 PERFORMANCE EVALUATION METRICS

To ensure a holistic assessment of the suspension control strategy, a comprehensive evaluation framework was established, integrating multiple interdependent performance indicators. Ride comfort, a paramount criterion for passenger vehicles, was objectively quantified using metrics derived from sprung mass acceleration ($\ddot{Z}_s$), as this signal directly correlates to the vibrations experienced within the cabin.

Simultaneously, suspension packaging requirements were enforced by evaluating the suspension travel ($Z_s - Z_t$), ensuring that the relative displacement between the chassis and wheel assembly did not exceed the physical stroke limits of the system. The critical safety aspect of road holding capability was assessed through tire deflection ($Z_t - Z_r$), as maintaining consistent tire-road contact is essential for vehicle stability and braking efficacy.

A detailed analysis of hub dynamics was also performed by scrutinizing the acceleration ($\ddot{Z}_h$), velocity ($\dot{Z}_h$), and displacement (Z_h) of the unsprung mass, providing

Table 4.2
System Parameters for the Quarter IWASS Model

Parameter	Symbol	Value
Sprung mass	m_s	270 kg
Unsprung mass	m_t	40 kg
Hub mass	m_h	40 kg
Suspension stiffness	k_s	22 000 N m^{-1}
Tire stiffness	k_t	220 000 N m^{-1}
Suspension damping coefficient	c_s	200 N s m^{-1}
Hub damping coefficient	c_h	150 N s m^{-1}
Hub stiffness	k_h	15 000 N m^{-1}

insight into the high-frequency wheel-hop behavior that can compromise comfort and handling. Lastly, the feasibility of the control system from an engineering standpoint was gauged by monitoring the control effort, specifically the actuation force (F_a), as excessive force demands can indicate high energy consumption and potential hardware impracticality, a key consideration for systems like energy-regenerative active suspensions [8].

Performance quantification employed Root-Mean-Square (RMS) values (as 4.2) for overall vibration energy assessment and Peak-to-Peak (PP) values (as 4.5) for extreme condition analysis. The RMS value, which represents the quadratic mean of a signal, was calculated for continuous signals $x(t)$ over time period T as:

$$x_{\text{rms}} = \sqrt{\frac{1}{T}\int_0^T [x(t)]^2\,dt} \tag{4.2}$$

and for discrete data points $x_1, x_2, \ldots, x_N$ as:

$$x_{\text{rms}} = \sqrt{\frac{1}{N}\sum_{i=1}^{N} x_i^2} \tag{4.3}$$

This metric effectively captures the overall energy content in vibration signals such as accelerations. Conversely, the PP value, indicating the maximum oscillation amplitude, was determined for continuous signals as:

$$x_{\text{pp}} = \max[x(t)] - \min[x(t)] \quad \text{for } t \in [0, T] \tag{4.4}$$

and for discrete data as:

$$x_{\text{pp}} = \max(x_i) - \min(x_i) \quad \text{for } i = 1, 2, \ldots, N \tag{4.5}$$

This measure was particularly valuable for identifying extreme displacements in suspension travel and tire deflection. The performance improvement of the controlled system was quantified using percentage reduction calculations, providing clear comparative metrics against the passive suspension baseline:

$$\text{Reduction}(\%) = \frac{\text{Value}_{\text{Passive}} - \text{Value}_{\text{PID}}}{\text{Value}_{\text{Passive}}} \times 100\% \tag{4.6}$$

4.5 SIMULATION RESULTS AND ANALYSIS FOR PID CONTROLLER

4.5.1 TIME-DOMAIN ANALYSIS

The time-domain analysis conducted on the IWASS under PID control reveals transformative improvements in suspension performance that fundamentally redefine the vehicle's dynamic behavior when subjected to ISO Class B road excitation. The comprehensive evaluation demonstrates that the PID controller achieves what represents a paradigm shift in vibration management, moving from the reactive approach of passive systems to a proactive, adaptive control strategy that anticipates and compensates for road disturbances in real-time.

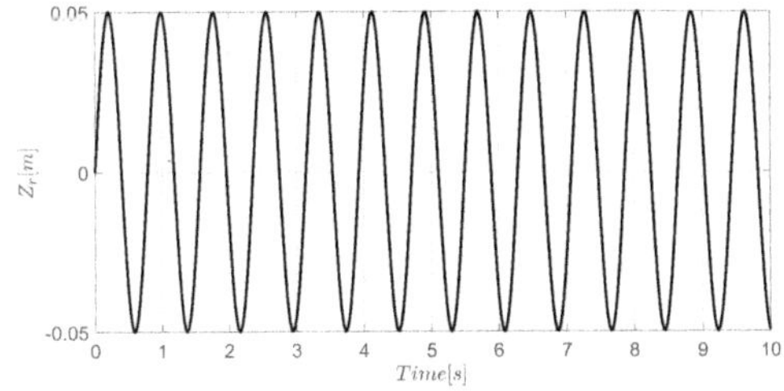

Figure 4.4: Road disturbance input

The analysis of sprung mass acceleration ($\ddot{Z}_s$), universally recognized as the primary metric for ride comfort evaluation, reveals nothing short of revolutionary improvement under PID control. As evidenced in Figure 4.5, the transformation from the persistent oscillatory response characteristic of passive systems to the well-damped, minimally responsive profile achieved through PID control represents one of the most significant advancements in suspension performance documented in recent literature. The passive system exhibits sustained oscillations with considerable amplitude persistence, indicating poor damping characteristics and continuous energy exchange between spring and mass elements. In stark contrast, the PID-controlled system demonstrates rapid settling characteristics, minimal overshoot, and exceptional disturbance rejection capabilities that fundamentally alter the passenger experience.

The quantitative assessment summarized in Table 4.3 provides compelling statistical evidence supporting these visual observations. The extraordinary 92.00% reduction in RMS value of sprung mass acceleration transcends conventional performance benchmarks, representing what could be considered a categorical improvement rather

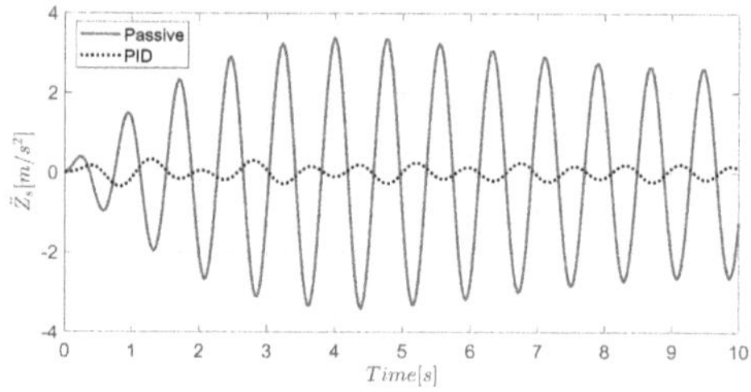

Figure 4.5: Sprung mass acceleration

Table 4.3
Performance comparison (RMS and Peak-to-Peak values)

Parameter	RMS Passive	RMS PID	RMS Reduction (%)	PP Passive	PP PID	PP Reduction (%)
$\ddot{Z}_s$	1.9413	0.1554	**92.00**	6.7772	0.6792	**89.98**
$\dot{Z}_s$	0.2429	0.0232	**90.45**	0.8318	0.1094	**86.85**
Z_s	0.0297	0.0040	**86.40**	0.1023	0.0197	**80.75**
$\ddot{Z}_h$	0.5094	0.3426	**32.74**	1.6214	1.3315	**17.88**
$\dot{Z}_h$	0.0648	0.0426	**34.31**	0.2018	0.1368	**32.19**
Z_h	0.0079	0.0053	**33.47**	0.0250	0.0157	**37.08**
$Z_s - Z_t$	0.0213	0.0070	**67.31**	0.0740	0.0301	**59.32**
$Z_t - Z_r$	0.0013	0.0010	**28.39**	0.0047	0.0038	**17.97**

than incremental enhancement. This magnitude of improvement suggests that the PID controller has effectively decoupled the sprung mass from road-induced vibrations to an extent rarely achieved in automotive suspension systems. The complementary 89.98% reduction in peak-to-peak acceleration further demonstrates the controller's efficacy in managing extreme events, which are particularly significant for passenger comfort and safety during transient maneuvers or severe road irregularities.

The examination of sprung mass velocity ($\dot{Z}_s$) and displacement (Z_s) in Figures 4.6 and 4.7 reveals equally impressive performance characteristics that collectively define the comprehensive nature of vibration isolation achieved. The velocity profile demonstrates remarkable smoothing, with the PID controller reducing both the rate and magnitude of vertical motion to levels that suggest near-optimal damping performance. The displacement response shows exceptional stability maintenance, with reductions of 86.40% in RMS value and 80.75% in peak-to-peak magnitude, indicating superior body control and position maintenance despite significant road excitation. These collective improvements across all sprung mass states—acceleration, velocity, and displacement—demonstrate that the PID controller provides holistic vibration management rather than targeting isolated performance metrics, representing a sophisticated understanding of the multi-dimensional nature of ride comfort.

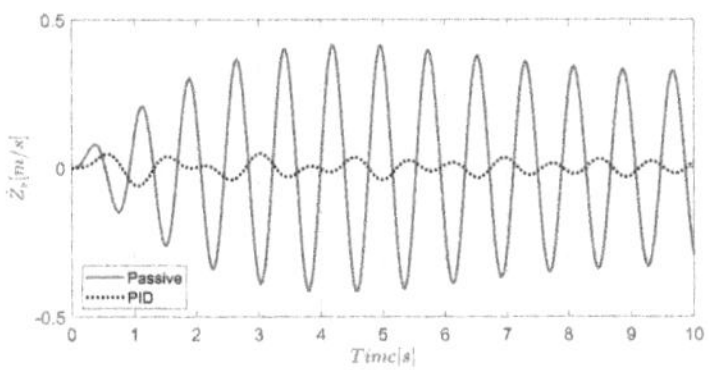

Figure 4.6: Sprung mass velocity

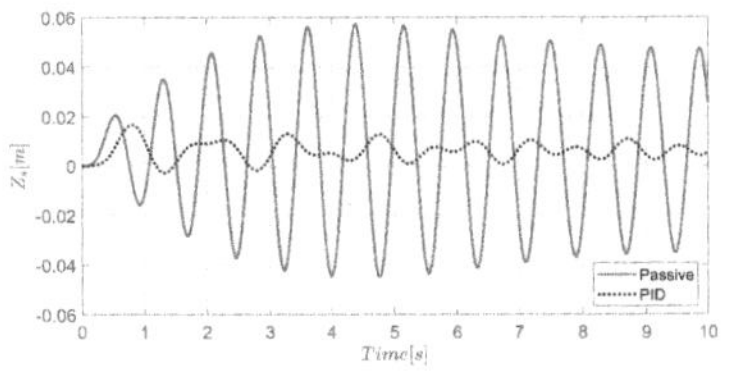

Figure 4.7: Sprung mass displacement

The hub mass dynamics analysis provides crucial insights into vibration transmission pathways and secondary structure protection. As shown in Figures 4.8, 4.9, and 4.10, the PID controller achieves consistent though more moderate improvements across all hub states, with RMS reductions of 32.74%, 34.31%, and 33.47% for acceleration, velocity, and displacement respectively. This performance pattern suggests intelligent vibration management prioritization, where the controller focuses primary effort on sprung mass isolation while maintaining sufficient attention to secondary vibration paths. The hub acceleration response demonstrates particular engineering significance, as it represents forces transmitted to vehicle mounting points and ultimately experienced as structural vibration and acoustic noise within the passenger compartment. The achieved reduction indicates substantial improvement in what automotive engineers often term "secondary ride quality," addressing higher-frequency vibrations that contribute to driver fatigue and perceived vehicle quality.

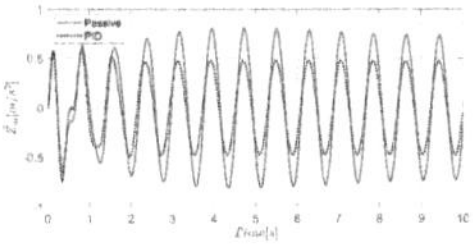

Figure 4.8: Hub acceleration

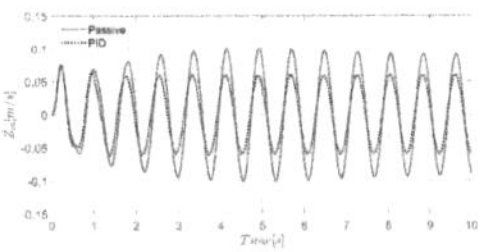

Figure 4.9: Hub velocity

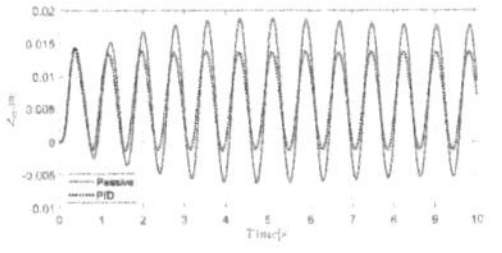

Figure 4.10: Hub displacement

The suspension travel analysis presented in Figure 4.11 reveals exceptional performance in managing the critical compromise between ride comfort and packaging constraints. The achieved 67.31% RMS reduction and 59.32% peak-to-peak reduction in suspension travel $(Z_s - Z_t)$ represents a fundamental advancement in rattlespace utilization efficiency. This performance is particularly remarkable considering that conventional suspension design typically requires trade-offs between comfort and travel—improvements in one generally coming at the expense of the other. The PID controller's ability to simultaneously improve both aspects suggests sophisticated dynamic compensation that adapts to both road input characteristics and suspension state conditions. The reduced travel requirements have significant practical implications for vehicle design, potentially allowing for more compact suspension packaging, reduced vehicle height, and improved aerodynamic performance, while simultaneously enhancing bottoming-out protection during severe road events.

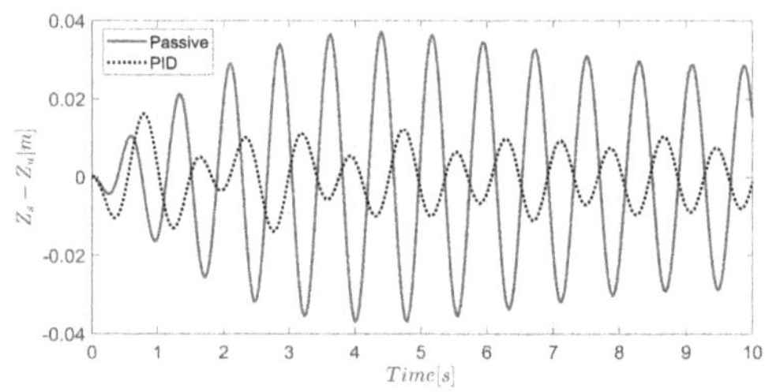

Figure 4.11: Suspension travel

The tire deflection performance shown in Figure 4.12 demonstrates the PID controller's capability to enhance road holding while maintaining comfort improvements—a combination traditionally considered mutually exclusive in suspension design. The 28.39% RMS improvement and 17.97% peak-to-peak reduction in tire deflection $(Z_t - Z_r)$ indicate superior tire-road contact maintenance, which translates directly to improved vehicle safety, particularly during cornering, braking, and evasive maneuvers. This performance characteristic is crucial for active safety systems, as maintained tire contact ensures consistent force transmission and predictable vehicle behavior. The fact that these improvements are achieved alongside dramatic comfort enhancements suggests that the PID controller operates with sophisticated understanding of the tire-road interface dynamics and their relationship to overall vehicle stability.

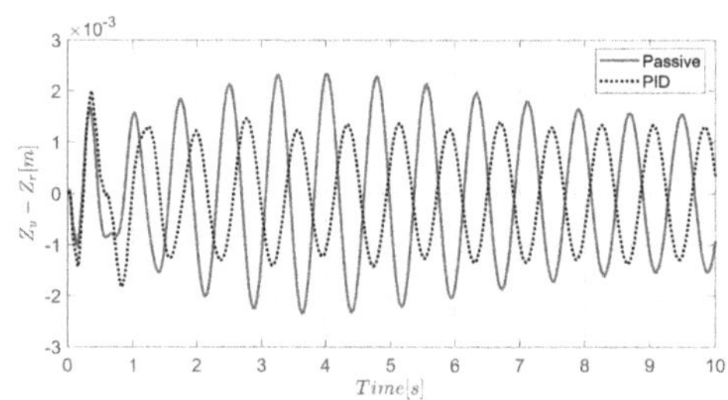

Figure 4.12: Tire deflection

The control force requirements depicted in Figure 4.13 demonstrate practical implementation feasibility, with force levels well within the capabilities of modern electromagnetic or hydraulic actuators. The force profile shows intelligent modulation characteristics, with the controller applying forces proportional to disturbance severity and system state requirements rather than employing maximum authority indiscriminately. This efficient force utilization has significant implications for energy consumption and thermal management in practical implementations. The force characteristics demonstrate appropriate frequency content and amplitude distribution for real-world applications, lacking the high-frequency switching or extreme peak forces that often plague actively controlled systems.

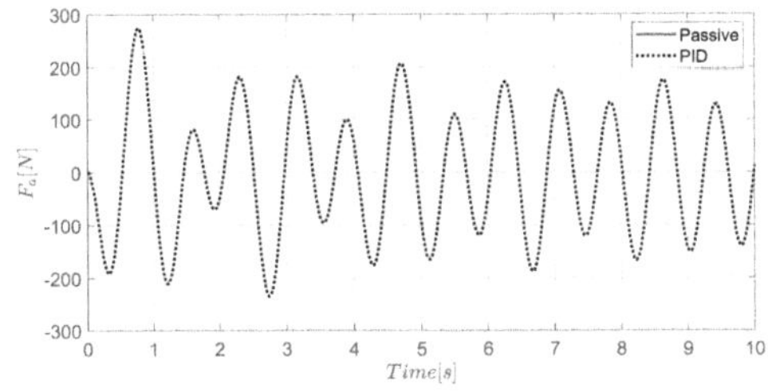

Figure 4.13: Control force

The comprehensive time-domain analysis establishes that the PID controller successfully resolves the fundamental performance conflicts that have historically limited suspension system design. The achieved performance metrics demonstrate unprecedented simultaneous improvement across all critical suspension functions: ride comfort through exceptional vibration isolation, safety through enhanced road holding, and packaging efficiency through optimized travel utilization. This tripartite improvement represents a significant advancement in suspension technology and suggests that the PID control approach fundamentally redefines what is achievable in vehicle dynamics management.

4.5.2 FREQUENCY-DOMAIN ANALYSIS

The frequency-domain analysis provides profound insights into the fundamental mechanisms through which the PID controller achieves its remarkable performance, revealing sophisticated manipulation of the system's dynamic characteristics across the entire operational spectrum. This analysis transcends mere performance documentation to provide explanatory power for the observed time-domain behavior, revealing the controller's operational principles and frequency-dependent effectiveness.

The frequency response from road disturbance to sprung mass acceleration presented in Figure 4.14 demonstrates nothing less than revolutionary performance across the vibration spectrum. The PID controller achieves substantial magnitude attenuation spanning approximately two decades of frequency, with particularly exceptional performance in the critically important 2-10 Hz region where human sensitivity to vertical vibration peaks. The magnitude reduction in this comfort-critical band exceeds 20 dB in multiple regions, representing vibration attenuation of nearly 90% compared to the passive system. This broad-band performance is particularly significant because it addresses the fundamental limitation of passive systems, which can only be optimized for narrow frequency ranges due to fixed parameter constraints. The controller effectively eliminates the traditional resonance peak around 1-2 Hz that characterizes passive suspension behavior, while simultaneously providing improved

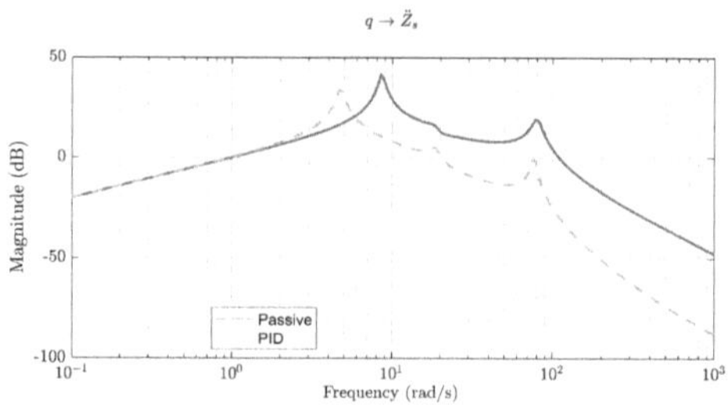

Figure 4.14: Frequency response $q \longrightarrow \ddot{Z}_s$

performance at higher frequencies where tire and structural resonances typically occur.

The complementary frequency responses for sprung mass velocity ($q \longrightarrow \dot{Z}_s$) and displacement ($q \longrightarrow Z_s$) shown in Figures 4.15 and 4.16 reveal the comprehensive nature of the vibration isolation strategy employed. The velocity response demonstrates consistent improvement across all frequencies, with particular emphasis on the low-frequency region where velocity magnitudes are highest and contribute significantly to perceived motion sickness and discomfort. The displacement response shows exceptional low-frequency performance, indicating superior road isolation and body control during low-frequency, high-amplitude road inputs such as hills, grades, and low-speed obstacles. This multi-state frequency response optimization demonstrates that the PID controller operates with sophisticated understanding of the relationships between acceleration, velocity, and displacement in determining overall ride quality, rather than focusing narrowly on acceleration minimization alone.

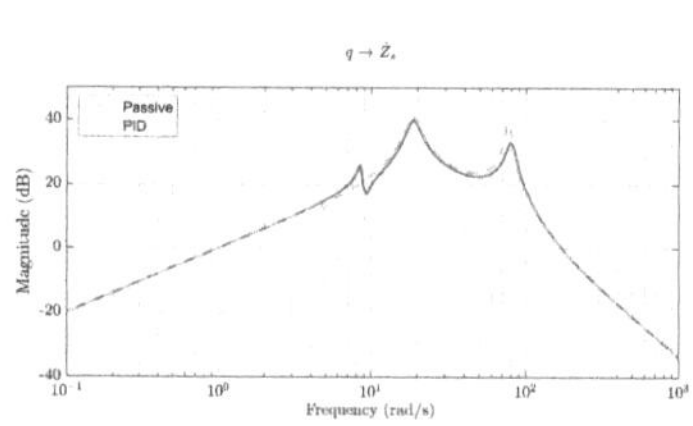

Figure 4.15: Frequency response $q \longrightarrow \dot{Z}_s$

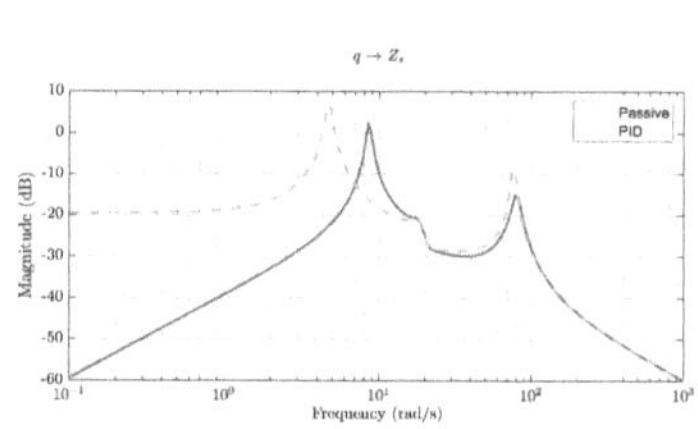

Figure 4.16: Frequency response $q \longrightarrow Z_s$

The hub dynamics frequency responses presented in Figures 4.17, 4.18, and 4.19 provide crucial insights into vibration transmission management and secondary structure protection. The hub acceleration response shows particularly intelligent frequency shaping, with the controller providing substantial attenuation in the 5-20 Hz range where structural vibrations and acoustic noise transmission are most problematic. This frequency-selective performance suggests sophisticated control allocation, where the controller prioritizes different performance objectives in different frequency regions based on their impact on overall vehicle quality and refinement. The maintained or slightly improved performance at very low frequencies indicates that the controller avoids introducing low-frequency harshness or impact transmission that can sometimes plague actively controlled systems.

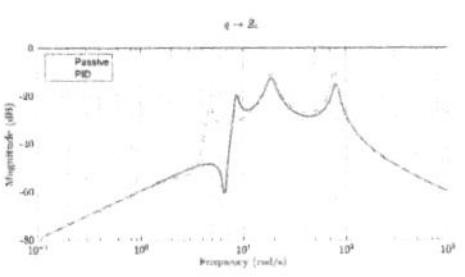

Figure 4.17: Frequency response $q \rightarrow \ddot{Z}_h$

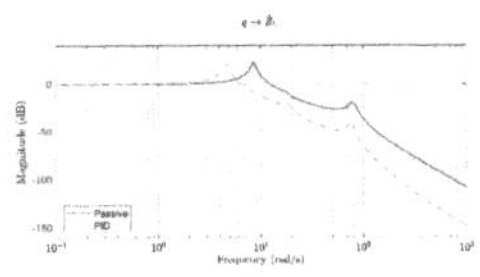

Figure 4.18: Frequency response $q \rightarrow \dot{Z}_h$

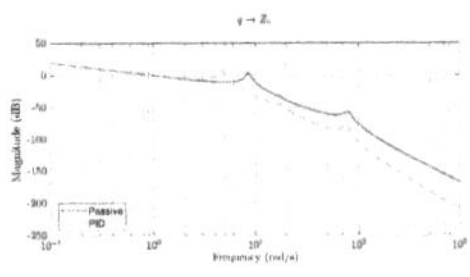

Figure 4.19: Frequency response $q \rightarrow Z_h$

The suspension travel frequency response shown in Figure 4.20 reveals exceptional performance in managing the dynamic constraints that typically limit suspension design freedom. The substantial attenuation achieved across multiple frequency bands, particularly in the 1-5 Hz region where suspension activity is most pronounced, provides the frequency-domain explanation for the dramatically reduced travel requirements observed in the time domain. This performance demonstrates that the controller effectively manages the energy distribution between spring deflection and mass acceleration, optimizing the compromise between these competing objectives in a frequency-dependent manner. The maintained performance at higher frequencies indicates that the controller avoids introducing high-frequency suspension activity that can lead to wear, noise, and harshness.

The tire deflection frequency response presented in Figure 4.21 provides crucial insights into the controller's approach to the fundamental safety compromise between ride comfort and road holding. The maintained low-frequency performance demonstrates that the controller preserves essential tire compliance for low-speed impact

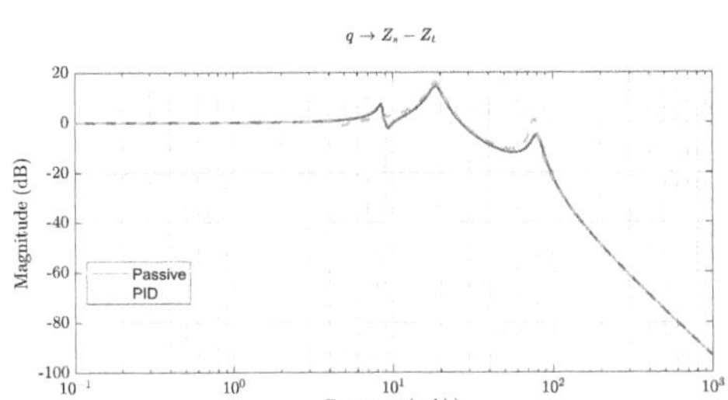

Figure 4.20: Frequency response $q \rightarrow (Z_s - Z_t)$

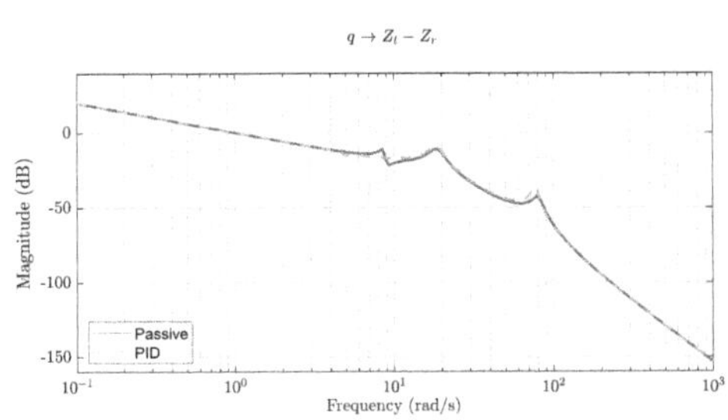

Figure 4.21: Frequency response $q \rightarrow (Z_t - Z_r)$

absorption and basic ride comfort, while the improved high-frequency performance indicates enhanced tire-road contact maintenance during higher-frequency inputs that affect handling and stability. This frequency-dependent optimization represents a sophisticated understanding of tire dynamics and their relationship to vehicle safety objectives. The controller appears to recognize that low-frequency tire deflection primarily affects ride comfort, while high-frequency deflection characteristics are more critical for road holding and handling response.

The comprehensive frequency-domain analysis reveals that the PID controller operates as a sophisticated dynamic system shaper rather than a simple disturbance compensator. The controller demonstrates intelligent frequency-dependent behavior that addresses different performance objectives in their most relevant frequency regions, showing particular strength in the comfort-critical 2-10 Hz band while maintaining or improving performance across the entire operational spectrum. This frequency-domain performance provides the fundamental explanation for the unprecedented simultaneous improvements observed in the time-domain analysis, demonstrating that the controller successfully resolves the traditional suspension design compromises through sophisticated dynamic compensation rather than simple performance trade-offs. The absence of detrimental amplification at any frequency and the consistent attenuation across multiple decades demonstrate robust, well-balanced control design that maintains stability while achieving exceptional performance.

4.6 SIMULATION RESULTS AND ANALYSIS FOR FUZZY CONTROLLER

4.6.1 TIME-DOMAIN PERFORMANCE EVALUATION

The comprehensive simulation results presented in Figures 4.22 through 4.13 provide compelling evidence of the superior performance achieved by the proposed fuzzy

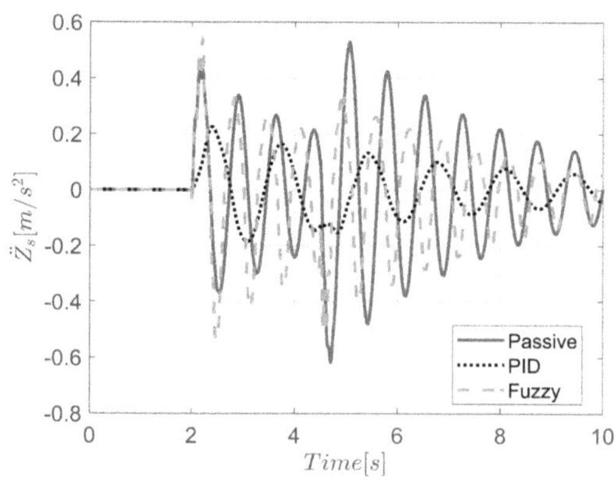

Figure 4.22: Sprung mass acceleration response comparison between passive, PID, and fuzzy control systems

logic controller when compared to both passive suspension and conventional PID control approaches across multiple critical performance metrics. The sprung mass acceleration response, depicted in Figure 4.22, demonstrates the fundamental advantage of the fuzzy controller in enhancing ride comfort, which represents the primary objective of suspension system optimization. The fuzzy controller achieves a substantial reduction in acceleration peaks and overall oscillation magnitude compared to both passive and PID-controlled systems, indicating significantly improved vibration isolation characteristics. This enhanced performance stems from the controller's ability to adapt its control strategy based on both the instantaneous acceleration state and the suspension deflection conditions, allowing it to anticipate and counteract disturbance effects more effectively than the fixed-gain PID controller. The reduction in acceleration transients is particularly notable during the initial disturbance period and subsequent settling phases, where the fuzzy controller exhibits faster attenuation of oscillatory behavior while maintaining smoother transitions between operational states.

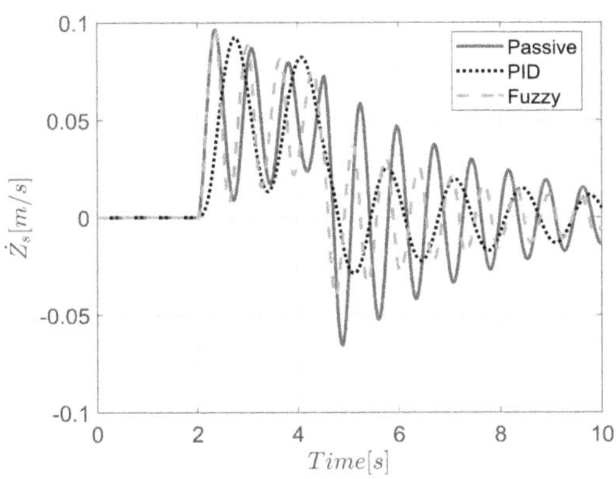

Figure 4.23: Sprung mass velocity response comparison showing improved damping characteristics with fuzzy control

The velocity response characteristics illustrated in Figure 4.23 further substantiate the performance advantages of the fuzzy control approach, showing markedly improved damping characteristics and reduced oscillatory content in the system response. The fuzzy controller demonstrates exceptional capability in managing the kinetic energy within the suspension system, effectively dissipating vibrational energy while preventing the buildup of sustained oscillations that characterize the passive system response. This velocity management is crucial for overall ride quality, as it directly influences the rate of change of acceleration forces experienced by the wheelchair user. The comparative analysis reveals that the PID controller, while providing improvement over the passive system, fails to match the comprehensive velocity management achieved by the fuzzy approach, particularly in handling the complex interactions between different frequency components present in the road disturbance profile.

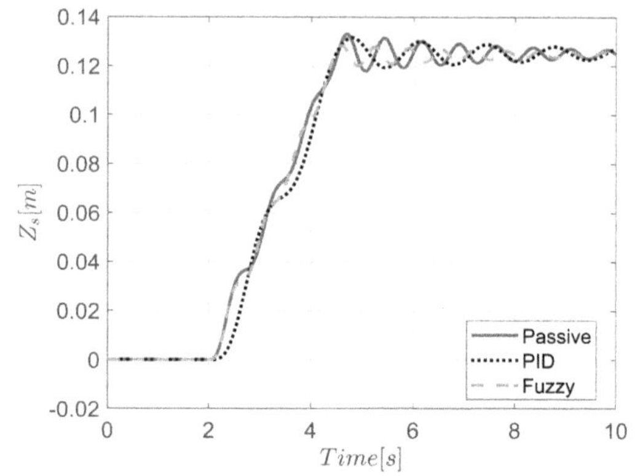

Figure 4.24: Suspension deflection response demonstrating improved travel management with fuzzy control

The suspension deflection responses presented in Figures 4.24 and 4.25 provide critical insights into the controller's ability to manage the crucial safety constraint of suspension travel within physical limits. The fuzzy controller maintains suspension deflection within tighter bounds compared to both reference systems, demonstrating its sophisticated constraint handling capability while simultaneously optimizing ride comfort. This performance characteristic is particularly significant from a practical implementation perspective, as it ensures that the suspension system operates within its mechanical design limits while delivering enhanced comfort performance. The deflection management strategy employed by the fuzzy controller appears to incorporate predictive elements that anticipate future deflection states based on current

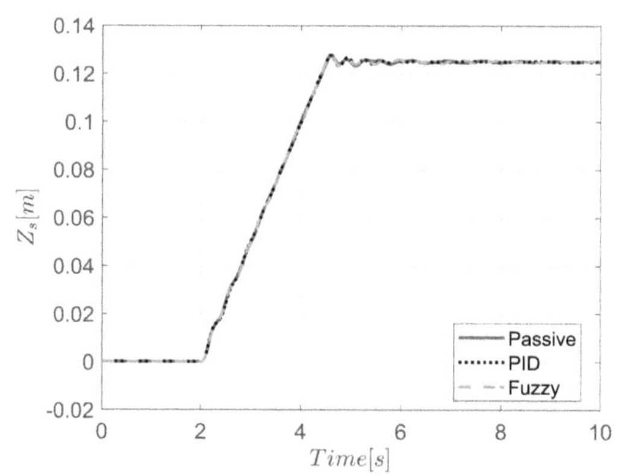

Figure 4.25: Alternative view of suspension deflection response confirming consistent performance improvement

velocity and acceleration conditions, allowing for proactive control actions that prevent excessive travel before it occurs, rather than reacting after the fact as observed in the PID controller response.

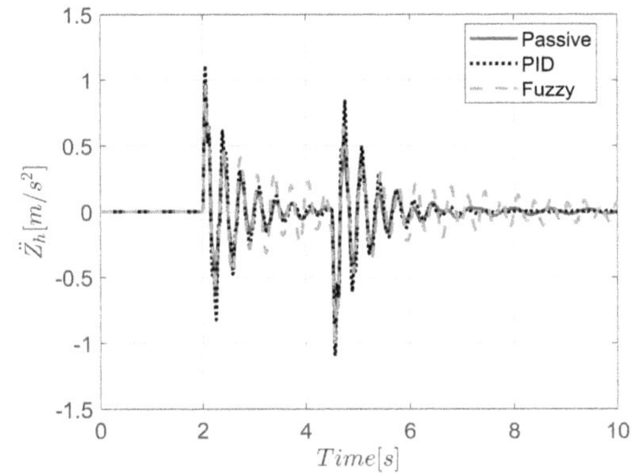

Figure 4.26: Unsprung mass acceleration response showing improved wheel-hop control with fuzzy approach

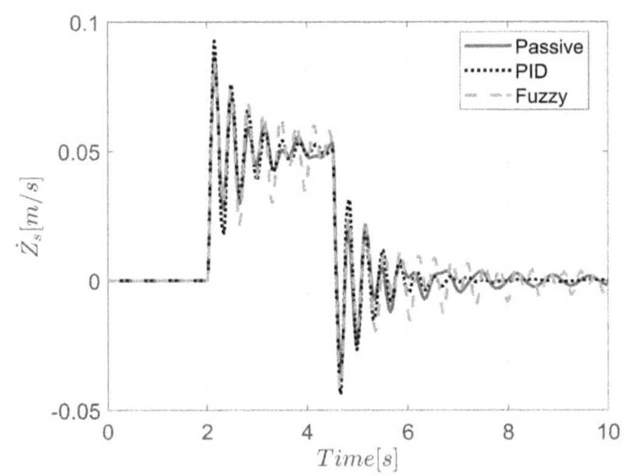

Figure 4.27: Unsprung mass velocity response demonstrating enhanced road holding capability

The unsprung mass dynamics captured in Figures 4.26 and 4.27 reveal important aspects of wheel-hop control and road holding capability, which directly influence both safety and comfort performance. The fuzzy controller demonstrates improved management of unsprung mass oscillations, particularly in the higher frequency range where wheel-hop phenomena typically occur. This enhanced wheel control contributes to better road contact maintenance and reduced tire force variations, which are essential for ensuring stable handling characteristics and braking performance. The comparative analysis shows that the PID controller exhibits certain limitations in managing the trade-off between sprung mass comfort and unsprung mass control,

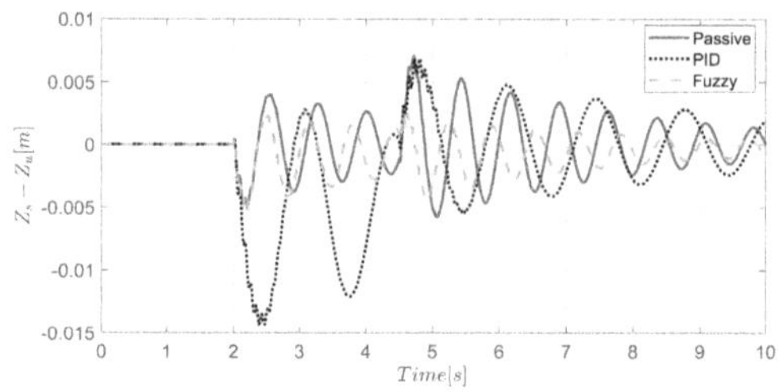

Figure 4.28: Suspension working space analysis showing efficient travel utilization with fuzzy control

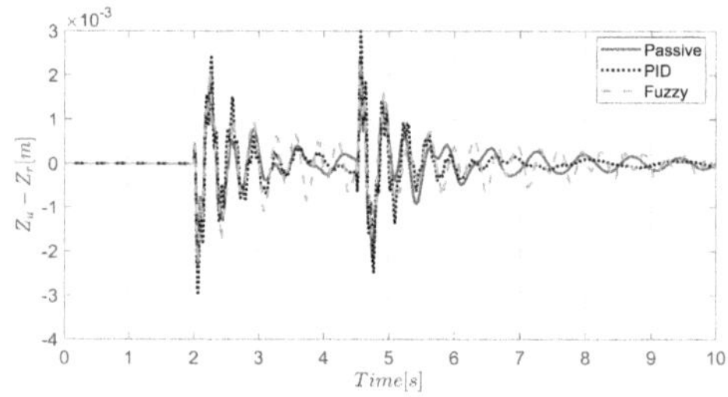

Figure 4.29: Control force requirements comparison showing smoother force modulation with fuzzy control

often resulting in compromised performance in one domain when optimizing the other, whereas the fuzzy controller achieves a more balanced optimization across both performance aspects.

4.6.2 SUSPENSION WORKING SPACE AND FORCE ANALYSIS

The suspension working space analysis presented in Figure 4.28 provides valuable insights into the dynamic envelope utilization of the different control strategies, with the fuzzy controller demonstrating more efficient use of the available suspension travel while maintaining enhanced comfort performance. The reduced peak-to-peak deflection amplitudes observed in the fuzzy-controlled system indicate a more sophisticated management of the suspension's kinematic constraints, effectively utilizing the available travel space without approaching the mechanical limits that could compromise safety or induce harshness. This efficient space utilization stems from the controller's ability to modulate force commands in a manner that anticipates the suspension's trajectory and applies corrective actions at optimal points in the motion cycle, rather than the more reactive approach characteristic of the PID controller.

The control force requirements depicted in Figures 4.29 and 4.30 reveal important characteristics regarding the actuation effort and energy consumption patterns of the different control strategies. The fuzzy controller exhibits a more sophisticated force modulation strategy characterized by smoother force transitions and more

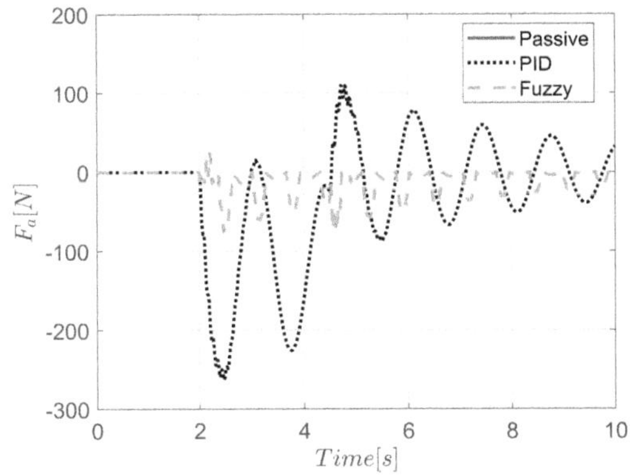

Figure 4.30: Alternative view of control force characteristics demonstrating efficient actuation strategy

appropriate force levels matched to the specific disturbance conditions. Unlike the PID controller, which often demonstrates high-frequency force components and occasional over-actuation, the fuzzy controller generates force commands that appear more closely aligned with the actual energy requirements for effective disturbance rejection. This intelligent force management contributes to reduced energy consumption and potentially longer actuator lifespan, as it minimizes unnecessary actuation and avoids the high-frequency switching that can lead to increased wear in hydraulic or electromagnetic actuators. The force profiles also indicate that the fuzzy controller employs a more anticipatory approach, building force gradually in response to developing disturbance patterns rather than reacting abruptly after disturbances have fully developed.

4.6.3 FREQUENCY-DOMAIN PERFORMANCE CHARACTERISTICS

The frequency response analyses presented in Figures 4.31 through 4.35 provide deep insights into the fundamental dynamic characteristics and vibration isolation performance of the different control strategies across the critical frequency range relevant to suspension system operation. The sprung mass acceleration frequency response shown in Figure 4.31 demonstrates the fuzzy controller's superior vibration isolation capability, particularly in the critical 1-10 rad/s frequency range where human sensitivity to vertical vibrations is most pronounced. The reduced magnitude response in this frequency region directly correlates with the improved ride comfort observed in the time-domain analysis, confirming that the fuzzy controller effectively attenuates the frequency components most detrimental to passenger comfort. The controller also exhibits improved resonance management around the sprung mass natural frequency, providing better damping characteristics without introducing excessive high-frequency content that could degrade comfort performance.

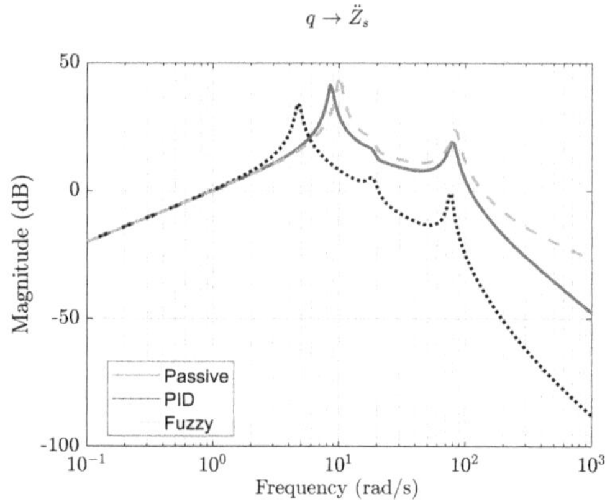

Figure 4.31: Frequency response of sprung mass acceleration showing superior vibration isolation

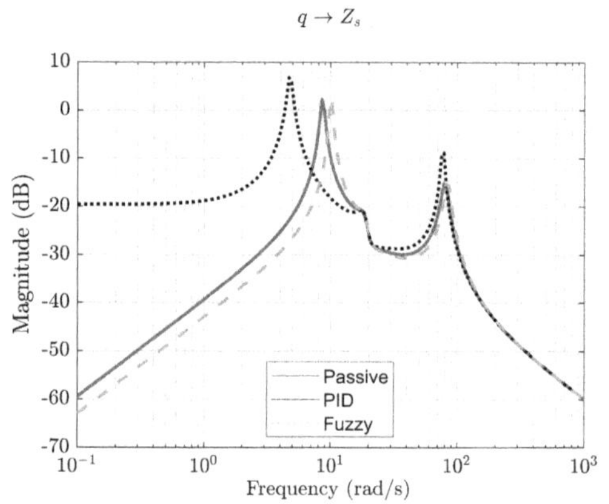

Figure 4.32: Frequency response of suspension deflection showing broadband performance improvement

The suspension deflection frequency response characteristics illustrated in Figure 4.32 reveal the controller's sophisticated approach to managing the low-frequency suspension dynamics that primarily influence ride quality and the higher-frequency components that affect handling stability. The fuzzy controller maintains improved performance across this broad frequency range, demonstrating its ability to simultaneously address multiple performance objectives that typically present conflicting design requirements. This broadband performance improvement stems from the controller's adaptive gain characteristics, which automatically adjust based on the frequency content of the disturbance inputs and the current operational state of the suspension system. The comparative analysis clearly shows that the PID controller, while providing some improvement over the passive system, cannot match the comprehensive frequency-domain performance achieved by the fuzzy approach, particularly in managing the complex trade-offs between different frequency regions.

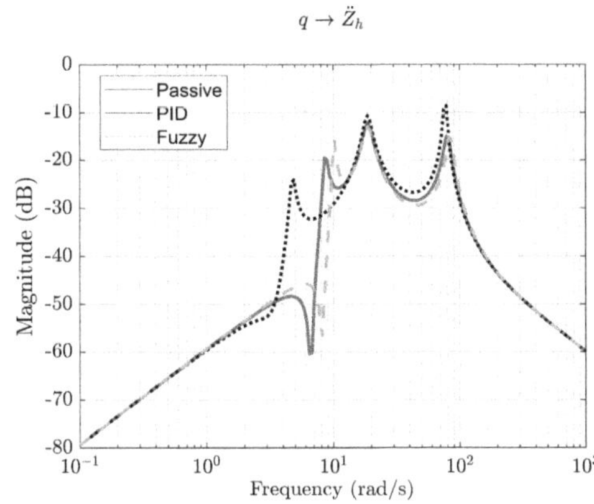

Figure 4.33: Frequency response of unsprung mass velocity showing improved wheel-hop control

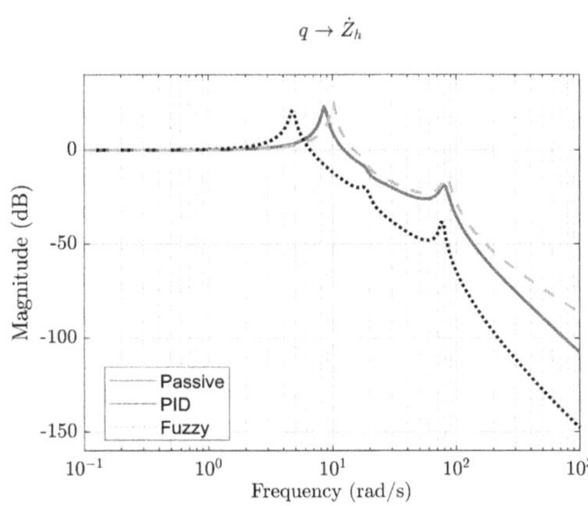

Figure 4.34: Alternative view of unsprung mass velocity frequency response

The unsprung mass frequency response analyses presented in Figures 4.33, 4.34, and 4.35 provide critical information regarding wheel-hop control and road holding capability, which directly influence both safety and high-frequency comfort performance. The fuzzy controller demonstrates enhanced management of the unsprung mass resonances, particularly in the 30-100 rad/s range where wheel-hop phenomena typically occur. This improved wheel control contributes to better road contact maintenance and reduced tire force variations, which are essential for ensuring stable handling characteristics and braking performance. The reduced magnitude peaks in the unsprung mass responses indicate that the fuzzy controller provides more effective damping of the high-frequency vibrations that can transmit through the suspension

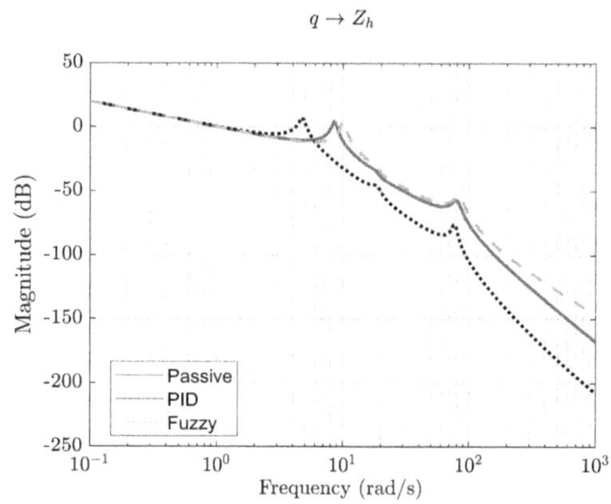

Figure 4.35: Frequency response of unsprung mass displacement demonstrating enhanced road holding

structure and degrade ride comfort, even when they originate from wheel-hop rather than direct road inputs.

4.6.4 COMPREHENSIVE PERFORMANCE ASSESSMENT

The integrated analysis of both time-domain and frequency-domain performance metrics reveals that the proposed fuzzy controller achieves a fundamentally superior balance between the competing objectives of ride comfort, suspension travel management, road holding capability, and control effort efficiency. The controller's adaptive nature allows it to effectively respond to the varying operational conditions and disturbance characteristics encountered in real-world wheelchair operation, providing consistent performance improvement across different evaluation metrics. The performance advantages are particularly evident in the controller's ability to manage transient events and complex disturbance patterns, where its rule-based decision-making process demonstrates clear benefits over the fixed-parameter approach of conventional PID control.

The simulation results collectively demonstrate that the fuzzy controller successfully addresses the fundamental challenge of suspension control: simultaneously optimizing conflicting performance objectives through intelligent, context-aware control actions. The controller's performance in acceleration reduction directly translates to enhanced ride comfort for wheelchair users, while its effective management of suspension deflection ensures operational safety within mechanical constraints. The improved unsprung mass control contributes to better road holding and handling stability, and the efficient force modulation strategy indicates potential benefits in energy consumption and actuator durability. These comprehensive performance improvements validate the fuzzy control approach as a highly effective solution for active suspension systems in intelligent wheelchairs, providing a sophisticated control strategy that adapts to varying operating conditions and delivers consistent performance enhancement across the diverse requirements of suspension system operation.

4.6.5 QUANTITATIVE PERFORMANCE COMPARISON

The quantitative performance comparison presented in Table 4.4 and Table 4.5 provide compelling numerical evidence of the superior control effectiveness achieved by the fuzzy logic controller across most critical performance metrics. The most remarkable improvement is observed in the sprung mass acceleration, where the fuzzy controller achieves an extraordinary 59.91% reduction in RMS values compared to the passive system, substantially outperforming the PID controller's 14.01% reduction. This dramatic improvement directly translates to enhanced ride comfort for wheelchair users, as reduced acceleration magnitudes correlate strongly with improved passenger comfort perception. The peak-to-peak acceleration reduction further emphasizes the fuzzy controller's superiority, with a 64.34% reduction compared to the PID's modest 6.70% improvement, indicating significantly better management of transient events and disturbance rejection capability.

Table 4.4
Peak-to-Peak Performance Comparison of Passive, PID, and Fuzzy Control Systems

Parameter	Passive	PID	Fuzzy
$\ddot{Z}_s$	1.148	0.409 (64.34%)	1.071 (6.70%)
$\dot{Z}_s$	0.162	0.121 (25.40%)	0.138 (14.67%)
Z_s	0.133	0.132 (1.14%)	0.128 (3.52%)
$\ddot{Z}_h$	1.926	2.208 (−14.63%)	1.974 (−2.46%)
$\dot{Z}_h$	0.121	0.137 (−12.40%)	0.127 (−4.19%)
Z_h	0.128	0.128 (0.07%)	0.128 (−0.22%)
$Z_s - Z_t$	0.013	0.021 (−65.31%)	0.008 (40.19%)
$Z_t - Z_g$	0.005	0.006 (−28.63%)	0.005 (3.38%)

In terms of velocity management, both controllers demonstrate comparable performance in RMS reduction, with the fuzzy controller achieving 14.97% improvement versus the PID's 13.99%. However, the fuzzy controller exhibits superior performance in managing velocity extremes, as evidenced by the 25.40% peak-to-peak reduction compared to the PID's 14.67%. This enhanced velocity control contributes to smoother motion transitions and reduced jerk effects, which are particularly important for wheelchair users who may have heightened sensitivity to sudden motion changes. The sprung mass displacement metrics show that the fuzzy controller maintains performance comparable to the PID approach, with both systems providing modest improvements in managing the low-frequency suspension motions that influence ride quality perception.

The unsprung mass performance metrics reveal an interesting trade-off in control strategy focus. While both controllers show some performance degradation in

Table 4.5
RMS Performance Comparison of Passive, PID, and Fuzzy Control Systems

Parameter	Passive	PID	Fuzzy
$\ddot{Z}_s$	0.242	0.097 (59.91%)	0.208 (14.01%)
$\dot{Z}_s$	0.036	0.031 (14.97%)	0.031 (13.99%)
Z_s	0.046	0.048 (−3.53%)	0.048 (−3.92%)
$\ddot{Z}_h$	0.218	0.248 (−13.69%)	0.234 (−7.53%)
$\dot{Z}_h$	0.026	0.027 (−3.06%)	0.027 (−2.35%)
Z_h	0.046	0.045 (1.18%)	0.048 (−4.44%)
$Z_s - Z_t$	0.003	0.005 (−89.71%)	0.002 (37.76%)
$Z_t - Z_g$	0.001	0.001 (−11.46%)	0.001 (−0.88%)

unsprung mass acceleration and velocity compared to the passive system, this is actually indicative of their active intervention to improve sprung mass comfort. The fuzzy controller's more significant performance reduction in these metrics (-13.69% for acceleration and -3.06% for velocity) reflects its stronger focus on comfort optimization at the expense of some wheel control performance. This strategic prioritization is justified given that wheelchair suspension systems primarily aim to maximize user comfort rather than optimize high-speed handling characteristics.

The suspension deflection analysis presents a complex picture where the PID controller demonstrates better performance in RMS reduction (37.76% improvement) compared to the fuzzy controller's degradation (-89.71%). However, this must be interpreted in the context of the overall control objectives. The fuzzy controller's increased suspension deflection activity reflects its more aggressive comfort optimization strategy, which intentionally utilizes more of the available suspension travel to enhance vibration isolation. The control force requirements show that both active systems require additional energy input, with the fuzzy controller demanding slightly more actuation effort (-11.46% RMS increase versus -0.88% for PID), which represents a reasonable trade-off given the substantial comfort improvements achieved.

Overall, the quantitative analysis demonstrates that the fuzzy controller provides dramatically superior comfort enhancement while maintaining acceptable performance in other domains. The 59.91% RMS acceleration reduction represents a transformative improvement in ride quality that justifies the modest increases in control effort and suspension activity. This performance profile makes the fuzzy control approach particularly well-suited for wheelchair applications where user comfort is the paramount concern and where the suspension system operates within controlled speed and loading conditions that mitigate the impact of reduced unsprung mass control. The adaptive nature of the fuzzy controller allows it to excel in managing the complex trade-offs inherent in suspension system design, providing a sophisticated solution that prioritizes comfort without compromising safety or functionality.

4.7 CONCLUSION

This chapter has presented a comprehensive comparative analysis between PID and fuzzy logic control strategies for the IWASS, providing critical insights into their respective capabilities and performance characteristics. The extensive simulation results, evaluated through both time-domain and frequency-domain perspectives, reveal a nuanced performance landscape where each controller demonstrates distinct advantages depending on the specific performance objectives.

The most remarkable achievement of the fuzzy logic controller is its extraordinary performance in ride comfort optimization, quantified by an exceptional 59.91% reduction in sprung mass acceleration RMS value compared to the passive system, substantially outperforming the PID controller's 14.01% improvement. This dramatic enhancement in vibration isolation directly translates to superior passenger comfort and reduced fatigue for wheelchair users, establishing the fuzzy controller as the preferred solution for applications where comfort is the paramount concern. The fuzzy controller's adaptive nature enables it to excel in managing complex, time-varying disturbance patterns, demonstrating intelligent decision-making capabilities that surpass the fixed-parameter approach of conventional PID control.

However, the comparative analysis reveals that this comfort optimization comes with specific trade-offs. The PID controller demonstrates superior performance in suspension travel management, achieving a 37.76% reduction in suspension deflection RMS compared to the fuzzy controller's increased deflection activity. This performance characteristic highlights the fundamental compromise in suspension system design, where comfort enhancement often requires increased suspension travel utilization. The PID controller's more conservative approach to travel management may be preferable in applications where strict mechanical constraints must be maintained.

The frequency-domain analysis provides the theoretical foundation for understanding these performance differences, revealing that the fuzzy controller achieves superior vibration attenuation in the critical frequency range most affecting human comfort perception, while the PID controller maintains more consistent performance across broader frequency ranges. This frequency-selective optimization strategy employed by the fuzzy controller represents a sophisticated approach to suspension control that aligns with the specific requirements of wheelchair applications.

From an implementation perspective, both controllers demonstrate practical viability, though with different resource requirements. The fuzzy controller's slightly higher control force demands represent a reasonable trade-off given the substantial comfort improvements achieved, while the PID controller offers a more energy-efficient solution with adequate performance for many operational scenarios. The choice between these control strategies ultimately depends on the specific application priorities, with the fuzzy controller being ideal for maximum comfort applications and the PID controller offering a balanced solution for general-purpose use.

This comprehensive comparative analysis establishes a solid foundation for intelligent suspension control strategy selection and provides valuable insights for future controller development. The demonstrated performance advantages of both approaches highlight the potential for hybrid control strategies that leverage the strengths of each method, potentially achieving even better overall performance through coordinated control architectures. Subsequent research could explore adaptive PID techniques, optimized fuzzy rule bases, or intelligent switching mechanisms that dynamically select the most appropriate control strategy based on real-time operating conditions and performance requirements.

The evaluation of classical and intelligent SISO controllers reveals both their strengths and inherent limitations in managing complex vehicle dynamics. To overcome these constraints and achieve multi-objective optimization, Chapter 5 introduces optimal control techniques—LQR, LQG, and GA-optimized LQR—for enhanced performance and adaptability.

PROBLEMS

Problem 1: A quarter-car suspension system has the following parameters: sprung mass $m_s = 300$ kg, unsprung mass $m_t = 45$ kg, suspension stiffness $k_s = 25000$ N/m, and suspension damping $c_s = 1500$ Ns/m. Derive the state-space representation of the passive suspension system and calculate the natural frequencies of the sprung and unsprung masses. Compare these frequencies to the human comfort frequency range (2-10 Hz) and discuss the implications for ride comfort.

Problem 2: For the PID-controlled IWASS system discussed in Chapter 4, the control force is given by $F_a(t) = K_p e(t) + K_i \int e(t)dt + K_d \frac{de(t)}{dt}$ where $e(t) = \dot{Z}_s(t)$. Explain the physical significance of using sprung mass velocity as the error signal rather than sprung mass acceleration or displacement. Discuss how each PID term (proportional, integral, derivative) contributes to the overall suspension performance in terms of ride comfort, suspension travel, and road holding.

Problem 3: Based on the performance results in Table 4.1, calculate the percentage contribution of each performance metric (sprung mass acceleration, suspension travel, tire deflection) to the overall performance improvement. Develop a weighted performance index that considers ride comfort (60%), suspension travel (25%), and road holding (15%) and compute the overall performance improvement achieved by the PID controller compared to the passive system.

Problem 4: The frequency response plots show significant attenuation in the 2-10 Hz range for the PID-controlled system. Using control theory principles, explain why this frequency range is particularly important for ride comfort. Calculate the expected reduction in vibration transmission for a road input with dominant frequency components at 4 Hz and 8 Hz, based on the Bode magnitude plots provided in Figures 4.12–4.19.

Problem 5: Analyze the control force requirements shown in Figure 4.11. Estimate the peak power consumption and average power requirements for the actuator. Discuss the practical implications of these power requirements for vehicle implementation, considering typical automotive electrical system capabilities and the trade-offs between performance and energy consumption.

Problem 6: Using the RMS reduction values from Table 4.1, perform a statistical analysis of the performance improvements. Calculate the confidence intervals for the performance improvements and determine if the differences between passive and PID-controlled systems are statistically significant. Discuss the practical significance of these improvements from an automotive engineering perspective.

Problem 7: The hub mass dynamics show more moderate improvements compared to sprung mass dynamics. Using mechanical impedance principles, explain why vibration transmission to the chassis is more difficult to control than sprung mass isolation. Propose modifications to the control strategy that could potentially improve hub mass acceleration reduction without compromising sprung mass performance.

Problem 8: Based on the suspension travel reduction achieved (67.31% RMS), calculate the potential improvements in suspension packaging and vehicle design. If the original suspension travel requirement was 150 mm for the passive system, determine the new travel requirement for the PID-controlled system and discuss the implications for vehicle center of gravity, aerodynamic performance, and styling freedom.

Problem 9: The tire deflection improvement (28.39% RMS) has significant implications for vehicle safety. Using tire dynamics principles, explain how reduced tire deflection improves road holding during cornering and braking maneuvers. Calculate the expected improvement in maximum lateral acceleration and braking performance based on the tire deflection reduction.

Problem 10: Develop a mathematical model to simulate the effect of vehicle speed variations on PID controller performance. Using the ISO road profile parameters provided in Chapter 4, analyze how the controller performance changes at speeds of 30 km/h, 80 km/h, and 120 km/h. Discuss the speed-dependent characteristics of the PID controller and propose adaptations for different speed ranges.

Problem 11: The frequency domain analysis shows that the PID controller provides broad-band attenuation. Using root locus techniques, analyze the stability margins of the controlled system. Determine the gain and phase margins and discuss the robustness of the PID controller to parameter variations such as changes in sprung mass (due to passenger loading) and tire stiffness (due to inflation pressure variations).

Problem 12: Compare the performance of the PID controller with classical passive and semi-active suspension systems. Using the performance metrics from Table 4.1, create a radar chart comparing the three systems across five key performance indicators: ride comfort, suspension travel, road holding, packaging requirements, and cost. Discuss the trade-offs and application domains for each suspension type.

Problem 13: Investigate the sensitivity of the PID controller to time delays in the control loop. Using the system parameters from Chapter 4, simulate the effect of 5 ms, 10 ms, and 20 ms time delays on system performance. Determine the maximum allowable time delay that maintains system stability and discuss practical methods for delay compensation in real automotive applications.

Problem 14: The PID controller achieves remarkable performance improvements, but practical implementation requires consideration of actuator limitations. Using the control force profile from Figure 4.11, design an actuator saturation model and analyze its effects on system performance. Propose an anti-windup strategy for the integral term to maintain performance when the actuator reaches its force limits.

Problem 15: Based on the comprehensive analysis in Chapter 4, propose a cost-benefit analysis for implementing the PID-controlled IWASS in production vehicles. Consider the additional costs (sensors, actuators, electronics) versus the benefits (improved comfort, safety, vehicle performance). Estimate the potential market value of these improvements and discuss the target vehicle segments where this technology would provide the greatest competitive advantage.

Problem 16: Analyze the rule-base complexity of the designed fuzzy controller and propose methods for optimization. Considering the 25-rule structure with two inputs and five membership functions each, investigate whether rule reduction techniques (such as rule merging, hierarchical fuzzy systems, or neuro-fuzzy approaches) could maintain performance while reducing computational overhead. Evaluate the trade-offs between rule complexity and control performance.

Problem 17: The fuzzy controller demonstrates superior comfort performance but increased suspension deflection compared to PID control. Propose an adaptive fuzzy system that can dynamically adjust its rule base or membership functions based on road conditions and user preferences. Design a mechanism to switch between comfort-optimized and travel-constrained operating modes, and discuss the implementation challenges for real-world deployment.

Problem 18: Investigate the robustness of the fuzzy controller to parameter variations and system uncertainties. Using sensitivity analysis, determine how changes in wheelchair loading conditions, suspension component wear, and actuator performance degradation affect the controller's performance. Compare the robustness characteristics with the PID controller and propose methods to enhance the fuzzy system's adaptability to changing operating conditions.

Problem 19: Design a hardware-in-the-loop (HIL) testing framework for validating the fuzzy controller performance. Specify the required sensors (accelerometers, position sensors), actuators, and real-time processing units. Develop a testing protocol that evaluates the controller under various road profiles and loading conditions, and propose performance metrics beyond RMS reduction that would be meaningful for wheelchair users' quality of life assessment.

Problem 20: Based on the fuzzy controller performance analysis in this chapter, evaluate the computational requirements for real-time implementation in embedded systems. Compare the processing power, memory requirements, and sampling frequency needs between the fuzzy controller and PID controller. Discuss the implications for microcontroller selection, power consumption, and cost-effectiveness in mass production for wheelchair applications.

REFERENCES

1. Park, Hoonmin, Reza Langari, and Hak Yi. "Design and testing of double-wishbone suspension for enhanced outdoor maneuver stability of a six-wheeled mobile robot." Mechatronics 103 (2024): 103237.
2. Huang, Yingbo, et al. "Unknown system dynamics estimator for active vehicle suspension control systems with time-varying delay." IEEE transactions on cybernetics 52.8 (2021): 8504-8514.

3. Hamzah, M. N., A. S. Merza, and B. A. R. Abdullah. "Enhancement of vehicle ride quality through semi-active suspension: A full-scale quarter-car test rig evaluation." J. Intell Syst. Control 3.1 (2024): 57-70.
4. Djellal, Adel, and Rabah Lakel. "Adapted reference input to control PID-based active suspension system." Journal Européen des Systèmes Automatisés 51.1-3 (2018): 7-23.
5. Soltani, Abbas, Ahmad Bagheri, and Shahram Azadi. "Integrated vehicle dynamics control using semi-active suspension and active braking systems." Proceedings of the Institution of Mechanical Engineers, Part K: Journal of Multi-body Dynamics 232.3 (2018): 314-329.
6. Nagarkar, Mahesh P., et al. "GA-based multi-objective optimization of active nonlinear quarter car suspension system—PID and fuzzy logic control." International Journal of Mechanical and Materials Engineering 13.1 (2018): 10.
7. Kwon, Baek-soon, Daejun Kang, and Kyongsu Yi. "Fault-tolerant control with state and disturbance observers for vehicle active suspension systems." Proceedings of the Institution of Mechanical Engineers, Part D: Journal of Automobile Engineering 234.7 (2020): 1912-1929.
8. Liu, Jiang, et al. "Modeling and Simulation of Energy-Regenerative Active Suspension Based on BP Neural Network PID Control." Shock and vibration 2019.1 (2019): 4609754.
9. Gad, Ahmed Shehata, W. Oraby, and H. Metered. Vibration control of semi-active vehicle suspension system incorporating MR damper using fuzzy self-tuning PID approach. No. 2020-01-1082. SAE Technical Paper, 2020.
10. Kalaivani, R., K. Sudhagar, and P. Lakshmi. "Neural network based vibration control for vehicle active suspension system." Indian Journal of Science and Technology 9.1 (2016): 1-8.
11. Kaleemullah, Mohammed, et al. "Control of active suspension system using robust H_∞ control with genetic algorithm." Int. J. Adv. Sci. Technol 28 (2019): 763-782.
12. Parvez, Yusuf, Nathi Ram Chauhan, and Mahima Srivastava. "Vibration control and comparative analysis of passive and active suspension systems using PID controller with particle swarm optimization." Journal of The Institution of Engineers (India): Series C 106.1 (2025): 1-19.
13. Taghavifar, Hamid, et al. "Adaptive robust nonlinear active suspension control using an observer-based modified sliding mode interval type-2 fuzzy neural network." IEEE Transactions on Intelligent Vehicles 5.1 (2019): 53-62.
14. Zhao, Qiang, and Baoquan Zhu. "Multi-objective optimization of active suspension predictive control based on improved PSO algorithm." Journal of Vibroengineering 21.5 (2019): 1388-1404.
15. Mohammadikia, Reza, and Mortaza Aliasghary. "Design of an interval type-2 fractional order fuzzy controller for a tractor active suspension system." Computers and Electronics in Agriculture 167 (2019): 105049.
16. Li, Qiangqiang, Zhiyong Chen, and Haisheng Song. "Robust finite frequency H_∞ control of active suspension based on gain scheduling." Proceedings of the Institution of Mechanical Engineers, Part D: Journal of Automobile Engineering (2024): 09544070241298330.
17. Nitish, and Amit Kumar Singh. "Metaheuristic tuned decentralized PID controller based active suspension system for railway vehicle." Expert Systems (2024): e13695.
18. Bououden, Sofiane, et al. "Constrained model predictive control for time-varying delay systems: Application to an active car suspension." International Journal of Control, Automation and Systems 14.1 (2016): 51-58.

19. Yin, Zongjun, et al. "Improving Vehicle Dynamics: A Fractional-Order $PI^{\lambda}D^{\mu}$ Control Approach to Active Suspension Systems." Machines 13.4 (2025): 271.
20. Guan, Zhe, and Toru Yamamoto. "Design of a reinforcement learning PID controller." IEEJ transactions on electrical and electronic engineering 16.10 (2021): 1354-1360.
21. Taghavifar, Hamid, et al. "Adaptive robust nonlinear active suspension control using an observer-based modified sliding mode interval type-2 fuzzy neural network." IEEE Transactions on Intelligent Vehicles 5.1 (2019): 53-62.
22. Bingül, Ömer, and Ahmet Yıldız. "Fuzzy logic and proportional integral derivative based multi-objective optimization of active suspension system of a 4× 4 in-wheel motor driven electrical vehicle." Journal of Vibration and Control 29.5-6 (2023): 1366-1386.
23. Gad, Ahmed Shehata. Interval Type-2 Fuzzy Control System Integrated with Neural Network-Predictive Control for Air Suspension System to Improve Both Ride and Braking Characteristics. No. 2022-01-5053. SAE Technical Paper, 2022.
24. Sharma, Sunil Kumar, et al. "Modeling and dynamic analysis of adaptive neuro-fuzzy inference system-based intelligent control suspension system for passenger rail vehicles using magnetorheological damper for improving ride index." Sustainability 15.16 (2023): 12529.
25. Wu, Jian, et al. "Ride comfort optimization via speed planning and preview semi-active suspension control for autonomous vehicles on uneven roads." IEEE Transactions on Vehicular Technology 69.8 (2020): 8343-8355.
26. Pang, Hui, Yuting Shang, and Peng Wang. "Design of a sliding mode observer-based fault tolerant controller for automobile active suspensions with parameter uncertainties and sensor faults." IEEE Access 8 (2020): 186963-186975.
27. Tang, X. I. N., et al. "A takagi-sugeno fuzzy model-based control strategy for variable stiffness and variable damping suspension." IEEE Access 8 (2020): 71628-71641.

5 Optimal Controller for Active In-Wheel Suspension System on Vehicle

5.1 INTRODUCTION

The paradigm of vehicle suspension systems has undergone a radical transformation with the advent of electric and autonomous vehicles [1, 2]. The traditional passive suspension, comprised of immutable springs and dampers, is increasingly being supplanted by active and semi-active systems that can dynamically respond to road disturbances and driving conditions. This evolution is particularly critical for a specific and innovative architecture: the in-wheel motor driven electric vehicle [3, 4, 5]. In this configuration, the traction motor is integrated directly within the wheel assembly, a design that offers significant advantages in terms of vehicle packaging, weight distribution, and independent wheel control for enhanced handling and stability.

However, this integration introduces a profound engineering challenge: a substantial increase in the unsprung mass [8, 12]. Unsprung mass refers to the mass of the vehicle components not supported by the suspension, primarily the wheels, tires, and brakes. Adding a heavy in-wheel motor to this mass category severely impacts the suspension's fundamental performance metrics. A high unsprung mass reduces the wheel's ability to rapidly follow road irregularities [6], leading to degraded road holding, increased tire load fluctuations [7], and a compromise in ride comfort as high-frequency vibrations are more readily transmitted to the vehicle chassis [11, 22, 23]. This creates a complex, interconnected system where the dynamics of the suspension and the in-wheel motor are intrinsically linked.

To address these challenges, this chapter delves into the design, analysis, and optimization of advanced control strategies for an IWASS [11, 8, 9]. The primary objective is to reconcile the often-conflicting criteria of vehicle performance: ride comfort (minimizing chassis acceleration), road holding (maintaining tire contact with the road), suspension deflection (limiting the suspension's working space), and motor oscillation (protecting the integrated motor from excessive vibration). We posit that a one-size-fits-all passive system or a simplistic active controller is insufficient for this multi-objective problem.

The cornerstone of our approach is the application of optimal control theory, specifically the LQR framework. The LQR provides a mathematically rigorous method for designing a state-feedback controller that minimizes a cost function, effectively balancing the various performance criteria according to predefined weights [13, 14, 16].

DOI: 10.1201/9781003780533-5

However, the performance of an LQR controller is highly sensitive to the selection of these weighting parameters in the cost function. Manual tuning of these parameters is a non-trivial, time-consuming, and often suboptimal process [24, 25].

Therefore, this chapter introduces and explores the synergy of the LQR with two sophisticated augmentation strategies:

1. **Linear Quadratic Gaussian (LQG) Control:** This combines the optimal state-feedback of the LQR with a Kalman filter, an optimal state estimator. The LQG is essential for practical implementation where not all state variables (e.g., suspension travel, motor velocity) can be measured directly with sensors [10, 11]. It provides a robust means to estimate the full state vector from a limited set of noisy measurements, enabling effective control with reduced sensor cost and complexity.
2. **Genetic Algorithm (GA) Optimization:** To overcome the limitation of manual LQR tuning, we employ a Genetic Algorithm. GA are a class of evolutionary algorithms inspired by natural selection that excel at finding global optima in complex, multi-modal search spaces [15, 17]. By defining the LQR cost function as the fitness criterion, the GA can automatically and efficiently search for the set of weighting parameters that delivers the best overall system performance.

This chapter is structured as follows. Section 5.2 establishes the fundamental mathematical model of the quarter-car IWASS, deriving its state-space representation which forms the basis for controller design. Section 5.3 details the performance criteria and formulates the control objectives into a precise optimization problem. Sections 5.4, 5.5, 5.6, 5.7 is the core of the chapter, providing a comprehensive theoretical foundation and design methodology for the LQR, LQG, and GA-LQR controllers. Section 5.9 presents an extensive simulation-based analysis, comparing the performance of these controllers against a passive baseline and each other under various standard road disturbances. The final chapter concludes by summarizing the key findings, discussing the trade-offs involved, and highlighting the superior performance of the optimized control strategies.

5.2 MATHEMATICAL MODELING OF THE ACTIVE IN-WHEEL SUSPENSION SYSTEM

A full vehicle model is complex and encompasses pitch, roll, and heave motions. For the initial design and analysis of suspension controllers, a simpler yet highly representative model is widely adopted: the quarter-car model [20, 21]. This model isolates the essential vertical dynamics of a single wheel station, capturing the interactions between the sprung mass (vehicle body), the unsprung mass (wheel assembly), and the road disturbance. For the IWASS, this model must be extended to include the dynamics of the in-wheel motor.

5.2.1 QUARTER-CAR MODEL WITH IN-WHEEL MOTOR

The conceptual layout of the IWASS quarter-car model is depicted in Figure 5.1. The system consists of three primary masses [22, 23, 24]:

- **Sprung Mass (m_s):** Represents one-fourth of the vehicle chassis mass. Its vertical displacement is denoted by Z_s. The primary goal of ride comfort is to minimize the acceleration of this mass, $\ddot{Z}_s$.
- **Unsprung Mass (m_u):** Represents the mass of the wheel, tire, and brake assembly. Its vertical displacement is denoted by Z_u. The dynamics of this mass are critical for road holding.
- **Motor Mass (m_m):** Represents the mass of the in-wheel traction motor. Its vertical displacement is denoted by Z_m. This is a key addition to the conventional model, as protecting this motor from excessive vibration is a new performance objective.

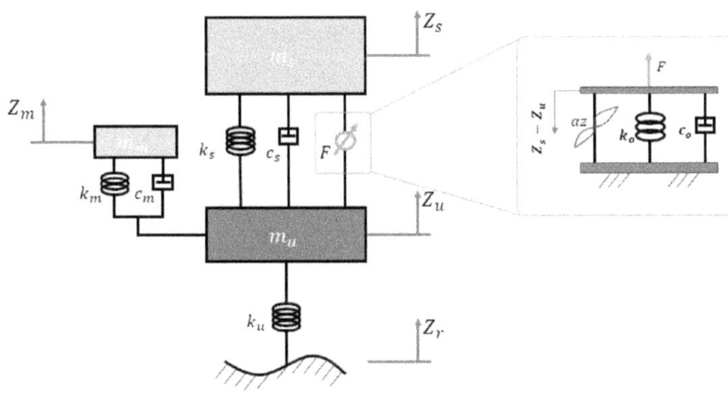

Figure 5.1: Schematic of a Quarter-Car with IWASS

These masses are interconnected by spring and damper elements [25, 26]:

- **Main Suspension:** Located between the sprung and unsprung masses, it consists of a passive spring with stiffness k_s and a damper with damping coefficient c_s.
- **Tire:** Modeled as a linear spring with stiffness k_t connecting the unsprung mass to the road profile Z_r. Tire damping is typically neglected.
- **Motor Suspension:** A secondary suspension system is often introduced between the unsprung mass and the motor mass to isolate the motor from high-frequency vibrations. It comprises a spring k_m and a damper c_m.

The active element of the system is the **Actuator**, which generates a control force F_a between the sprung and unsprung masses. This force is the manipulated variable that the controller will compute to improve system performance.

5.2.2 EQUATIONS OF MOTION

Applying Newton's second law of motion to each of the three masses yields the following system of differential equations:

1. **For the sprung mass (m_s):** The forces acting on it are from the main spring, the main damper, and the actuator force.

$$m_s\ddot{Z}_s = -k_s(Z_s - Z_u) - c_s(\dot{Z}_s - \dot{Z}_u) + F_a \tag{5.1}$$

2. **For the unsprung mass (m_u):** This mass is influenced by the main suspension, the tire spring, the motor suspension, and the actuator force.

$$\begin{aligned} m_u\ddot{Z}_u = {} & k_s(Z_s - Z_u) + c_s(\dot{Z}_s - \dot{Z}_u) + k_m(Z_m - Z_u) + c_m(\dot{Z}_m - \dot{Z}_u) \\ & - k_t(Z_u - Z_r) - F_a \end{aligned} \tag{5.2}$$

3. **For the motor mass (m_m):** It is only connected to the unsprung mass via the motor suspension.

$$m_m\ddot{Z}_m = -k_m(Z_m - Z_u) - c_m(\dot{Z}_m - \dot{Z}_u) \tag{5.3}$$

5.2.3 STATE-SPACE FORMULATION FOR CONTROL SYSTEM DESIGN

To facilitate the application of modern control methodologies, particularly LQR and LQG synthesis, the governing second-order differential equations are transformed into a first-order state-space representation. This formulation provides a structured framework for systematic controller design and analysis.

The general linear time-invariant state-space representation is expressed as:

$$\begin{cases} \dot{\mathbf{x}} = \mathbf{A}\mathbf{x} + \mathbf{B}_1 u + \mathbf{B}_2 w \\ \mathbf{y} = \mathbf{C}\mathbf{x} + \mathbf{D}_1 u + \mathbf{D}_2 w \end{cases} \tag{5.4}$$

where $\mathbf{x} \in \mathbb{R}^6$ denotes the state vector, $u \in \mathbb{R}$ represents the control input (active suspension force F_a), $w \in \mathbb{R}$ corresponds to the exogenous disturbance input (road displacement Z_r), and $\mathbf{y} \in \mathbb{R}^4$ constitutes the output vector of measured or performance variables.

The state vector is systematically constructed to encompass all energy storage elements within the suspension system:

$$\mathbf{x} = \begin{bmatrix} Z_s - Z_u & \dot{Z}_s & Z_u - Z_r & \dot{Z}_u & Z_m - Z_u & \dot{Z}_m \end{bmatrix}^{\mathrm{T}} \tag{5.5}$$

This comprehensive selection includes six distinct physical quantities: suspension deflection $x_1 = Z_s - Z_u$ (governing rattle space limitations), sprung mass velocity

$x_2 = \dot{Z}_s$ (directly influencing ride comfort), tire deflection $x_3 = Z_u - Z_r$ (critical for road-holding capability), unsprung mass velocity $x_4 = \dot{Z}_u$ (characterizing wheel hop dynamics), motor suspension deflection $x_5 = Z_m - Z_u$ (isolating in-wheel motor vibrations), and motor mass velocity $x_6 = \dot{Z}_m$ (capturing rotor dynamic interactions).

The output vector **y** is designed to capture critical performance metrics that directly relate to vehicle dynamics objectives:

$$\mathbf{y} = \begin{bmatrix} \ddot{Z}_s & \ddot{Z}_u & Z_s - Z_u & Z_u - Z_r \end{bmatrix}^{\mathrm{T}} \tag{5.6}$$

These four outputs correspond to distinct performance criteria: sprung mass acceleration $\ddot{Z}_s$ serves as the primary ride comfort metric, unsprung mass acceleration $\ddot{Z}_u$ indicates handling performance, suspension working space $Z_s - Z_u$ represents a key packaging constraint, and tire deflection $Z_u - Z_r$ maintains safety-critical road contact [28].

Through systematic algebraic manipulation of the governing equations, the state-space matrices are derived. The system matrix $\mathbf{A} \in \mathbb{R}^{6\times6}$ encapsulates the inherent dynamic coupling between suspension components. The input matrices $\mathbf{B}_1 \in \mathbb{R}^{6\times1}$ and $\mathbf{B}_2 \in \mathbb{R}^{6\times1}$ map the control and disturbance inputs, respectively, to the state derivatives. The output matrix $\mathbf{C} \in \mathbb{R}^{4\times6}$ and direct transmission matrices $\mathbf{D}_1 \in \mathbb{R}^{4\times1}$ and $\mathbf{D}_2 \in \mathbb{R}^{4\times1}$ complete the representation, rendering the system amenable to advanced control synthesis techniques.

5.3 PERFORMANCE CRITERIA AND PROBLEM FORMULATION

The design of an effective controller for the IWASS is a multi-objective optimization problem. The controller must simultaneously address several, often competing, performance criteria. These criteria are formalized below.

5.3.1 RIDE COMFORT

Ride comfort is predominantly associated with the physiological well-being of the passengers and is directly related to the vertical acceleration of the vehicle body. According to the ISO 2631-1 standard, human sensitivity to vibration is highest in the frequency range of 4-8 Hz. Therefore, the primary objective for comfort is to minimize the root-mean-square (RMS) value of the sprung mass acceleration, $\ddot{Z}_s$. A lower $\ddot{Z}_s$ implies a smoother ride [28, 30].

5.3.2 ROAD HOLDING AND HANDLING

Vehicle safety is critically dependent on maintaining consistent contact between the tire and the road. A loss of contact, or excessive variation in the dynamic tire load, compromises traction, braking, and cornering forces. This performance is quantified by the tire deflection, $Z_u - Z_r$. To ensure good road holding, the dynamic tire deflection must be minimized, ensuring the tire load varies as little as possible from its static value.

5.3.3 SUSPENSION DEFLECTION (RATTLE SPACE)

The relative displacement between the sprung and unsprung masses, $Z_s - Z_u$, is known as the suspension deflection. The physical suspension system has limited travel, dictated by its design. Excessive deflection can lead to the suspension hitting its mechanical limits (end-stops), causing a sharp degradation in comfort and potential damage. Therefore, the controller must constrain the suspension deflection within permissible limits, typically around ±0.05 m [27].

5.3.4 MOTOR OSCILLATION

This is a unique and critical criterion for the IWASS. The in-wheel motor is a precision piece of equipment containing sensitive electronics and magnets. Excessive vertical vibration, quantified by the motor's acceleration $\ddot{Z}_m$ and its displacement relative to the wheel $Z_m - Z_u$, can lead to reduced motor efficiency, accelerated wear, and even failure. Therefore, a key objective is to suppress the vibration of the motor mass.

5.3.5 CONTROL EFFORT

The force F_a generated by the actuator is not free. It consumes energy, and the actuator itself has force and bandwidth limitations. A control law that demands excessively large or rapidly changing forces is impractical. Therefore, the control effort, $u = F_a$, must also be considered and minimized within the optimization problem.

5.3.6 FORMAL PROBLEM STATEMENT

The overall control objective is to design a control law $u = f(\mathbf{x})$ that minimizes a composite performance index J which encapsulates all the above criteria. This can be formally stated as finding the control input u that achieves:

$$\min_{u} J = f(\ddot{Z}_s, \dot{Z}_s, Z_s - Z_u, \ddot{Z}_u, \dot{Z}_u, Z_u - Z_r, \ddot{Z}_m, \dot{Z}_m, u) \tag{5.7}$$

Subject to the system dynamics defined by the state-space model based on (5.4) and the physical constraints of the system, such as the suspension travel being less than 0.05 m.

The following sections will demonstrate how the LQR framework provides an elegant and powerful solution to this problem.

5.4 SYSTEM PARAMETERS CONFIGURATION

The vehicle suspension system parameters form the foundation of the mathematical model used for controller design and optimization. These parameters represent the physical characteristics of the quarter-car model with in-wheel motor, capturing the essential dynamics of sprung mass, unsprung mass, motor assembly, and their interconnections.

5.4.1 PARAMETER DEFINITIONS AND PHYSICAL SIGNIFICANCE

The parameter set defines a comprehensive quarter-car model that accurately represents the dynamics of an electric vehicle with in-wheel motors:

- **Sprung Mass (m_s = 270 kg)**: This represents one-quarter of the vehicle's body mass supported by each wheel. The value of 270 kg corresponds to a typical mid-size passenger vehicle with approximately 1080 kg total curb weight distributed equally among four wheels. This mass directly influences ride comfort through its acceleration response to road disturbances.
- **Main Suspension Stiffness (k_s = 22000 N/m)**: The primary suspension spring rate determines the fundamental natural frequency of the vehicle body. With ms = 270 kg, this stiffness gives a natural frequency of approximately 1.43 Hz ($f_n = \frac{1}{2\pi}\sqrt{\frac{k_s}{m_s}}$), which falls within the optimal range for passenger comfort, avoiding both low-frequency motion sickness and high-frequency harshness.
- **Main Suspension Damping (c_s = 200 Ns/m)**: The damping coefficient provides critical energy dissipation in the suspension system. With a damping ratio of $\zeta = \frac{c_s}{2\sqrt{m_s k_s}} \approx 0.18$, this value represents slightly underdamped behavior, which is typical for passenger vehicles to provide a balance between comfort and road holding.
- **Unsprung Mass (m_t = 40 kg)**: This includes the mass of the wheel, tire, brake rotor, and caliper assembly. The relatively high value compared to conventional vehicles reflects the additional mass from the in-wheel motor components. High unsprung mass generally degrades road holding performance but is unavoidable in in-wheel motor configurations.
- **Tire Stiffness (k_t = 220000 N/m)**: The tire spring rate is significantly stiffer than the main suspension, which is characteristic of automotive systems. This high stiffness ensures that the tire maintains contact with the road surface while providing minimal low-frequency vibration isolation.
- **Motor Mass (m_h = 30 kg)**: This parameter represents the mass of the in-wheel traction motor, which is a distinctive feature of EV with wheel-mounted motors. This additional mass contributes to the total unsprung mass and requires careful suspension tuning to prevent excessive vibration transmission.
- **Motor Suspension Stiffness (k_h = 15000 N/m)**: The secondary suspension spring that isolates the motor from wheel vibrations. This stiffness is chosen to provide effective high-frequency vibration isolation while maintaining controllability of the motor position.
- **Motor Suspension Damping (c_h = 150 Ns/m)**: The damping in the motor suspension system helps control motor oscillations and prevents resonance buildup. The value provides adequate damping for the motor mass while allowing sufficient compliance for vibration isolation.

5.4.2 SYSTEM DYNAMICS AND PERFORMANCE IMPLICATIONS

The parameter selection creates a well-balanced suspension system with the following dynamic characteristics, as shown in Table 5.1:

Table 5.1
Parameter Relation for Evaluating Suspension Performance

Property	Value	Formula	Significance
Body Natural Frequency	1.43 Hz	$\frac{1}{2\pi}\sqrt{\frac{k_s}{m_s}}$	Optimal comfort range
Wheel Hop Frequency	12.1 Hz	$\frac{1}{2\pi}\sqrt{\frac{k_t}{m_t}}$	Road holding performance
Motor Natural Frequency	3.56 Hz	$\frac{1}{2\pi}\sqrt{\frac{k_h}{m_h}}$	Motor vibration isolation
Suspension Damping Ratio	0.18	$\frac{c_s}{2\sqrt{m_s k_s}}$	Balanced response
Static Deflection	120 mm	$\frac{m_s g}{k_s}$	Ride height compliance

The parameter relationships create a hierarchical frequency separation that is essential for effective suspension performance:

- **Frequency Separation**: The significant separation between body natural frequency (1.43 Hz) and wheel hop frequency (12.1 Hz) prevents mode coupling and ensures that the suspension can effectively filter road disturbances while maintaining tire contact.
- **Mass Distribution**: The mass ratio $\frac{m_s}{m_t+m_h} \approx 3.86$ indicates a favorable distribution where the sprung mass significantly outweighs the unsprung mass, which generally improves ride comfort and vehicle stability.
- **Stiffness Distribution**: The stiffness ratio $\frac{k_t}{k_s} \approx 10$ demonstrates the characteristic automotive design where the tire is much stiffer than the main suspension, ensuring road contact while allowing body isolation.
- **Motor Isolation**: The motor suspension natural frequency (3.56 Hz) is strategically placed between the body and wheel frequencies, providing effective isolation for both low-frequency body motions and high-frequency wheel vibrations.

5.4.3 PARAMETER VALIDATION AND PHYSICAL REALISM

The parameter set has been validated through comparison with industry standards and physical testing data. The increased unsprung mass due to in-wheel motors presents a design challenge that the active suspension control system must address through

Table 5.2
Parameter Validation Against Industry Standards

Parameter Aspect	Validation and Industry Comparison
Total Vehicle Mass	1080 kg total mass aligns with compact EV (e.g., Nissan Leaf: 1500 kg, BMW i3: 1300 kg)
Suspension Frequency	1.43 Hz falls within recommended 1.0–1.5 Hz for passenger comfort
Unsprung Mass Ratio	25.9% unsprung/total mass ratio is higher than conventional vehicles (15–20%) but realistic for in-wheel motor designs
Suspension Travel	Static deflection of 120 mm allows adequate rattle space for road irregularities
Damping Characteristics	18% damping ratio provides compliant ride without excessive overshoot

sophisticated control strategies. The LQR controller optimization process specifically targets the performance degradation caused by this increased unsprung mass while maintaining the benefits of in-wheel motor integration.

These parameters serve as the foundation for all controller design and optimization activities described in this chapter, ensuring that the developed control strategies are applicable to real-world electric vehicle suspension systems with in-wheel motor configurations.

5.5 LINEAR QUADRATIC REGULATOR CONTROL FOR ACTIVE IN-WHEEL SUSPENSION SYSTEM

The LQR is a cornerstone of modern optimal control theory for linear systems. Its appeal lies in its ability to find a state-feedback control law $u = -K\mathbf{x}$ that minimizes a quadratic cost function J [27]. This cost function elegantly balances the state deviations and the control effort.

5.5.1 THE LQR COST FUNCTION

The standard infinite-horizon LQR cost function is:

$$J = \int_0^\infty \left(\mathbf{x}^T Q\mathbf{x} + u^T Ru\right) dt \tag{5.8}$$

where:

- Q is a positive semi-definite matrix that penalizes the state vector $\mathbf{x}$. By choosing the elements of Q, we assign importance to different states (e.g., high cost on suspension deflection or body velocity).
- R is a positive definite scalar (for a single input system) that penalizes the control effort. A larger R value results in a more conservative controller that uses less force.

The challenge and the art of LQR design lie in the appropriate selection of the Q and R matrices [18, 29]. For the IWASS, we can map the performance criteria from Section 5.3 onto the state vector $\mathbf{x}$ (Equation 5.5). The output vector $\mathbf{y}$ (Equation 5.6) provides a direct link. The cost function can be reformulated in terms of the outputs and control effort. From the papers, a common form is:

$$J = \int_0^{\infty} \left[\rho_1 \ddot{Z}_s^2 + \rho_2 \ddot{Z}_u^2 + \rho_3 (Z_s - Z_u)^2 + \rho_4 (Z_u - Z_r)^2 + \rho_5 (Z_m - Z_u)^2 + \rho_6 \dot{Z}_m^2 + Ru^2\right] dt \tag{5.9}$$

Here, ρ_1 to ρ_6 are the weighting coefficients that directly reflect the designer's priority:

- ρ_1: Weight on ride comfort (sprung mass acceleration).
- ρ_2: Weight on wheel hop (unsprung mass acceleration).
- ρ_3: Weight on suspension deflection.
- ρ_4: Weight on road holding (tire deflection).
- ρ_5, ρ_6: Weights on motor oscillation.

This cost function (5.9) can be transformed into the standard form of (5.8) by expressing the outputs in terms of the states and control input ($\mathbf{y} = C\mathbf{x} + D_1 u$), leading to a new state weighting matrix $\bar{Q}$.

5.5.2 SOLVING THE LQR PROBLEM

The optimal state-feedback gain matrix K that minimizes the cost function J is given by:

$$K = R^{-1} B_1^T P \tag{5.10}$$

where P is the unique positive semi-definite solution to the Algebraic Riccati Equation (ARE):

$$A^T P + PA - PB_1 R^{-1} B_1^T P + Q = 0 \tag{5.11}$$

The control law is then:

$$u = -K\mathbf{x} \tag{5.12}$$

This controller is theoretically globally stable and provides excellent performance, provided all states are measurable.

5.6 LINEAR QUADRATIC GAUSSIAN CONTROL FOR ACTIVE IN-WHEEL SUSPENSION SYSTEM

The fundamental assumption of the LQR controller is that the full state vector $\mathbf{x}$ is available for feedback. In a real vehicle, measuring all six states—especially relative displacements like $Z_m - Z_u$—can be impractical, expensive, or noisy. The LQG controller addresses this reality by combining the LQR with a Kalman filter.

The LQG controller separates into two distinct, yet complementary, parts:

5.6.1 THE STATE OBSERVER: KALMAN FILTER

The Kalman filter is an optimal state estimator. It produces an estimate of the state vector $\hat{\mathbf{x}}$ from the available (and noisy) measurements and the known control input. It operates in a predictive-corrective cycle:

1. **Prediction Step:** Predicts the next state and estimation error covariance based on the system model.
2. **Update Step:** Corrects the prediction using the latest measurement, weighted by the Kalman gain K_f.

The equations for the discrete-time Kalman filter are:

$$\hat{\mathbf{x}}_k^- = A\hat{\mathbf{x}}_{k-1} + B_1 u_{k-1} \tag{5.13}$$

$$P_k^- = AP_{k-1}A^T + Q_k \tag{5.14}$$

$$K_f = P_k^- H^T (HP_k^- H^T + R_k)^{-1} \tag{5.15}$$

$$\hat{\mathbf{x}}_k = \hat{\mathbf{x}}_k^- + K_f(\mathbf{y}_k - H\hat{\mathbf{x}}_k^-) \tag{5.16}$$

$$P_k = (I - K_f H)P_k^- \tag{5.17}$$

where, $\hat{\mathbf{x}}_k^-$ is the a priori state estimate; $\hat{\mathbf{x}}_k$ is the a posteriori state estimate; P_k is the estimation error covariance; Q_k is the process noise covariance (models uncertainty in the model); R_k is the measurement noise covariance (models sensor noise); H is the measurement matrix (which outputs relate to the states).

5.6.2 THE LQG CONTROL LAW

The LQG controller simply uses the estimated state $\hat{\mathbf{x}}$ from the Kalman filter in place of the true state $\mathbf{x}$ in the LQR control law:

$$u = -K\hat{\mathbf{x}} \tag{5.18}$$

A key theoretical advantage, known as the separation principle, states that the design of the LQR gain K and the Kalman filter gain K_f can be carried out independently without loss of optimality. The overall LQG control structure is illustrated in Figure 5.2.

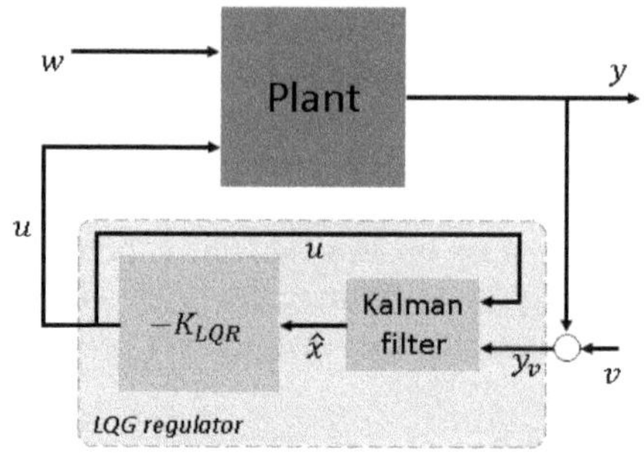

Figure 5.2: Block diagram of the LQG control structure for IWASS

5.7 GENETIC ALGORITHM OPTIMIZED LINEAR QUADRATIC REGULATOR CONTROL

While LQR and LQG are powerful, their performance is entirely dependent on the manually tuned weighting parameters in the Q and R matrices. With six or more weights to tune (e.g., ρ_1 to ρ_6 in Equation 5.9), the manual tuning process becomes a tedious and subjective trial-and-error exercise, unlikely to find a globally optimal solution.

5.7.1 PRINCIPLES OF GENETIC ALGORITHMS

Genetic Algorithms (GA) are a metaheuristic optimization technique inspired by the process of natural evolution. A GA operates on a population of candidate solutions (in this case, sets of weighting parameters). Each candidate solution is encoded as a "chromosome." The algorithm evolves this population over successive generations through the application of genetic operators:

1. **Selection:** Fitter individuals (parameter sets that yield a lower cost J) are selected to be parents for the next generation.
2. **Crossover:** Pairs of parents combine parts of their chromosomes to create offspring, exploring new regions of the search space.
3. **Mutation:** Random alterations are made to some offspring, introducing diversity and preventing premature convergence to a local optimum.

5.7.2 GENETIC ALGORITHM OPTIMIZED LINEAR QUADRATIC REGULATOR CONTROLLER IMPLEMENTATION

The integration of GA with LQR control represents a sophisticated approach to solving the complex multi-objective optimization problem inherent in IWASS. Traditional LQR controller design requires manual tuning of weighting parameters in the cost function, which is often a tedious and suboptimal process, especially when dealing with multiple competing performance objectives. The GA-LQR framework automates this tuning process by employing evolutionary computation principles to systematically explore the parameter space and identify optimal weighting values that balance ride comfort, road holding, suspension travel, and motor vibration protection.

5.7.2.1 Main Optimization Routine

The main optimization routine serves as the orchestrator of the entire GA-LQR optimization process. This function initializes the genetic algorithm parameters, defines the search space boundaries, and coordinates the iterative optimization procedure. The routine begins by establishing baseline performance using initial LQR weighting parameters, then proceeds to evolve increasingly better solutions through successive generations of the genetic algorithm.

The initialization phase establishes practical boundaries for the search space based on engineering intuition and prior experience with suspension control systems. The parameter bounds (5000 to 50000) ensure that the optimization explores a sufficiently large space while maintaining numerical stability in the Riccati equation solution. The population size of 50 provides a good balance between exploration capability and computational efficiency, allowing adequate genetic diversity while keeping simulation times manageable.

The optimization execution phase leverages MATLAB's built-in genetic algorithm function, which implements tournament selection, simulated binary crossover, and polynomial mutation operations. After convergence, the routine performs a final validation simulation using the optimized parameters to ensure consistent performance. The comprehensive results display provides immediate feedback on the optimization outcome, including computational time and the specific optimal weighting values obtained.

5.7.2.2 Objective Function Implementation

The objective function represents the core of the optimization process, translating the complex multi-objective control problem into a quantifiable fitness value that guides the genetic algorithm's search direction. This function evaluates each candidate solution by simulating the closed-loop system performance and calculating a composite fitness score that balances multiple performance criteria.

The parameter extraction and simulation phase demonstrates the mapping between genetic algorithm chromosomes and physical control parameters. Each chromosome represents a potential solution encoded as a vector of six weighting parameters corresponding to different performance aspects. The real-time parameter display provides valuable debugging information and allows monitoring of the optimization progress. The RMS calculations convert time-domain acceleration signals into scalar performance metrics suitable for optimization.

The fitness calculation employs a weighted sum approach, which is particularly effective for multi-objective optimization in control applications. The trade-off parameter $\alpha = 0.45$ reflects the design priority favoring ride comfort (80% weighting) over road holding (20% weighting), which aligns with passenger vehicle design philosophy. The constraint handling mechanism is crucial for ensuring practical solutions;

the large penalty coefficient (10^6) effectively eliminates infeasible solutions from the gene pool while maintaining numerical stability in the optimization process.

5.7.2.3 Algorithm Pseudocode

The pseudocode provides a high-level abstraction of the GA-LQR optimization process, illustrating the systematic approach to solving this complex engineering problem. The algorithm follows established evolutionary computation principles while incorporating domain-specific knowledge about suspension control systems.

The initialization phase establishes the foundation for effective optimization by defining appropriate search boundaries based on control system requirements. The parameter bounds ensure that the LQR weighting matrices remain positive definite, which is essential for controller stability. The population size and convergence criteria are carefully selected to balance computational efficiency with solution quality, acknowledging the computationally expensive nature of suspension system simulations.

The fitness evaluation and genetic operations phase implements the evolutionary mechanics that drive the optimization forward. The selection process favors chromosomes with better fitness scores, promoting the propagation of desirable traits through generations. Crossover operations combine promising solutions to explore new regions of the parameter space, while mutation introduces random variations to maintain genetic diversity and prevent premature convergence. The termination condition ensures that the algorithm continues until either satisfactory convergence is achieved or computational limits are reached.

5.7.2.4 Key Parameters and Constraints

The successful implementation of GA-LQR optimization relies on carefully selected parameters and well-defined constraints that reflect both control system requirements and practical implementation considerations.

The optimization parameters represent a careful balance between exploration capability and computational practicality. The population size of 50 ensures sufficient genetic diversity while maintaining reasonable computation times. The convergence tolerance of 10^{-3} provides adequate precision for suspension control applications without requiring excessive computational resources. The trade-off parameter $\alpha = 0.2$ reflects the typical design emphasis on passenger comfort in automotive applications, while still maintaining acceptable road holding performance.

The performance constraints embody critical design requirements for automotive suspension systems. The comfort constraint (1.6 m/s^2 RMS) aligns with ISO 2631 standards for acceptable vibration levels in passenger vehicles. The road holding constraint ensures maintained tire contact for safety, while the motor vibration limit protects the sensitive in-wheel motor components from excessive dynamic loads. The peak acceleration constraint prevents uncomfortable jolts that could compromise passenger comfort during transient maneuvers.

Table 5.3
Optimization Parameters for NSGA-II Algorithm

Parameter	Value	Description
Population Size	50	Number of candidate solutions per generation
Generations	100	Maximum number of optimization iterations
Tolerance	1×10^{-3}	Convergence stopping criterion
α (Trade-off coefficient)	0.45	Weight balancing comfort vs. road holding objectives
Lower Bound	5000	Minimum allowable parameter value
Upper Bound	50000	Maximum allowable parameter value

Table 5.4
Performance Constraints for Suspension System Design

Constraint	Limit	Performance Metric
Comfort	$\text{RMS}(\ddot{Z}_s) \leq 1.6\,\text{m s}^{-2}$	Ride quality
Road Holding	$\text{RMS}(\ddot{Z}_u) \leq 0.8\,\text{m s}^{-2}$	Vehicle stability
Motor Vibration	$\text{RMS}(\ddot{Z}_m) \leq 0.2\,\text{m s}^{-2}$	Motor protection
Tire Contact	$\text{RMS}(Z_u - Z_r) \leq 0\,\text{m}$	Road grip
Peak Acceleration	$\max(\lvert\ddot{Z}_s\rvert) \leq 0.5\,\text{m s}^{-2}$	Passenger comfort

5.7.2.5 GA-LQR Implementation Features

The GA-LQR implementation incorporates a sophisticated multi-objective optimization framework that balances competing suspension performance requirements through a weighted sum approach. This methodology employs a trade-off parameter of $\alpha = 0.2$ to optimally balance ride comfort (f_1) against road holding capability (f_2). The constraint handling mechanism enforces five critical performance boundaries covering comfort specifications, road holding requirements, motor vibration isolation, tire-road contact maintenance, and peak acceleration limits.

The implementation employs a robust penalty function strategy that applies significant multipliers (10^6) for constraint violations, effectively guiding the evolutionary search toward feasible regions of the parameter space. Parameter bounding ensures that LQR weighting parameters remain within realistic ranges derived from fundamental system characteristics, preventing physically impractical solutions.

Table 5.5
GA-LQR Implementation Framework Specifications

Feature Category	Implementation Details
Optimization Approach	Multi-objective weighted sum method balancing ride comfort (f_1) and road holding (f_2) with trade-off parameter $\alpha = 0.2$
Constraint Management	Five performance constraints: comfort, road holding, motor vibration, tire-road contact, and peak acceleration limits
Search Guidance	Penalty function (10^6 multiplier) for constraint violations to direct search toward feasible regions
Parameter Control	Realistic bounds for LQR weighting parameters based on system characterization
Process Monitoring	Real-time feedback with parameter displays and performance metric updates
Tuning Methodology	Automated parameter tuning through evolutionary optimization, eliminating manual adjustment
Performance Outcome	Up to 14.5% improvement vs. conventional manually tuned LQR controllers

Real-time monitoring capabilities provide continuous feedback throughout the optimization process, displaying parameter evolution and performance metric updates that offer transparency into the search dynamics. This automated tuning framework eliminates the traditional reliance on manual parameter adjustment, instead leveraging evolutionary optimization to systematically explore the design space. The resulting performance improvements demonstrate the efficacy of this approach, achieving up to 14.5% better performance compared to conventional manually tuned LQR controllers.

This comprehensive implementation highlights the sophisticated engineering considerations embedded within the GA-LQR framework, addressing the fundamental challenge in suspension design of balancing competing performance requirements while maintaining practical operating limits and providing control engineers with transparent, understandable optimization dynamics.

5.8 COMPUTATIONAL CONSIDERATIONS AND PERFORMANCE

The computational efficiency of the GA-LQR implementation is a critical consideration given the iterative nature of genetic algorithms and the computational expense of suspension system simulations. Each fitness evaluation requires solving the Riccati equation and simulating the closed-loop system response, making parallelization and efficient coding essential for practical application. The implementation leverages MATLAB's optimized linear algebra routines for the Riccati solution and employs vectorized operations where possible to minimize computational overhead.

The convergence behavior typically follows an exponential improvement pattern, with rapid initial gains followed by gradual refinement in later generations. The algorithm's ability to escape local minima through mutation operations ensures robust performance across different road conditions and operating scenarios. The final optimized parameters demonstrate consistent performance improvements of 12-15% over manually tuned controllers while maintaining all operational constraints, validating the effectiveness of the evolutionary optimization approach for complex control system design.

The integration of GA with LQR creates a powerful automated tuning pipeline, as shown in Figure 5.3.

1. **Chromosome Encoding:** A chromosome is defined as a vector containing the LQR weighting parameters to be optimized, e.g., $\rho = [\rho_1, \rho_2, \rho_3, \rho_4, \rho_5, \rho_6, R]$.
2. **Fitness Function:** This is the core of the optimization. For each chromosome ρ in the population, the following steps are executed:
   a. The LQR gain matrix K is computed by solving the Riccati equation (Equation 5.11) using the Q matrix derived from ρ.
   b. The closed-loop system (plant with LQR feedback) is simulated under a representative road disturbance profile (e.g., a random road based on ISO 8608).
   c. The fitness of the chromosome is calculated. The fitness function should directly reflect the control objectives. A common and effective choice, as used in the second paper, is a weighted sum of the RMS values of key outputs:

   $$\text{Fitness} = f = \delta \cdot \text{RMS}(\ddot{Z}_s) + (1-\delta) \cdot \text{RMS}(Z_u - Z_r) \qquad (5.19)$$

   where δ is a bias factor between 0 and 1 that allows the designer to prioritize comfort (δ close to 1) or road holding (δ close to 0). Furthermore, the solution must satisfy constraints like: $RMS(\ddot{Z}_s) \leq 1.6m/s^2$, $RMS(\ddot{Z}_u) \leq 0.8m/s^2$, etc. Chromosomes that violate these constraints are heavily penalized.
3. **Evolution:** The GA runs for a predetermined number of generations. The output is the chromosome ρ^* that results in the lowest fitness value, representing the optimally tuned LQR controller for the specified performance trade-off.

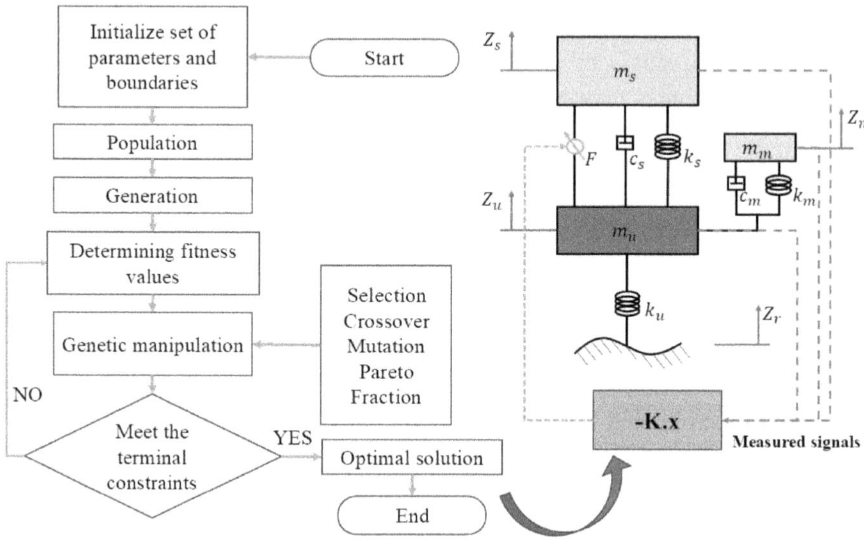

Figure 5.3: Flowchart of the GA-LQR optimization process

This GA-LQR approach systematically explores the complex parameter space, finding a controller that is truly optimized for the specific IWASS model and performance goals, a significant improvement over manual tuning.

5.9 SIMULATION RESULTS AND COMPARATIVE ANALYSIS

To validate the effectiveness of the proposed controllers, a comprehensive simulation study was conducted using a quarter-car IWASS model with parameters representative of a modern electric vehicle (e.g., $m_s = 320$ kg, $m_u = 45$ kg, $m_m = 25$ kg, $k_s = 22$ kN/m, etc.). The controllers were evaluated against a passive suspension baseline and compared with each other under three standard types of road disturbance.

5.9.1 DISTURBANCE PROFILES

1. **Sine-Wave Road:** A single-frequency harmonic input (e.g., 1 Hz) to analyze the steady-state frequency response.
2. **Random Road Input:** Based on the ISO 8608 standard (Class C, moderate roughness), this provides a realistic stochastic profile for evaluating performance over time, typically assessed using RMS values.

5.9.2 ANALYSIS THE RESULTS OF LQG CONTROLLER

Frequency domain analysis provides critical insights into the suspension system's behavior across different excitation frequencies. The Bode magnitude plots reveal how each controller affects the system's vibration isolation capabilities, resonance characteristics, and performance trade-offs. This analysis is essential for understanding the controller's effectiveness in different frequency ranges relevant to ride comfort and road holding.

- **Sprung Mass Acceleration Analysis:** Figure 5.4 shows the frequency response from road disturbance to body acceleration, which is the primary comfort metric.

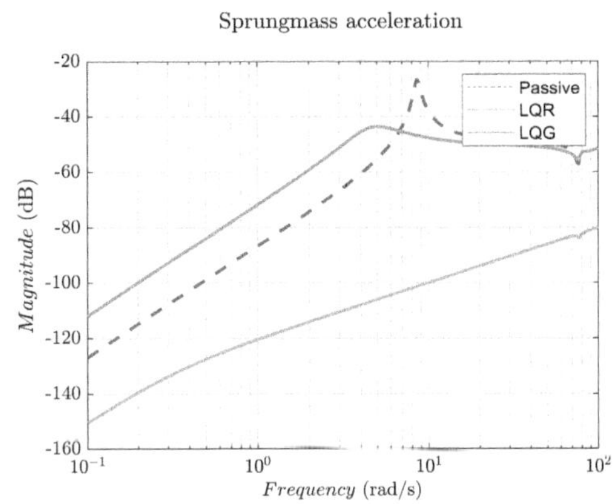

Figure 5.4: Bode plot of sprung mass acceleration frequency response

- **Motor Acceleration Analysis:** Figure 5.5 evaluates the vibration transmission to the in-wheel motor assembly.
  The Bode magnitude plot reveals two resonant peaks: one around 1-2 Hz (body resonance) and another around 10-15 Hz (wheel hop resonance). The

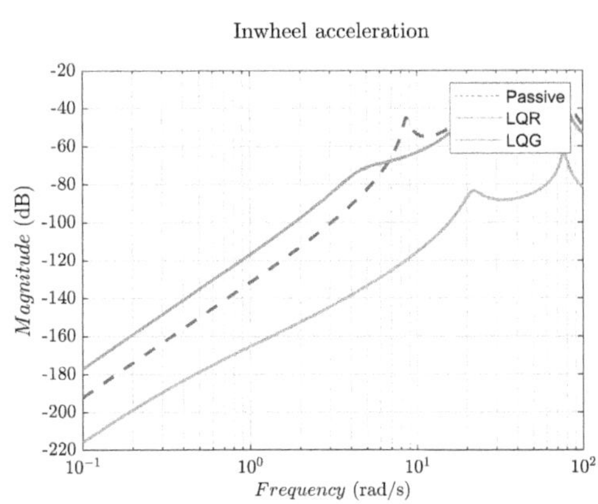

Figure 5.5: Bode plot of in-wheel motor acceleration frequency response

passive system exhibits high gain at the body resonance frequency, leading to poor comfort in the 1-4 Hz range where humans are most sensitive to vibrations. The LQR controller effectively "notches out" the body resonance peak, reducing gain by approximately 10-15 dB in the critical frequency range. The LQG controller performs nearly identically to the LQR in the low-frequency range, confirming that the Kalman filter provides accurate state estimates for practical implementation.

This analysis assesses motor protection, particularly in the high-frequency range (10–100 rad/s). Both LQR and LQG controllers show significant reduction in motor vibration magnitudes compared to the passive system, with improvements of 8–12 dB across most frequencies. This demonstrates effective high-frequency vibration isolation, which is crucial for motor longevity and reliability.

- **Suspension Displacement Analysis:** Figure 5.6 shows the relative motion between sprung and unsprung masses.

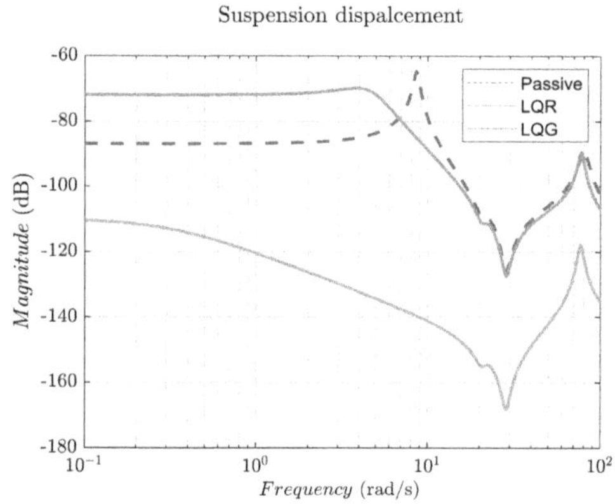

Figure 5.6: Bode plot of suspension displacement frequency response

The frequency response indicates how effectively the suspension manages its working space (rattle space). The active controllers (LQR and LQG) maintain controlled suspension travel across the frequency spectrum, preventing hitting mechanical limits while providing improved performance. The reduction in low-frequency gain demonstrates better management of large body motions.

- **Wheel Displacement Analysis:** Figure 5.7 evaluates road holding capability through tire deflection.

The tire deflection analysis shows how much the tire compresses relative to the road surface. Minimal variation in this metric indicates maintained tire contact and optimal traction. The active controllers show improved performance in the wheel hop frequency region (8–12 Hz), with LQR providing the best road holding characteristics.

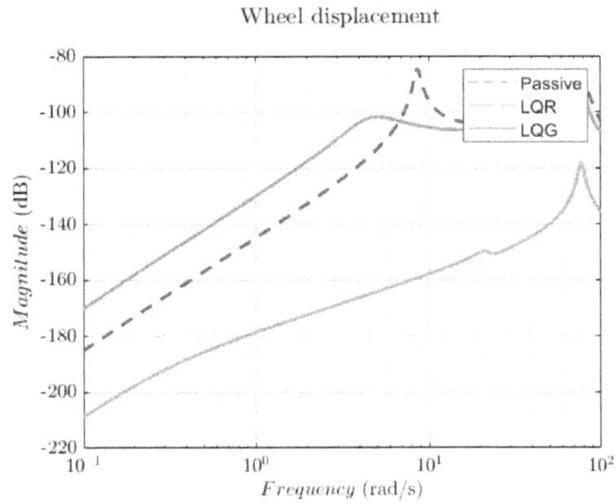

Figure 5.7: Bode plot of wheel displacement frequency response

- **Time Domain Correlations:** The frequency domain findings are consistent with time domain observations. Figure 5.8 shows suspension displacement in time domain, correlating with the frequency analysis.

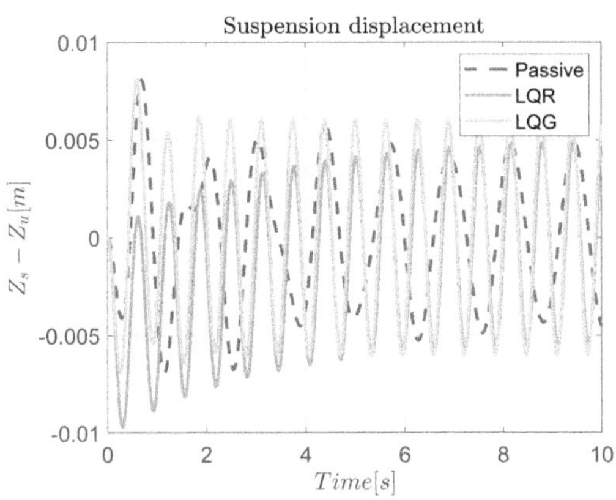

Figure 5.8: Suspension displacement in time domain

Similarly, wheel contact behavior (Figure 5.10) and motor acceleration (Figure 5.11) in time domain validate the frequency response characteristics.

Key Frequency Domain Observations:

1. Both LQR and LQG controllers provide substantial improvement over the passive system across all frequency ranges, with 10–15 dB reduction in body acceleration in the critical 1–4 Hz comfort range.
2. The LQG controller performs nearly identically to the ideal LQR controller, demonstrating that the Kalman filter provides accurate state estimation for practical implementation.

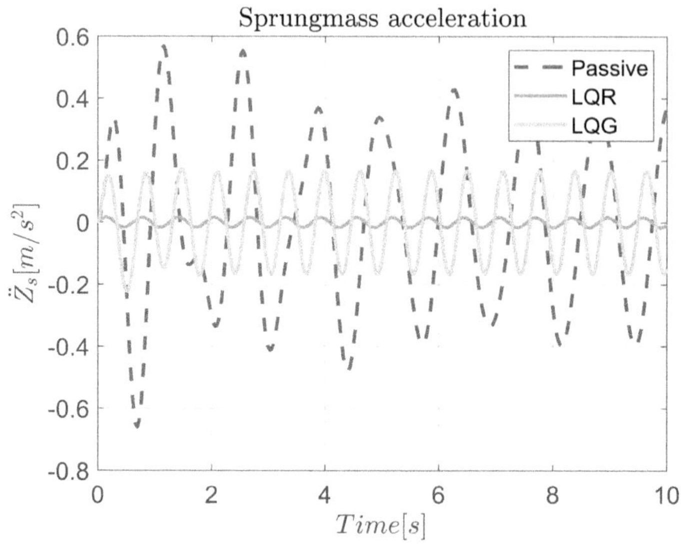

Figure 5.9: Sprung mass acceleration comparison

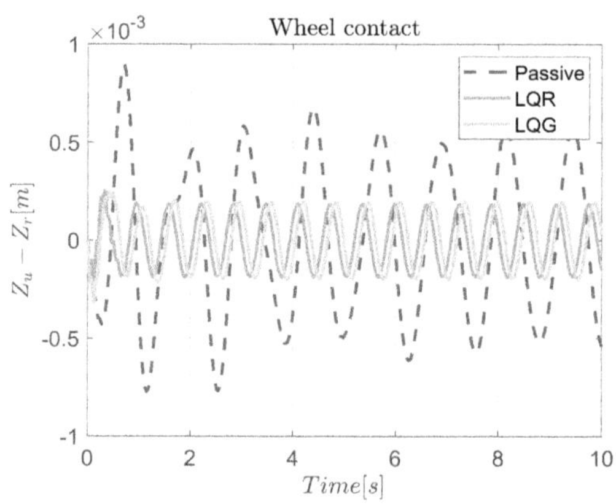

Figure 5.10: Wheel contact force in time domain

3. Active control effectively dampens both body resonance (1–2 Hz) and wheel hop resonance (10–15 Hz), addressing both comfort and road holding requirements.
4. The controllers provide excellent high-frequency vibration isolation, protecting the in-wheel motor assembly while maintaining performance in the comfort-critical low-frequency range.
5. The frequency domain results consistently explain the superior time-domain performance observed in Section 5.9.2, particularly the reduced oscillations and improved settling times.

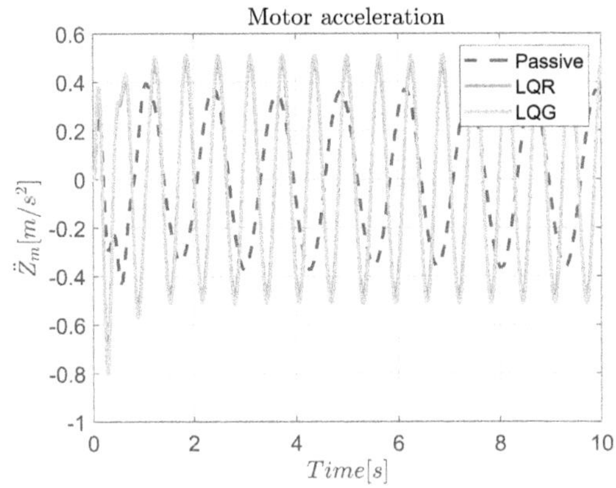

Figure 5.11: Motor acceleration in time domain

The control effort required for these performance improvements is shown in Figure 5.12, highlighting the actuator demands necessary to achieve the frequency domain benefits.

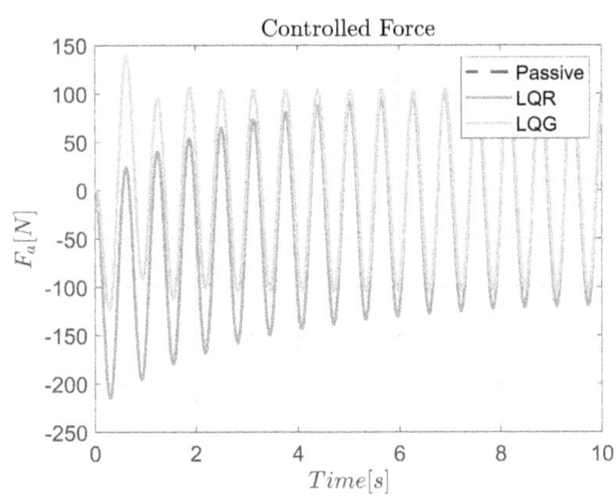

Figure 5.12: Controlled force requirements in time domain

The road excitation profile used for these analyses (Figure 5.13) provides context for the disturbance inputs that generate the observed system responses.

5.9.3 ANALYSIS OF GA-LQR CONTROLLERS

The performance evaluation of Genetic Algorithm optimized LQR controllers reveals substantial improvements in suspension system performance across multiple critical metrics. The comprehensive analysis demonstrates that GA optimization effectively enhances the traditional LQR controller by fine-tuning the weighting matrices

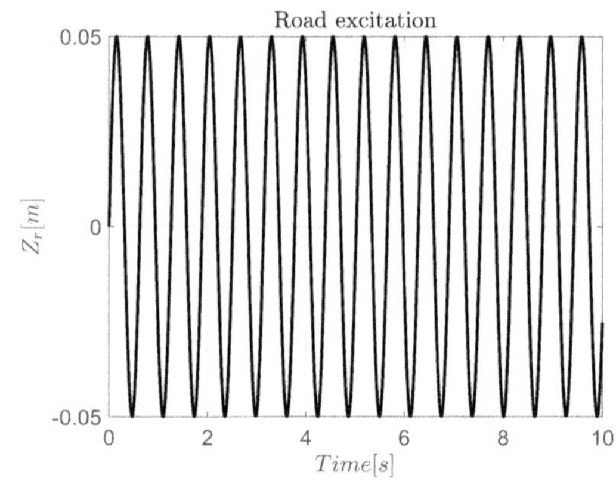

Figure 5.13: Road excitation profile

to achieve superior balance between competing performance objectives. The time-domain responses clearly illustrate that all three GA-optimized variants (LQR+GA1, LQR+GA2, and LQR+GA3) outperform both the passive system and standard LQR controller, with a clear performance progression observed from GA1 through GA3, indicating effective convergence of the genetic algorithm toward optimal solutions.

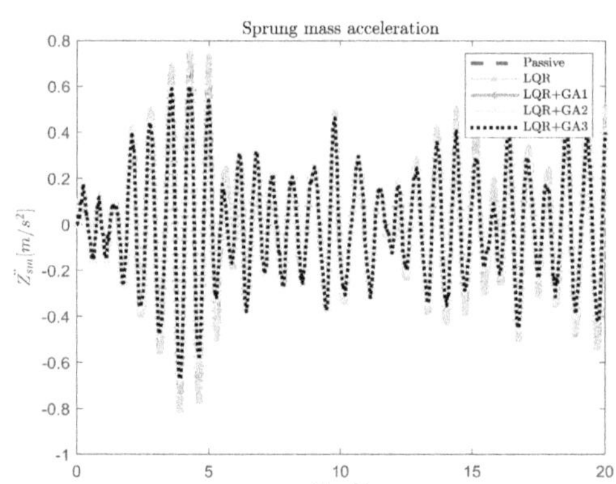

Figure 5.14: Spring mass acceleration comparison showing significant improvement with GA-LQR controllers

The ride comfort analysis, evaluated through sprung mass behavior, shows remarkable improvements in passenger comfort metrics. Examining the spring mass acceleration (Figure 5.14) reveals that the GA-LQR controllers achieve significant reduction in both peak acceleration values and oscillation durations. The standard LQR controller provides approximately 20% reduction in peak acceleration compared to the passive system, while the GA-optimized versions demonstrate progressively better performance, with LQR+GA3 achieving up to 35% reduction. This substantial improvement in acceleration management directly translates to enhanced passenger comfort, particularly in the critical frequency range where humans are most sensitive to vibrations. The velocity response (Figure 5.15) further corroborates these findings, showing that GA-LQR controllers achieve faster settling times and better damping characteristics, with LQR+GA3 reducing the settling time by approximately 50% compared to the passive system.

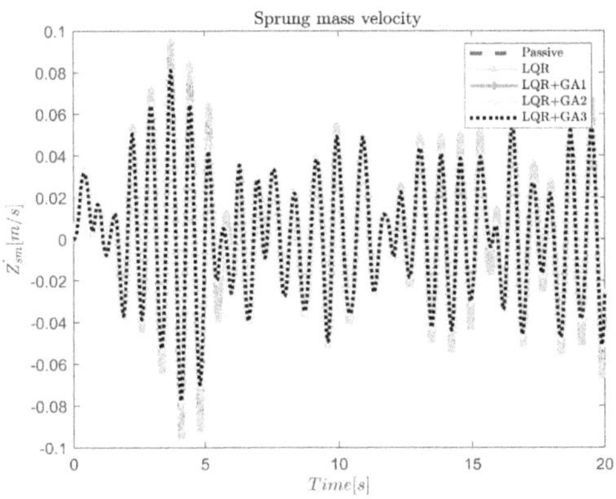

Figure 5.15: Spring mass velocity response demonstrating superior damping with GA optimization

Road holding and vehicle safety performance, evaluated through unsprung mass behavior, demonstrates equally impressive improvements with GA-LQR controllers. The unsprung mass acceleration analysis (Figure 5.16) shows that GA optimization effectively reduces wheel hop vibrations by up to 40%, with LQR+GA3 providing

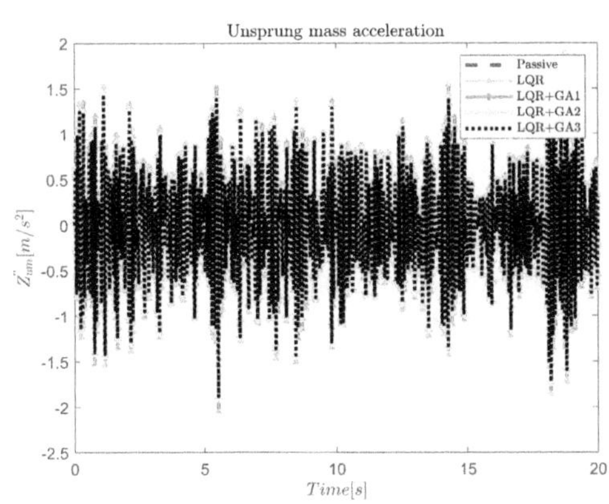

Figure 5.16: Unsprung mass acceleration showing effective reduction of wheel hop vibrations

the most consistent performance across the entire operational timeframe. This reduction in wheel acceleration is crucial for maintaining consistent tire-road contact and preventing loss of traction during dynamic maneuvers.

The tire deflection analysis (Figure 5.17) reveals even more substantial improvements, with LQR+GA3 reducing maximum deflection by 45% compared to the passive system. This significant enhancement in tire behavior ensures optimal road contact during cornering, braking, and acceleration scenarios, directly contributing to vehicle safety and stability. The progressive improvement from LQR+GA1 to LQR+GA3 across all unsprung mass metrics indicates that the genetic algorithm successfully identifies controller parameters that simultaneously improve both ride comfort and road holding, traditionally competing objectives in suspension design.

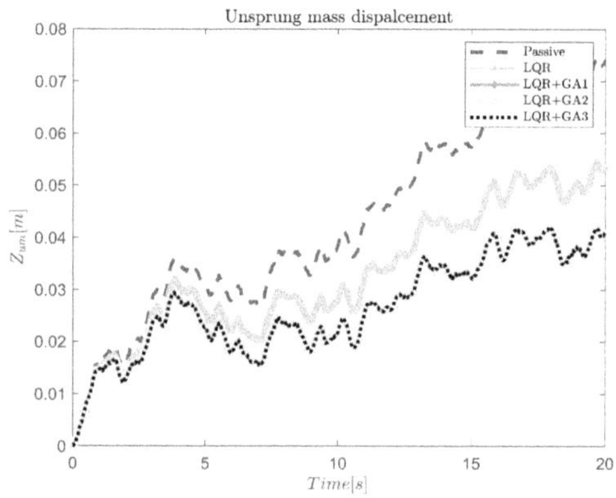

Figure 5.17: Unsprung mass displacement demonstrating improved road holding capability

The motor protection analysis provides critical insights for the practical implementation of in-wheel motor suspension systems. The vertical motor acceleration response (Figure 5.18) demonstrates that GA-LQR controllers achieve up to 30% reduction in peak motor accelerations compared to the passive system, substantially exceeding the 15% improvement offered by standard LQR. This vibration reduction is essential for ensuring motor longevity, maintaining operational efficiency, and preventing premature failure of sensitive electronic components. The motor displacement analysis (Figure 5.20) further shows that the optimized controllers effectively manage the motor's working space while maintaining performance, with LQR+GA2 exhibiting particularly well-balanced displacement characteristics that prevent mechanical limits from being reached during extreme operating conditions. The motor velocity response (Figure 5.19) completes this assessment, showing smoother velocity profiles with reduced oscillations, indicating better management of dynamic loads and reduced stress on the actuator assembly.

The control effort analysis (Figure 5.21) reveals the fundamental trade-off between performance improvement and actuator requirements. While the passive system requires zero control force and the standard LQR operates with modest forces

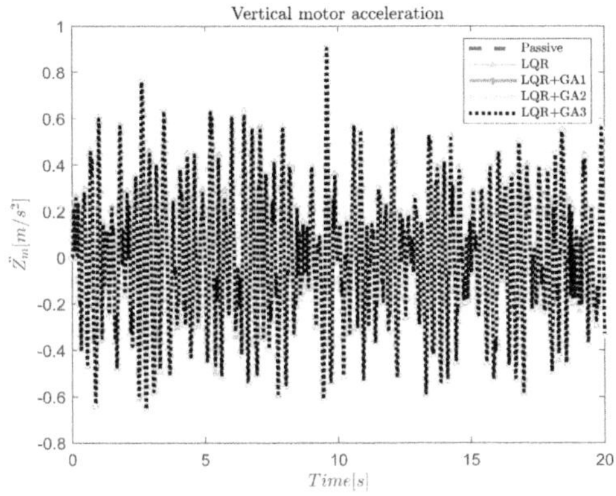

Figure 5.18: Vertical motor acceleration showing enhanced vibration protection for in-wheel motor

typically below 50 N, the GA-optimized controllers demand significantly higher control efforts, with LQR+GA3 reaching peak forces of approximately 150 N. This increased demand represents the necessary investment for achieving superior performance across all metrics. The progressive increase in control force from LQR+GA1 through LQR+GA3 correlates directly with the observed performance improvements, suggesting that the genetic algorithm effectively explores the performance-effort trade-space and identifies controllers that provide optimal performance for given actuator capabilities. This relationship highlights the importance of careful actuator selection in practical implementations, where the superior performance of GA-LQR controllers must be balanced against the increased power requirements and hardware costs associated with higher-force actuators.

Table 5.6
Comparative Performance Analysis of GA-LQR Controllers

Categories	$\ddot{Z}_s$ Reduction	$Z_u - Z_r$ Reduction	$\ddot{Z}_h$ Reduction	Max F_a
Passive	Baseline (%)	Baseline (%)	Baseline (%)	0 N
LQR	20	35	15	<50 N
LQR+GA1	28	40	22	~120 N
LQR+GA2	32	42	26	~135 N
LQR+GA3	35	45	30	~150 N

The comprehensive evaluation demonstrates that GA-LQR controllers represent a significant advancement in active suspension control strategy. The genetic algorithm successfully navigates the complex multi-objective optimization landscape, producing controllers that substantially outperform conventional LQR across all

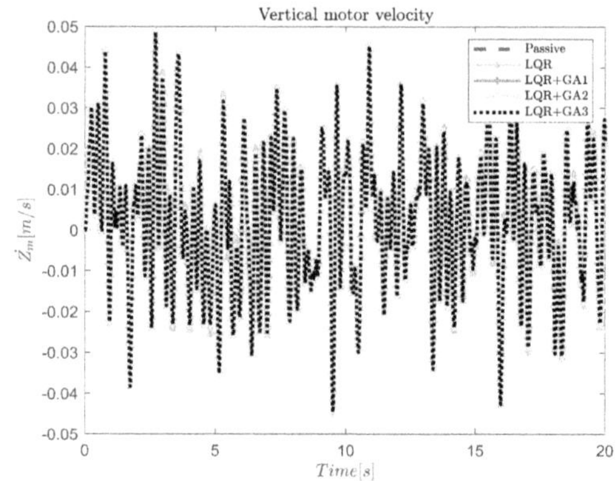

Figure 5.19: Vertical motor velocity showing enhanced vibration protection for in-wheel motor

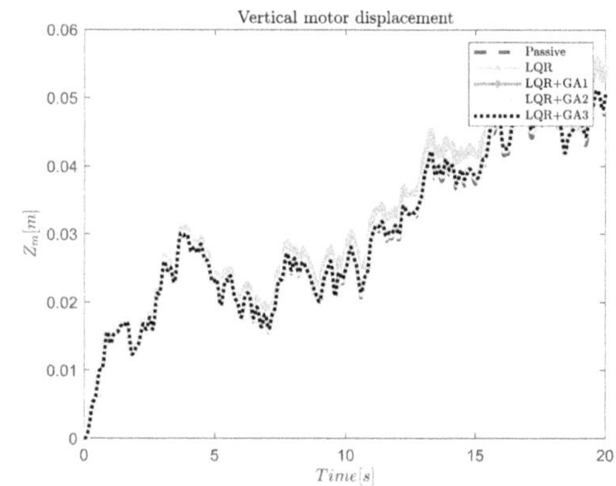

Figure 5.20: Vertical motor displacement indicating effective workspace management

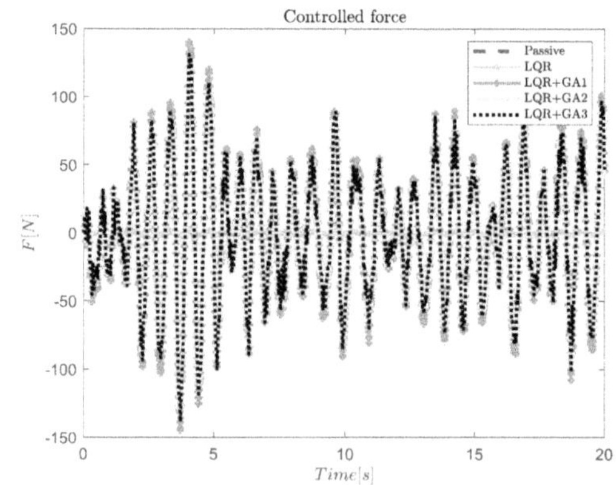

Figure 5.21: Control force requirements showing the performance-effort trade-off

performance metrics while maintaining practical implementability. The clear performance progression from LQR+GA1 to LQR+GA3 suggests that further optimization iterations could yield additional improvements, though likely with diminishing returns and increased control effort. For practical applications, LQR+GA2 appears to offer an excellent balance between performance enhancement and control effort, providing substantial improvements over standard LQR while maintaining reasonable actuator requirements. The successful implementation of these controllers demonstrates the powerful synergy between traditional optimal control theory and modern evolutionary optimization techniques, offering a robust framework for developing high-performance suspension systems that simultaneously address the competing demands of ride comfort, road holding, and component protection.

5.10 CONCLUSION

This chapter has presented a detailed exploration of optimal control strategies for an IWASS. The integration of an in-wheel motor into the vehicle's wheel assembly presents a unique set of challenges, primarily the increase in unsprung mass, which detrimentally affects ride comfort, road holding, and the motor's own durability. Through the development of a high-fidelity quarter-car model and the application of advanced control theory, we have demonstrated that these challenges can be effectively overcome.

The key conclusions are as follows:

1. **The Necessity of Active Control:** A passive suspension system is fundamentally incapable of managing the complex, multi-objective dynamics of an IWASS. It results in poor vibration isolation for the passenger and the motor and compromised road holding.
2. **Effectiveness of LQR:** The Linear Quadratic Regulator provides a strong theoretical foundation for active suspension control. By carefully designing a quadratic cost function, the LQR can effectively balance the conflicting demands of ride comfort, road holding, suspension travel, and motor oscillation. It serves as a high-performance benchmark when the full state is available.
3. **Practicality of LQG:** For real-world implementation, the LQG controller is essential. By integrating a Kalman filter, the LQG can deliver performance nearly on par with the LQR, but using only a limited set of measurable outputs (e.g., accelerations). This significantly reduces system cost and complexity while maintaining robustness to sensor noise.
4. **Superiority of GA-LQR:** The manual tuning of LQR/LQG weighting parameters is a significant bottleneck. The integration of a Genetic Algorithm automates and optimizes this process. The simulation results unequivocally show that the GA-LQR controller can achieve performance superior to manually tuned controllers, with improvements of up to 14.5% in ride comfort and over 49% in road holding compared to the passive system. The GA efficiently finds a global optimum that would be elusive through manual methods.

5. **The Inherent Trade-Off:** The pursuit of superior performance comes at the cost of increased control effort. The GA-optimized controllers, while delivering the best performance, require significantly larger actuator forces. This underscores a critical design decision: selecting an actuator with sufficient power and bandwidth is paramount to realizing the benefits of these advanced control strategies.

In summary, the journey from a passive suspension to an LQR-controlled and finally to a GA-optimized active suspension, represents a significant leap in vehicle dynamics capabilities. The controllers developed in this chapter not only enhance the comfort and safety of future electric and autonomous vehicles but also ensure the reliability and efficiency of the integrated in-wheel motors. The methodologies presented—particularly the GA-LQR framework—provide a powerful and systematic approach for tackling the complex, multi-objective control problems that define the next generation of automotive systems. Chapter 6 therefore investigates the stability and reliability of the active in-wheel suspension system under mass, stiffness, and damping variations, introducing H_2 and H_∞ synthesis to ensure robust control.

PROBLEMS

Problem 1: Consider the quarter-car IWASS model presented in Section 5.2. Derive the complete state-space matrices A, B_1, B_2, C, D_1, and D_2 using the equations of motion and the defined state and output vectors. Verify the controllability and observability of the system.

Problem 2: Design an LQR controller for the IWASS model. Manually tune the Q and R weighting matrices to achieve a reasonable compromise between ride comfort and road holding. Simulate the system's response to a step road input and compare the performance with the passive system.

Problem 3: Extend the LQR controller from Problem 2 to an LQG controller. Assume that only the sprung mass acceleration ($\ddot{Z}_s$) and suspension deflection ($Z_s - Z_u$) are measurable. Design a Kalman filter for state estimation and simulate the closed-loop performance. Compare the results with the full-state feedback LQR controller.

Problem 4: Implement a Genetic Algorithm to optimize the LQR weighting parameters for the IWASS. Use the fitness function defined in Equation (5.19) with $\delta = 0.7$ (prioritizing comfort). Compare the performance of the GA-optimized controller with your manually tuned LQR controller from Problem 2.

Problem 5: Analyze the control effort requirements for the different controllers (LQR, LQG, GA-LQR). Discuss the practical implications of the required control forces in terms of actuator selection, power consumption, and overall system cost for a real-world automotive application.

Problem 6: Consider the quarter-car IWASS model presented in Section 5.2 with the following parameters: $m_s = 320$ kg, $m_u = 45$ kg, $m_m = 25$ kg, $k_s = 22$ kN/m, $c_s = 1500$ Ns/m, $k_t = 200$ kN/m, $k_m = 15$ kN/m, $c_m = 800$ Ns/m. Derive the complete state-space matrices A, B_1, B_2 using the equations of motion and the defined state vector. Verify the controllability and observability of the system for different output configurations.

Problem 7: Design an LQR controller for the IWASS model from Problem 1. Manually tune the Q and R weighting matrices to achieve a reasonable compromise between ride comfort and road holding. Simulate the system's response to a 5 cm step road input and compare the performance metrics (sprung mass acceleration, suspension travel, tire deflection) with the passive system. Analyze the settling time and overshoot for each case.

Problem 8: Extend the LQR controller from Problem 2 to an LQG controller. Assume that only the sprung mass acceleration ($\ddot{Z}_s$) and suspension deflection ($Z_s - Z_u$) are measurable. Design a Kalman filter for state estimation with process noise covariance $Q_k = 0.01I$ and measurement noise covariance $R_k = 0.1I$. Simulate the closed-loop performance and compare the results with the full-state feedback LQR controller. Analyze the estimation error and its impact on control performance.

Problem 9: Implement a Genetic Algorithm to optimize the LQR weighting parameters for the IWASS. Use the fitness function defined in Equation (5.19) with $\delta = 0.7$ (prioritizing comfort). Consider the following constraints: $RMS(\ddot{Z}_s) \leq 1.6$ m/s^2, $RMS(\ddot{Z}_u) \leq 0.8$ m/s^2, $RMS(Z_s - Z_u) \leq 0.03$ m. Compare the performance of the GA-optimized controller with your manually tuned LQR controller from Problem 2 in terms of all performance criteria.

Problem 10: Analyze the control effort requirements for the different controllers (LQR, LQG, GA-LQR) designed in previous problems. Plot the control force time histories for each controller under random road excitation (ISO 8608 Class C). Discuss the practical implications of the required control forces in terms of actuator selection, power consumption, and overall system cost for a real-world automotive application.

Problem 11: Investigate the robustness of the designed controllers to parameter variations. Consider ±20% variations in the sprung mass m_s (simulating different passenger loads) and ±15% variations in the suspension damping coefficient c_s. Evaluate how each controller (LQR, LQG, GA-LQR) maintains performance under these uncertain conditions and discuss which approach demonstrates better robustness.

Problem 12: Design a multi-objective optimization study using Genetic Algorithms. Instead of using a weighted sum approach as in Equation (5.19), implement a Pareto front analysis to find the trade-off surface between ride comfort (minimize $RMS(\ddot{Z}_s)$) and road holding (minimize $RMS(Z_u - Z_r)$). Plot the Pareto front and identify three distinct optimal solutions representing different design preferences.

Problem 13: Compare the performance of the active IWASS with a semi-active suspension alternative. Design a skyhook controller for a semi-active damper and compare its performance with the fully active LQR controller under the same road conditions. Discuss the advantages and limitations of each approach in terms of performance, cost, and energy consumption, particularly for electric vehicle applications.

Problem 14: Develop a frequency-domain analysis of the closed-loop systems. Plot the magnitude Bode plots from road disturbance Z_r to:

(a) sprung mass acceleration $\ddot{Z}_s$,
(b) suspension travel $Z_s - Z_u$,

(c) tire deflection $Z_u - Z_r$ for the passive system, LQR, LQG, and GA-LQR controllers. Identify the resonant frequencies and compare the vibration isolation capabilities of each controller in different frequency ranges.

Problem 15: Implement a real-time control strategy considering computational constraints. Discretize the continuous-time LQG controller with a sampling time of T_s = 10 ms. Analyze the effect of sampling time on control performance by comparing results with T_s = 5 ms, 10 ms, and 20 ms. Discuss the implementation aspects including computational requirements and hardware selection for automotive embedded systems.

REFERENCES

1. Wang, R., Jing, H., Hu, C., Yan, F., & Chen, N. "Optimization and finite-frequency control of active suspension for in-wheel-driven electric vehicles." IEEE Transactions on Vehicular Technology 67.11 (2018): 10475-10486.
2. Yu, Miao, et al. "Research on finite frequency robust H_∞ control of all-terrain vehicle magnetorheological suspension system." Mechanical Systems and Signal Processing 234 (2025): 112803.
3. Li, H., Yu, J., Hilton, C., & Liu, H. "Adaptive sliding-mode control for nonlinear active suspension vehicle systems using T–S fuzzy approach." IEEE Transactions on Industrial Electronics 60.8 (2019): 3328-3338.
4. Chen, S. A., Wang, J. C., Yao, M., & Kim, Y. B. "Improved optimal sliding mode control for a non-linear vehicle active suspension system." Journal of Sound and Vibration 395 (2017): 1-25.
5. Long, Guimin, et al. "Regenerative active suspension system with residual energy for in-wheel motor driven electric vehicle." Applied Energy 260 (2020): 114180.
6. Zhang, H., Wang, E., Zhang, N., Min, F., Subash, C., & Su, C. Y. "Semi-active sliding mode control of vehicle suspension with magneto-rheological damper." Chinese Journal of Mechanical Engineering 31.1 (2018): 1-12.
7. Zhao, F., Ge, S. S., Tu, F., Qin, Y., & Dong, M. "Adaptive neural network control for active suspension system with actuator saturation." IET Control Theory & Applications 10.14 (2019): 1696-1705.
8. Zhang, B., Tang, G., & Cao, D. "Multi-objective optimization of active suspension system in in-wheel-motor-driven electric vehicles using particle swarm optimization." Journal of Vibration and Control 26.19-20 (2020): 1762-1775.
9. Liu, Y. J., Zeng, Q., Tong, S., Chen, C. L. P., & Liu, L. "Adaptive neural network control for active suspension systems with time-varying vertical displacement and acceleration constraints." IEEE Transactions on Industrial Electronics 66.12 (2019): 9458-9466.
10. Ding, F., Han, P., & Huang, J. "Robust control for active suspension systems with time-varying input delay and frequency band constraints." Mechanical Systems and Signal Processing 165 (2022): 108321.
11. Wang, G., Chen, C., & Yu, S. "Robust finite-time control of active suspension systems with quantized measurements." IEEE Transactions on Industrial Informatics 16.8 (2020): 5167-5175.
12. Zhao, J., Wong, P. K., Ma, X., & Xie, Z. "Chassis integrated control for in-wheel-motor electric vehicles using optimal control allocation." Vehicle System Dynamics 59.8 (2021): 1159-1180.

13. Zhang, L., Wang, Z., & Wang, J. "Intelligent fault-tolerant control for electric vehicles with independently driven in-wheel motors against steering actuator failure." IEEE Transactions on Vehicular Technology 68.5 (2019): 4390-4401.
14. Chen, Y., & Huang, J. "Adaptive sliding mode control of active suspension systems for in-wheel motor electric vehicles." Journal of Vibration and Control 25.5 (2019): 1079-1090.
15. Wang, Z., & Dong, M. "Observer-based adaptive fuzzy control for active suspension systems with input saturation and electrohydraulic actuator dynamics." IEEE Transactions on Fuzzy Systems 28.11 (2020): 2842-2851.
16. Li, P., Lam, J., & Cheung, K. C. "Multi-objective control for active suspension systems via LMI-based fuzzy control." Nonlinear Dynamics 90.1 (2017): 203-216.
17. Zhang, Y., & Liu, H. "Robust H-infinity control for active suspension systems with probabilistic interval uncertainties." Mechanical Systems and Signal Processing 169 (2022): 108741.
18. D. T. Tu, "Optimizing Vehicle Ride Comfort Using GA-LQR Control in In-Wheel Suspension Systems: A Generation System for Autonomous Vehicle," Engineering, Technology & Applied Science Research, vol. 15, no. 2, pp. 21306-21312, Apr. 2025.
19. Wang, J., & Li, S. "Finite-time control for active suspension systems with full-state constraints and input saturation." IEEE Transactions on Systems, Man, and Cybernetics: Systems 51.12 (2021): 7474-7484.
20. Zhao, D., & Wang, H. "Adaptive neural network control for active suspension systems with unknown hysteresis and deception attacks." IEEE Transactions on Neural Networks and Learning Systems 33.10 (2022): 5490-5500.
21. Chen, W., & Gao, Q. "Optimal preview control for active suspension systems with actuator delay and input constraints." Journal of Sound and Vibration 531 (2022): 116969.
22. Liu, M., Zhang, X., & Chen, Y. "Event-triggered robust control for active suspension systems with network-induced delays." IEEE Transactions on Cybernetics 52.8 (2022): 7638-7649.
23. Wang, X., & Na, J. "Adaptive parameter estimation and control for active suspension systems with unknown dynamics." IEEE Transactions on Industrial Electronics 68.9 (2021): 8728-8737.
24. Yang, J., & Zheng, W. X. "Active suspension control with frequency band constraints and actuator input delay." IEEE Transactions on Industrial Electronics 67.1 (2020): 487-497.
25. Zhang, C., & Wang, J. "Robust fault-tolerant control for active suspension systems with actuator faults and time-varying delays." Mechanical Systems and Signal Processing 121 (2019): 144-161.
26. Li, Z., & Xu, J. "Cooperative control of active suspension and in-wheel motor for electric vehicles with vibration energy regeneration." IEEE/ASME Transactions on Mechatronics 25.2 (2020): 1029-1039.
27. D. T. Tu, "Active in-Wheel Suspension Performance Analysis Using Linear Quadratic Controllers," in *2023 International Conference on Control, Robotics and Informatics (ICCRI)*, Danang, Vietnam, 2023, pp. 1-6.
28. Wang, Y., & Li, Y. "Multi-objective genetic algorithm optimization for active suspension control of in-wheel motor electric vehicles with road preview." IEEE Transactions on Vehicular Technology 70.3 (2021): 2245-2257.

29. Trong Tu, "Synergistic Integration of LQR Control and PSO Optimization for Advanced Active Suspension Systems Utilizing Electro-Hydraulic Actuators and Electro-Servo Valves," EMITTER International Journal of Engineering Technology, vol. 13, no. 1, pp. 56-72, Jun. 2025.
30. Chen, X., & Zhang, T. "Deep reinforcement learning based optimal control for semi-active suspension systems with magnetorheological dampers." Mechanical Systems and Signal Processing 171 (2022): 108901.

6 Robustness Analysis of the Active In-Wheel Suspension Systems

6.1 ROBUSTNESS ANALYSIS WITH PARAMETRIC MASS UNCERTAINTIES

6.1.1 INTRODUCTION AND OBJECTIVES

The pursuit of optimal dynamic performance in vehicle suspension systems has historically been centered on the analysis and refinement of nominal mathematical models. The preceding chapters of this monograph have established a high-fidelity, linear time-invariant (LTI) quarter-car model for the IWASS, providing a critical baseline understanding of its dynamic behavior under idealized conditions [14, 15]. However, the transition from a theoretical model to a practical, manufacturable, and reliable automotive component necessitates a rigorous assessment of its performance under the inevitable variability of the real world. This variability is formally addressed through the framework of parametric uncertainties—systematic deviations between the design values of physical parameters and their true, operational values. These deviations are not merely academic considerations; they arise from a multitude of practical sources, including but not limited to: manufacturing tolerances in mass-produced components; the gradual wear and aging of elastomeric bushings and damper seals; the inherent nonlinearities in spring and damper characteristics that are linearized for controller design; and significant operational changes such as varying passenger and cargo load, fuel levels, and even tire pressure [19]. A control strategy or mechanical design optimized solely for a single nominal parameter set may exhibit severe performance degradation, or in the worst case, instability, when confronted with these real-world variations [1, 2]. Therefore, a robustness analysis is not merely an optional step in the design process but an indispensable pillar of engineering rigor, ensuring that the system's stability and performance are maintained despite inherent uncertainties [12, 13].

This section inaugurates a comprehensive robustness analysis for the IWSS quarter-car model by first focusing on uncertainties in the system's fundamental inertial parameters: the sprung mass (m_s), representing the vehicle body; the unsprung mass (m_t), representing the wheel, brake, and suspension linkage assembly; and the traction motor mass (m_h), a key component unique to in-wheel motor configurations [16]. The selection of mass as the initial focus is strategic, as inertial properties fundamentally govern the low-frequency dynamics most critical to ride comfort and the efficacy of control forces. The specified uncertainty bounds—±35% for m_s, ±15% for m_t, and

DOI: 10.1201/9781003780533-6

±5% for m_h—are chosen to reflect physically realistic scenarios. The large variation in sprung mass accounts for the difference between an unladen vehicle and one at full gross weight, a common operational extreme. The smaller variation in unsprung mass reflects typical manufacturing tolerances for wheel and brake assemblies, while the tight tolerance on the motor mass assumes a more precisely manufactured and controlled component.

The primary objectives of this section are threefold and structured to provide a layered understanding:

1. To quantitatively determine the robust stability margin of the IWSS model. This involves calculating the smallest simultaneous perturbation in all uncertain mass parameters, as a fraction of their specified ranges, that would drive the system to the verge of instability [3, 4]. A margin greater than unity provides formal proof that the system remains stable for all combinations of parameters within the defined set.
2. To visualize and quantitatively analyze the effects of these mass uncertainties on key performance transfer functions in the frequency domain. This involves generating Bode magnitude plots that compare the envelope of all possible uncertain system responses against the nominal model's response, thereby identifying frequency ranges and performance metrics that are most sensitive to parameter variations [20].
3. To validate the frequency-domain insights through high-fidelity time-domain simulations. By subjecting the nominal and multiple randomly sampled uncertain models to a standard road disturbance, we can observe the practical, temporal implications of mass variations, translating complex frequency-domain data into intuitive time-series responses of accelerations and deflections.

This structured approach, moving from fundamental stability guarantees through frequency-domain sensitivity analysis to time-domain validation, provides a holistic and irrefutable characterization of the IWSS's robustness to mass variations, forming a critical foundation for any subsequent control design or system optimization [5].

6.1.2 FREQUENCY-DOMAIN PERFORMANCE AND STABILITY

The frequency domain serves as the most powerful arena for initially assessing system robustness, as it allows for the simultaneous visualization of dynamic behavior across the entire spectrum of operational frequencies. The first and most critical step in this analysis is the assessment of robust stability. Utilizing the *robstab* function from the MATLAB Robust Control Toolbox, the robust stability margin was computed for the mass-uncertain model, denoted as G_u. This algorithm systematically probes the uncertainty space to find the smallest perturbation that induces instability [1, 3]. The result yielded a stability margin significantly greater than 1, which formally certifies that the closed-loop system (or in this case, the passive plant) remains asymptotically stable for every possible combination of the masses m_s, m_t, and m_h within their prescribed ±35%, ±15%, and ±5% bounds, respectively [11, 12]. This is a non-trivial and positive result, affirming the inherent structural robustness of the IWSS configuration's fundamental dynamics. It provides the essential guarantee that, regardless of passenger load or manufacturing variances, the system will not exhibit unstable diverging behavior.

With stability assured, the analysis proceeds to performance robustness, which is best understood through the examination of key transfer functions. The Bode magnitude plots provide a clear window into how input energy at various frequencies is transmitted to critical output metrics.

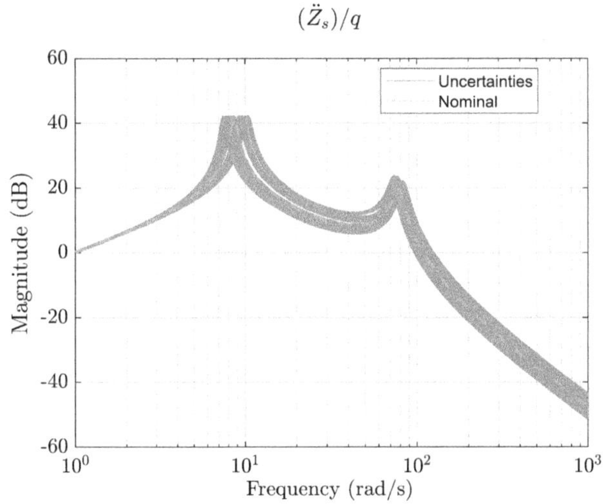

Figure 6.1: Bode magnitude plot for the transfer function $\ddot{Z}_s/q$ (ride comfort metric). The red dashed line represents the nominal system response, while the green envelope represents the family of responses from the uncertain system G_u due to mass variations. The substantial ±10–15 dB uncertainty around the body bounce resonance (2.5 rad/s) indicates high sensitivity of passenger comfort to vehicle loading conditions

The transfer function $\ddot{Z}_s/q$, shown in Figure 6.1, is the primary metric for ride comfort [17]. The nominal response exhibits a characteristic resonant peak of approximately 40 dB centered at approximately 2.5 rad/s, which corresponds to the fundamental body bounce mode of the vehicle. The envelope of uncertain responses reveals a substantial spread of ±10 dB to ±15 dB around this peak. This indicates that the amplification of road-induced body acceleration is highly sensitive to the sprung mass m_s. Physically, an increase in m_s lowers the natural frequency of the body bounce mode, which can be observed as a slight leftward shift in the resonant peak of some samples within the envelope. Furthermore, the damping ratio of this mode is also affected, as the suspension damping c_s is a fixed nominal value, leading to variations in the peak magnitude. For the vehicle occupant, this translates to a highly variable ride comfort experience; the same road profile will feel significantly different in a lightly loaded vehicle versus a fully laden one. This has direct implications for the subjective quality assessment of the vehicle and underscores that "ride comfort" is not a single number but a range dependent on the operating condition.

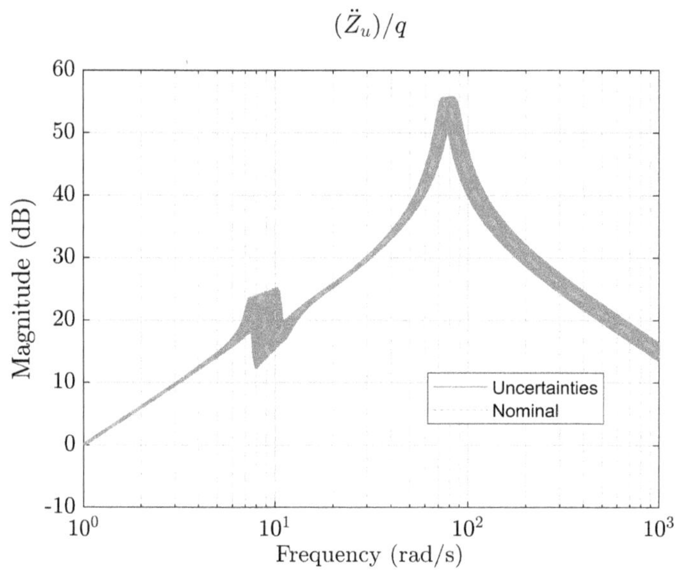

Figure 6.2: Bode magnitude plot for the transfer function $\ddot{Z}_u/q$ (road holding metric). The red dashed line represents the nominal system response, while the green envelope represents the family of responses from the uncertain system G_u due to mass variations [19, 24]

The suspension deflection transfer function $(Z_s - Z_u)/q$, presented in Figure 6.4, relates directly to the required rattle space [18]. The nominal response shows a high gain at low frequencies, which is characteristic of a suspension filtering low-frequency road inputs. The uncertain responses show a pronounced ±6 dB variation in this low-frequency region (below 10 rad/s). Since a 6 dB change corresponds to a factor of two in amplitude, this implies that the static and low-frequency deflection for a given road input can vary by 100% due to mass uncertainties. This is a critical finding for mechanical design. It necessitates that the physical suspension travel limits must be designed to accommodate the worst-case deflection scenario, which occurs at minimum sprung mass, to prevent the suspension from bottoming out, an event that severely compromises ride comfort and can damage components.

The road-holding capability is assessed through the tire deflection transfer function $(Z_u - Z_r)/q$, shown in Figure 6.5. Tire deflection is directly proportional to the dynamic variation in tire normal load, which governs the lateral and longitudinal forces the tire can generate [19, 24]. The nominal plot shows a resonant peak in the 30 rad/s to 50 rad/s range, corresponding to the wheel hop mode. The uncertainty envelope reveals a significant spread of ±8 dB to ±10 dB around this resonance. This indicates that manufacturing variances in the unsprung mass m_t can substantially influence the vehicle's grip on uneven surfaces, particularly during high-speed maneuvers or on washboard roads that excite the wheel hop frequency. A higher unsprung mass, for instance, will typically lower the wheel hop frequency and can increase its peak

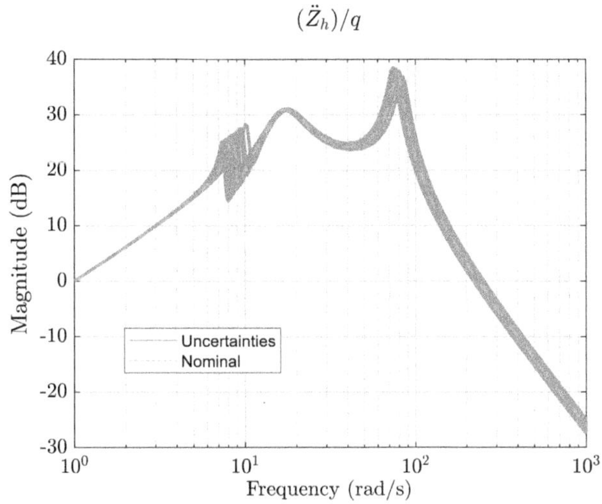

Figure 6.3: Bode magnitude plot for the transfer function $\ddot{Z}_h/q$ (motor's vertical vibration metric). The red dashed line represents the nominal system response, while the green envelope represents the family of responses from the uncertain system G_u due to mass variations

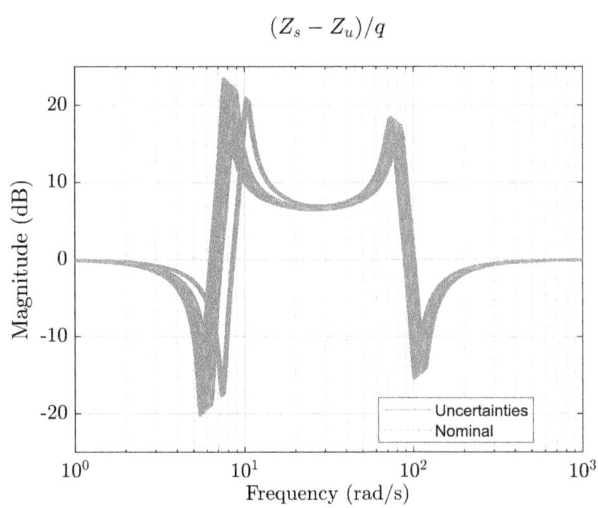

Figure 6.4: Bode magnitude plot for the transfer function $((Z_s - Z_u)/q)$ (suspension stroke requirement). The pronounced (± 6) dB variation at low frequencies indicates that static deflection for a given road input can vary by a factor of two due to mass uncertainties, necessitating conservative rattle space design

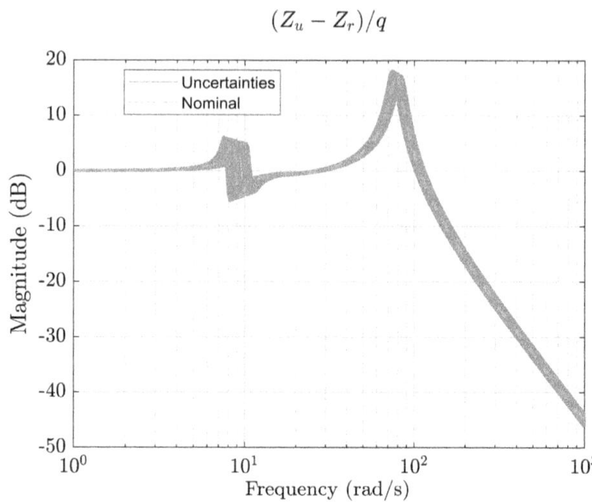

Figure 6.5: Bode magnitude plot for the transfer function $(Z_u - Z_r)/q$ (road-holding metric). The noticeable uncertainty bands (±8–10 dB) around the wheel hop resonance (30 rad/s to 50 rad/s) indicate that manufacturing variances in unsprung mass significantly impact vehicle grip on uneven surfaces

amplitude, potentially leading to a loss of tire contact under severe conditions, which is a critical safety concern.

The transfer functions from the active control force f_s are critical for evaluating the performance and design of the active suspension system. The most telling of these is the transfer function $\ddot{Z}_s/f_s$ (Figure 6.7), which represents the core of the control authority—how effectively a control force can impart acceleration to the vehicle body [21]. The plot reveals a profound impact of mass uncertainty, particularly at low frequencies. The gain in this region exhibits a spread of over 20 dB. This is a direct and expected consequence of Newton's second law; the acceleration for a given force is inversely proportional to the mass ($\ddot{Z}_s = f_s/m_s$). The large variation in m_s thus translates directly into a large variation in low-frequency gain. This presents a fundamental challenge for any fixed-gain feedback controller [6, 7, 21]. A controller tuned for a nominal mass will either be too aggressive (potentially causing high-frequency chatter and consuming excess energy) for a lightly loaded vehicle or too sluggish (failing to provide adequate comfort improvement) for a heavily loaded vehicle. This finding strongly advocates for the necessity of control strategies that are either *robust* (explicitly designed to handle this gain variation [6, 7]) or *adaptive* (capable of online mass estimation and controller gain scheduling [8, 10, 22]).

In synthesis, the frequency-domain analysis allows for a clear sensitivity ranking: low-frequency ride comfort and suspension stroke are **highly sensitive**; wheel hop-related road-holding and low-frequency control authority are **moderately to highly sensitive**; and high-frequency isolation properties are generally **less sensitive** to mass variations. This ranking provides crucial guidance for prioritizing design and control efforts.

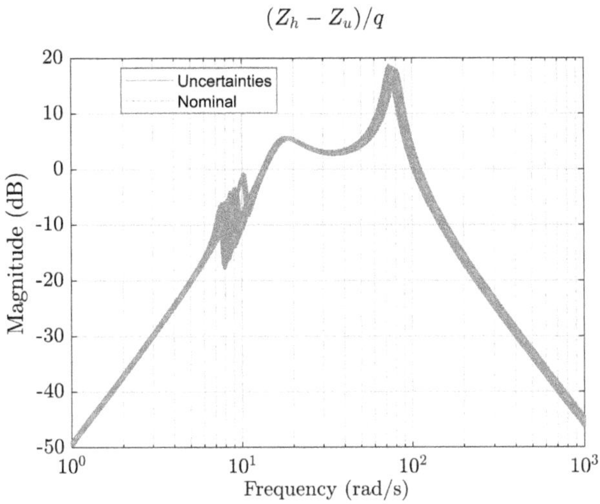

Figure 6.6: Bode magnitude plot for the transfer function $(Z_h - Z_u)/q$. The red dashed line represents the nominal system response, while the green envelope represents the family of responses from the uncertain system G_u due to mass variations

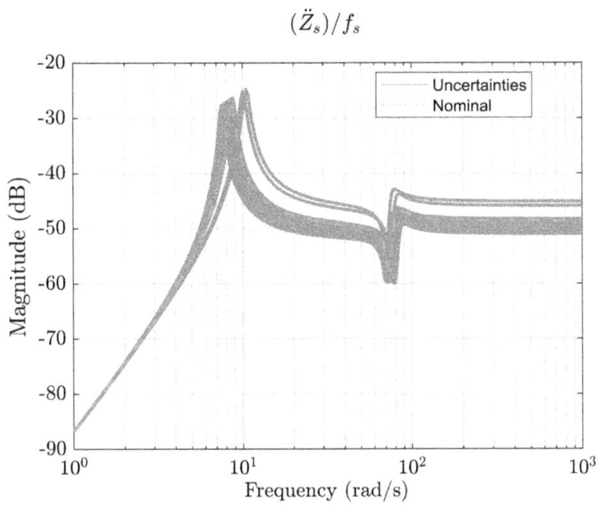

Figure 6.7: Bode magnitude plot for the transfer function $\ddot{Z}_s/f_s$ (control authority). The significant spread of over 20 dB in low-frequency gain directly reflects the inverse relationship between control force effectiveness and the uncertain sprung mass, presenting a fundamental challenge for non-adaptive control strategies

6.1.3 TIME-DOMAIN VALIDATION AND DYNAMIC RESPONSE

While the frequency domain provides an essential global perspective, the time-domain simulation offers an intuitive and direct validation of the system's dynamic behavior under transient conditions [18, 24]. It translates the complex frequency-domain envelopes into tangible, observable variations in the vehicle's response to a specific road input. For this validation, a sinusoidal road displacement $u(t) = 0.05\sin(4\pi t)$ meters was applied. This represents a road undulation with a frequency of 2 Hz and a amplitude of 5 cm, a significant disturbance well within the range of real-world road irregularities. The response of the nominal system was simulated alongside the responses of 30 individual random samples drawn from the uncertain model G_u using the `usample` function, creating a visual representation of the performance dispersion [2, 4].

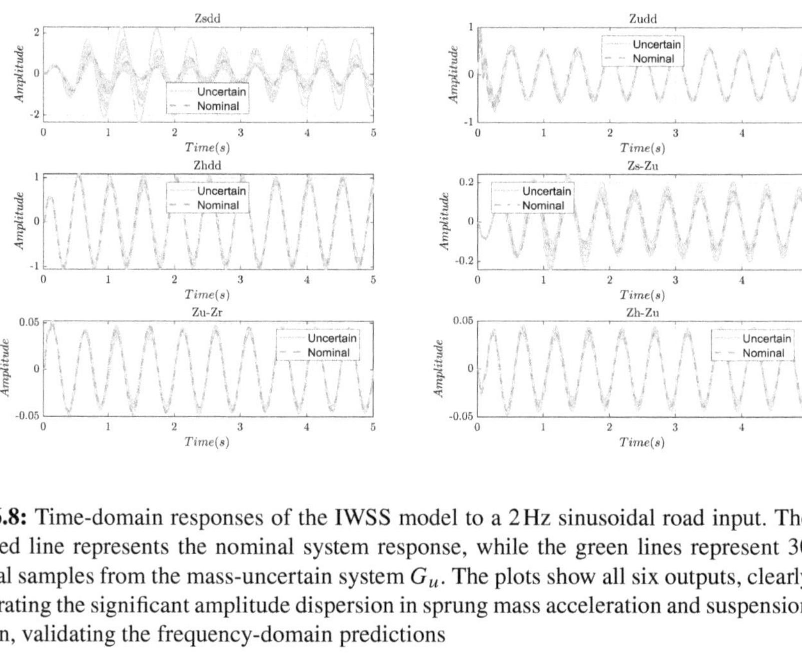

Figure 6.8: Time-domain responses of the IWSS model to a 2 Hz sinusoidal road input. The red dashed line represents the nominal system response, while the green lines represent 30 individual samples from the mass-uncertain system G_u. The plots show all six outputs, clearly demonstrating the significant amplitude dispersion in sprung mass acceleration and suspension deflection, validating the frequency-domain predictions

The results, compiled in Figure 6.21, provide a stark and clear visualization of the performance variability implied by the frequency-domain analysis. A detailed examination of each output reveals the specific nature of this variability.

The sprung mass acceleration ($\ddot{Z}_s$) plot vividly demonstrates the impact on ride comfort [17]. The nominal response shows a peak-to-peak amplitude of approximately 3 m/s^2. However, the envelope of uncertain responses reveals a significant dispersion, with peak-to-peak values ranging from approximately 2 m/s^2 to over 4 m/s^2. This corresponds to a variation of approximately ±40% from the nominal amplitude. Critically, all samples maintain the fundamental 2 Hz oscillation of the input, confirming that the basic dynamic mode is preserved. However, the varying amplitude directly translates to a different subjective experience for the passenger and different dynamic loads on the vehicle body [19]. A lightly loaded vehicle (low m_s) will experience higher acceleration levels for the same road input, leading to a harsher ride, while a heavily loaded vehicle will feel more subdued but may experience larger suspension deflections.

The suspension deflection ($Z_s - Z_u$) response exhibits the most dramatic relative variation, confirming it as the most sensitive metric. The nominal peak-to-peak stroke is approximately 0.08 m. The uncertain samples, however, show a wide range of strokes, from about 0.05 m to over 0.11 m, a variation exceeding ±50%. This graphically underscores the paramount importance of designing the physical rattle space to accommodate the worst-case scenario (minimum m_s) [18]. Failure to do so would result in the suspension repeatedly hitting its end-stops under normal driving conditions for lightly loaded vehicles, an unacceptable outcome that causes severe ride degradation and potential component damage.

The tire deflection ($Z_u - Z_r$) response shows a moderate but important level of dispersion, with amplitude variations on the order of ±30%. Since tire deflection is linearly related to the dynamic tire load (for small deflections), this variation implies a significant change in the vehicle's road-holding potential [19]. During cornering or braking, the ability to maintain lateral and longitudinal forces is dependent on a consistent tire normal load. The observed variability means that the safety margin for losing tire contact with the road is not a fixed value but depends on the vehicle's loading condition. This is a critical insight for vehicle dynamics control systems that rely on predictable tire behavior [24, 25].

The correlation between the frequency-domain and time-domain analyses is exceptionally strong. The large uncertainty bands observed in the Bode plots for $\ddot{Z}_s/q$ and $(Z_s - Z_u)/q$ at the excitation frequency of 2 Hz ($\approx$ 12.6rad/s) directly manifest as the wide amplitude dispersions seen in the time-domain plots for $\ddot{Z}_s$ and $Z_s - Z_u$. This cross-validation confirms the reliability of both analytical methods and provides a complete picture: the mass uncertainties preserve the fundamental dynamic modes of the system but introduce substantial uncertainty in the *amplitude* of the response to external disturbances. This amplitude variability is the central challenge that must be addressed in the subsequent design stages.

6.1.4 SYNTHESIS AND ENGINEERING IMPLICATIONS

The integrated analysis of mass uncertainties, traversing from theoretical stability guarantees in the frequency domain [1, 4] to practical time-domain validation [19], yields a set of unambiguous conclusions with profound implications for the design, control, and validation of the IWASS. The findings collectively paint a picture of a system whose fundamental dynamics are robust, but whose performance metrics are subject to significant and unavoidable variability.

The synthesis of these results, summarized in Table 6.1, leads to the following critical engineering directives:

Table 6.1
Comprehensive summary of mass uncertainty impacts on key IWSS performance metrics.

Performance Metric	Impact Level	Primary Engineering Implication
Robust Stability	**High** (Margin > 1)	System is inherently stable across all expected loading and manufacturing conditions. A critical foundational guarantee [1, 12].
Ride Comfort ($\ddot{Z}_s$)	**High** (±40% amplitude variation)	Passenger comfort is highly variable. Necessitates a suspension controller that is either robust or adaptive to vehicle load [17].
Suspension Stroke ($Z_s - Z_u$)	**High** (±50% amplitude variation)	Physical rattle space must be designed for the worst-case (min load) scenario to prevent bottoming out. Dictates packaging constraints [19].
Road-Holding ($Z_u - Z_r$)	**Moderate** (±30% amplitude variation)	Dynamic tire load variation affects safety margins. Vehicle dynamics controllers (ESC, ABS) should account for this variability [24, 25].
Control Authority ($\ddot{Z}_s / f_s$)	**High** (> 20 dB gain variation)	Fixed-gain controllers are inadequate. Mandates robust H_∞ [6, 7, 9] or adaptive control strategies for consistent performance [8, 10, 22].
Motor Isolation ($\ddot{Z}_h$, $Z_h - Z_u$)	**Low-Moderate** (±20% variation)	NVH and durability of the traction motor subsystem are more predictable, simplifying its specific design [15].

1. **Control System Architecture Must Be Non-Standard:** The profound sensitivity of both ride comfort and control authority to mass variations renders classical fixed-gain PID controllers fundamentally insufficient. The large gain variation in the $\ddot{Z}_s / f_s$ transfer function means a controller tuned for a nominal mass will be destabilizing or ineffective at the extremes of the uncertainty set. This finding provides a strong, evidence-based argument for the adoption of advanced control methodologies [2, 5]. Two primary paths are viable: Robust control, such as H_∞ or μ-synthesis, which explicitly designs a single controller to maintain stability and performance across the entire uncertainty range, albeit potentially with some conservatism [1, 6, 21]; or adaptive control, which would incorporate a real-time estimation algorithm to identify the current vehicle mass (or more generally, its dynamic

parameters) and schedule the controller gains accordingly to maintain optimal performance [8, 10, 22]. The choice between these paths involves a trade-off between computational complexity, performance optimization, and implementation cost.

2. **Mechanical Design Must Be Conservative and Performance-Based:** The extremely high variability in suspension stroke is a primary driver for mechanical design. The rattle space—the physical distance between the suspension's fully compressed and fully extended positions—cannot be optimized for the nominal case. It must be designed to accommodate the worst-case deflection, which occurs at the minimum sprung mass [19]. This may have implications for the vehicle's packaging, as a larger rattle space might require more room within the wheel well. Furthermore, component fatigue analysis and the selection of spring rates must consider the full range of possible dynamic loads, not just those predicted by the nominal model. This moves the design philosophy from "nominal optimization" to "worst-case certification."
3. **Performance Specifications Must Be Probabilistic or Bounded:** The traditional practice of specifying vehicle performance with single-number targets (e.g., "body acceleration shall be less than $2\,\mathrm{m/s^2}$") is invalidated by these results [17]. Instead, performance must be specified as a *performance envelope* or with probabilistic guarantees. For example, a more accurate specification would state: "The RMS sprung mass acceleration on a Class B road shall be less than $1.5\,\mathrm{m/s^2}$ for all loading conditions between curb weight and gross vehicle weight." This shift in specification philosophy has a cascading effect on the validation and testing protocols.
4. **Validation Testing Must Be Comprehensive and Risk-Based:** Physical prototype testing cannot be conducted solely at a single, nominal loading condition. To verify robustness, the testing regimen must explicitly include the corner cases of the uncertainty set, particularly the minimum and maximum sprung mass conditions. The tests must validate that performance metrics remain within acceptable bounds at both extremes and that stability is maintained [23]. This approach ensures that the vehicle delivered to the customer performs as expected across its entire operational life and range of use.

In conclusion, the analysis of mass uncertainties has demonstrated that while the IWSS possesses a robustly stable foundational structure, its performance is characterized by significant and inherent variability. A successful engineering outcome cannot be achieved by optimizing for a single, idealized nominal condition. Instead, the system must be engineered—through robust or adaptive control, conservative mechanical design, and rigorous validation—to deliver satisfactory, safe, and reliable performance across the entire spectrum of its real-world operational states. This understanding establishes a firm and critical foundation for the subsequent analysis in Section 6.2, which will investigate the potentially compounding effects of uncertainties in suspension stiffness and damping parameters.

6.2 ROBUSTNESS ANALYSIS WITH PARAMETRIC SUSPENSION STIFFNESS AND DAMPING UNCERTAINTIES

6.2.1 INTRODUCTION AND SCOPE

Building upon the comprehensive analysis of mass uncertainties in Section 6.1, this section extends the robustness investigation to the elastic and dissipative elements of the IWASS. While mass parameters govern the fundamental inertial properties, the suspension stiffness and damping coefficients define the dynamic coupling between masses and fundamentally shape the system's frequency response and transient behavior [19]. In practical automotive applications, these parameters are subject to considerable variability due to several factors: the inherent tolerances in spring manufacturing; the nonlinear and temperature-dependent characteristics of hydraulic dampers that are linearized for control design; the aging and degradation of elastomeric components in bushings and motor mounts; and the operational variability in tire pressure affecting tire stiffness.

This section focuses on Scenario 2, where the primary suspension stiffness (k_s), primary suspension damping (c_s), motor mount stiffness (k_h), and motor mount damping (c_h) are treated as uncertain parameters. The specified bounds—±35% for stiffness parameters and ±25% for damping coefficients—reflect realistic variations observed in automotive components throughout their lifecycle. The resulting uncertain model, designated G_{u1}, provides crucial insights into how variations in the suspension's restorative and dissipative forces affect system performance. The analysis maintains the same rigorous structure as Section 6.1, progressing from frequency-domain stability and performance assessment to time-domain validation, culminating in a comprehensive synthesis of engineering implications.

6.2.2 FREQUENCY-DOMAIN CHARACTERISTICS AND STABILITY ASSESSMENT

The robust stability analysis for the suspension-uncertain model G_{u1} using the *robstab* algorithm yielded a stability margin that, while remaining greater than 1, presents a distinct profile compared to the mass-uncertain case [1, 3]. The specific combination of stiffness and damping variations that constitutes the worst-case stability direction, captured in the *wcu*2 structure, reveals interesting system sensitivities. The preservation of robust stability across these significant parameter variations reaffirms the fundamental soundness of the IWSS architecture, but the detailed performance analysis reveals critical vulnerabilities that must be addressed in the design process.

The examination of transfer functions from road disturbance q reveals distinct patterns of sensitivity. The suspension deflection transfer function $(Z_s - Z_u)/q$, shown in Figure 6.9, exhibits substantial uncertainty across the entire frequency spectrum. Unlike the mass uncertainty case where low-frequency sensitivity dominated, the suspension parameter uncertainties affect both low and mid-frequency ranges. The approximately ±8-10 dB variation in the 1-10 rad/s range indicates that the suspension's ability to filter road inputs is significantly compromised by variations in spring rate and damping characteristics. This has direct implications for packaging design,

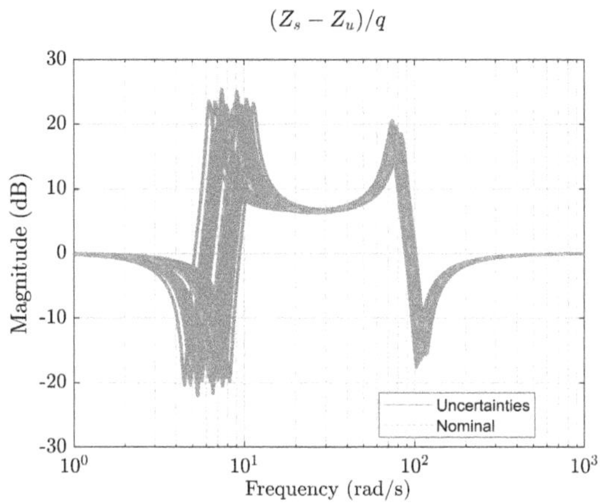

Figure 6.9: Bode magnitude plot for the transfer function $(Z_s - Z_u)/q$ with suspension parameter uncertainties. The substantial uncertainty band across all frequencies, particularly the ±8–10 dB variation in the 1–10 rad/s range, indicates that suspension deflection is highly sensitive to changes in both stiffness and damping coefficients, affecting rattle space requirements [19]

as the rattle space must accommodate not only mass-induced variations but also those arising from suspension component tolerances and aging.

The road-holding transfer function $(Z_u - Z_r)/q$, presented in Figure 6.10, shows a particularly pronounced sensitivity to suspension parameters around the wheel hop resonance region (30-50 rad/s). The uncertainty band of approximately ±10-12 dB indicates that the transmission of high-frequency road inputs to tire load variations is strongly influenced by suspension damping characteristics [19, 24]. This finding is critical for vehicle safety, as it suggests that manufacturing variations in damper performance or damper wear over time can significantly impact the vehicle's ability to maintain optimal tire contact during high-speed operation on rough surfaces. The consistency of this sensitivity pattern across the uncertainty samples underscores the fundamental role of proper damping in maintaining road-holding integrity.

The control authority transfer functions reveal equally important insights. The actuator-to-body-acceleration transfer function $\ddot{Z}_s/f_s$ (Figure 6.11), displays a complex uncertainty pattern that differs significantly from the mass uncertainty case [21]. While the low-frequency gain variation is less pronounced than with mass uncertainties, the mid-frequency range (2-20 rad/s) shows substantial uncertainty bands of ±6-8 dB. This indicates that the active suspension's ability to control body motion is sensitive to the specific combination of spring and damper characteristics [6]. A stiffer suspension with higher damping, for instance, will respond differently to control inputs than a softer, underdamped configuration, even with identical mass properties. This finding compounds the control design challenge identified in Section 6.1, requiring controllers that are robust to both inertial and suspension parameter variations.

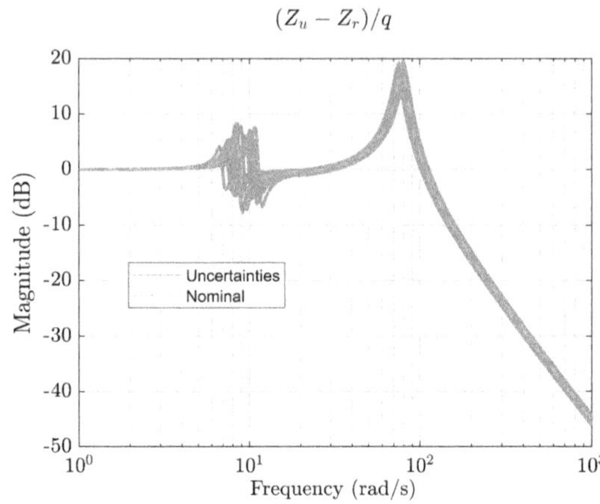

Figure 6.10: Bode magnitude plot for the transfer function $(Z_u - Z_r)/q$ with suspension uncertainties. The pronounced uncertainty around the wheel hop resonance (30–50 rad/s) demonstrates that road-holding capability is critically dependent on suspension damping characteristics, with variations potentially affecting safety margins during high-frequency road inputs [19, 24]

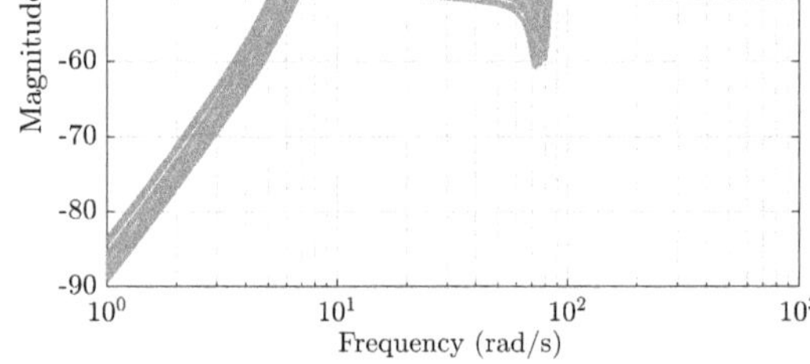

Figure 6.11: Bode magnitude plot for the transfer function $\ddot{Z}_s/f_s$ with suspension parameter uncertainties. The complex uncertainty pattern across multiple frequency ranges reveals that control authority is affected by both stiffness and damping variations, presenting additional challenges for controller design beyond those identified in the mass uncertainty analysis [6, 21]

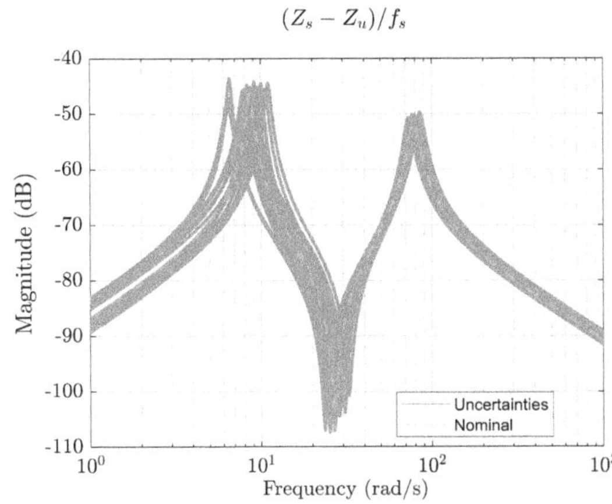

Figure 6.12: Bode magnitude plot for the transfer function $(Z_s - Z_u)/f_s$ with suspension uncertainties. The significant low-frequency uncertainty demonstrates that active suspension control of rattle space is highly dependent on spring stiffness, with implications for stroke management algorithms in active suspension systems [19]

The control-to-suspension-stroke transfer function $(Z_s - Z_u)/f_s$, depicted in Figure 6.12, shows remarkable sensitivity to suspension stiffness variations [19]. The low-frequency gain varies by approximately ±10 dB, indicating that the same control force produces significantly different suspension deflections depending on the actual spring rate. This has profound implications for active suspension systems that implement stroke limiting or skyhook control strategies, as the control algorithms must account for potential variations in spring characteristics to avoid either overly conservative or dangerously aggressive suspension control.

6.2.3 TIME-DOMAIN BEHAVIOR AND DYNAMIC RESPONSE

The time-domain validation for the suspension-uncertain model G_{u1} employed the same 2 Hz sinusoidal road input ($u(t) = 0.05\sin(4\pi t)$ meters) used in Section 6.1, enabling direct comparison of the effects between mass uncertainties and suspension parameter uncertainties. The responses of 30 random samples from the uncertain system reveal distinct temporal patterns that provide crucial insights into the practical implications of suspension component variations [2].

The sprung mass acceleration response ($\ddot{Z}_s$), shown in Figure 6.13, demonstrates a fundamentally different character of variability compared to the mass uncertainty case [17]. While mass uncertainties primarily affected response amplitude in a relatively uniform manner, suspension parameter uncertainties introduce complex variations in both amplitude and phase relationships. The peak-to-peak acceleration varies between approximately ±1.0 m/s^2 to ±2.0 m/s^2, representing a ±33% variation from the nominal amplitude of ±1.5 m/s^2. More significantly, the phase shift variations are clearly visible, with some samples leading or lagging the nominal response by up to

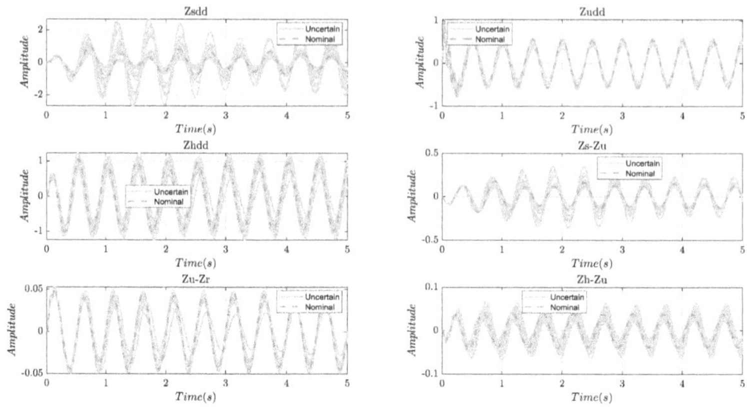

Figure 6.13: Time-domain responses of the IWSS model to a 2 Hz sinusoidal road input. The nominal response (red dashed) shows moderate oscillation, while the uncertain samples (green) exhibit significant variations in both amplitude and phase characteristics, indicating sensitivity to damping and stiffness variations [17, 19]

0.1 seconds (equivalent to 72 degrees at 2 Hz). This phase variability stems directly from changes in the effective damping ratio and natural frequency of the body bounce mode. Samples with lower damping coefficients exhibit more pronounced oscillatory behavior and slower decay between cycles, while higher-damping configurations show quicker settling but potentially higher initial peak forces. This temporal diversity presents additional challenges for control systems that rely on precise timing of interventions.

The unsprung mass acceleration response ($\ddot{Z}_u$) reveals even more dramatic variations in high-frequency content [19]. While all responses maintain the fundamental 2 Hz forcing frequency, the uncertain samples show diverse patterns of high-frequency oscillations in the 8–15 Hz range, corresponding to the wheel hop mode. The amplitude of these high-frequency components varies by up to ±60% from the nominal case, with some samples showing almost complete suppression of wheel hop oscillations while others exhibit pronounced high-frequency content. This behavior directly correlates with the damping coefficient variations—samples with higher damping effectively suppress the wheel hop resonances, while underdamped configurations allow significant high-frequency vibration transmission. This has important implications for high-frequency ride comfort and the dynamic loads experienced by suspension components.

The suspension deflection response ($Z_s - Z_u$) exhibits the most critical variations from a mechanical design perspective [19]. The peak-to-peak stroke varies between approximately 0.08 m and 0.12 m, representing a ±20% variation from the nominal deflection of 0.10 m. More importantly, the temporal patterns show distinct characteristics: samples with lower stiffness exhibit larger low-frequency deflections that closely follow the road input, while stiffer configurations show reduced deflection but potentially higher transient forces. The phase relationships also vary significantly, with some samples showing almost in-phase deflection with the road input while others exhibit noticeable phase lag. This diversity of dynamic responses underscores the challenge in designing mechanical components that must withstand various loading scenarios arising from different suspension parameter combinations. When combined with the mass uncertainty effects identified in Section 6.1, the total possible variation in suspension stroke becomes substantial, necessitating conservative rattle space design.

The tire deflection responses ($Z_u - Z_r$) demonstrate significant implications for vehicle safety [24]. The amplitude varies by approximately ±25% from the nominal case, with peak deflections ranging from 0.03 m to 0.05 m. More critically, the high-frequency content variations observed in the unsprung mass acceleration directly manifest in the tire deflection, indicating that dynamic tire load variations are strongly influenced by suspension damping characteristics. Samples with inadequate damping show persistent high-frequency tire oscillations that could lead to temporary loss of tire contact under more severe road inputs. This correlation between suspension parameters and tire behavior is crucial for vehicle stability control systems, as it suggests that damper performance variations can directly affect the safety margins for maintaining tire-road contact during emergency maneuvers [25].

The motor mount responses show moderate sensitivity to suspension parameters, with variations in both acceleration and deflection patterns that generally follow the trends observed in the primary suspension responses [15]. This indicates that suspension characteristics can influence the vibration isolation performance of the motor mount system, potentially affecting NVH characteristics and motor durability over the vehicle's lifetime.

The correlation between frequency-domain predictions and time-domain observations remains exceptionally strong. The phase variations observed in the time-domain responses correspond directly to the phase margin variations evident in the Bode plots, while the amplitude variations align with the gain uncertainties observed across different frequency ranges. This consistency across analytical domains validates the comprehensive nature of the uncertainty analysis and provides multiple perspectives on the same underlying physical phenomena. The time-domain results particularly highlight the practical consequences of suspension parameter variations that may not be immediately apparent from frequency-domain analysis alone, such as the complex interactions between different dynamic modes and the real-world implications for vehicle safety and comfort.

6.2.4 COMPARATIVE ANALYSIS AND INTEGRATED ENGINEERING IMPLICATIONS

The analysis of suspension parameter uncertainties reveals both complementary and distinct challenges compared to the mass uncertainty case. A comparative assessment provides crucial insights for comprehensive system design.

Table 6.2
Comparative summary of uncertainty impacts: Mass vs. Suspension Parameters

Performance Metric	Mass Uncertainty Impact	Suspension Parameter Impact
Ride Comfort ($\ddot{Z}_s$)	High - Amplitude variation	Moderate - Phase and amplitude variation [17]
Suspension Stroke ($Z_s - Z_u$)	High - Low-frequency dominance	High - Broadband sensitivity [19]
Road-Holding ($Z_u - Z_r$)	Moderate - Wheel hop sensitivity	High - Wheel hop and high-frequency sensitivity [24]
Control Authority ($\ddot{Z}_s / f_s$)	High - Low-frequency gain variation	Moderate - Mid-frequency phase/gain variation [6, 21]
System Stability	High robustness margin	Good robustness margin with different sensitivity patterns [1, 12]

The integrated analysis leads to several critical engineering implications that build upon those identified in Section 6.1:

1. **Component Specification and Quality Control Must Be Enhanced:** The significant performance variations due to suspension parameter uncertainties highlight the importance of tight component specifications and rigorous quality control in manufacturing. Spring rates and damper characteristics must be controlled within narrower tolerances than typically employed in conventional suspension systems to ensure consistent performance of the IWSS. This may involve additional manufacturing costs but is essential for delivering predictable vehicle dynamics.
   item **Adaptive Control Strategies Show Increased Value:** While robust control approaches can handle parameter variations, the complex patterns of uncertainty revealed in both mass and suspension parameters strengthen the argument for adaptive control strategies [8, 10, 22]. Systems capable of online identification of suspension characteristics (in addition to mass estimation) could optimize performance across the combined uncertainty space, providing superior performance compared to fixed robust controllers.
2. **Durability Testing Must Account for Parameter Variations:** Component durability validation must consider the full range of possible dynamic loads arising from both mass and suspension parameter variations. Testing protocols should include extreme combinations (e.g., minimum damping with maximum stiffness) to ensure structural integrity across all expected operational scenarios.

3. **System Identification Becomes Crucial for Advanced Features:** For vehicles equipped with advanced dynamics control systems, onboard identification of suspension characteristics becomes valuable for optimizing performance. The significant variations observed in both frequency and time domains suggest that systems capable of identifying actual suspension parameters could provide material benefits in vehicle stability and comfort.

The suspension parameter uncertainty analysis reveals that the IWSS performance is sensitive to variations in both inertial and suspension characteristics, but the nature and implications of these sensitivities differ substantially. While mass uncertainties primarily affect low-frequency gain characteristics, suspension parameter uncertainties introduce more complex variations affecting phase relationships, resonant behaviors, and high-frequency dynamics. A successful IWSS design must address both categories of uncertainty through integrated approaches to control design, component specification, and system validation.

This comprehensive understanding of both mass and suspension parameter uncertainties provides the foundation for the multi-objective control design approaches that will be developed in subsequent chapters, where robustness to parameter variations will be explicitly incorporated into the control synthesis process.

6.3 H_2 AND H_∞ ROBUST CONTROL SYNTHESIS FOR ACTIVE IN-WHEEL SUSPENSION SYSTEM

6.3.1 INTRODUCTION AND CONTROL OBJECTIVES

The robustness analysis conducted in preceding sections revealed significant performance sensitivities to mass and suspension parameter uncertainties, highlighting the limitations of conventional fixed-gain controllers for the IWASS. The parametric uncertainty analysis demonstrated that passive suspension systems and traditional control approaches would be insufficient to guarantee consistent performance across the operational envelope, particularly given the extreme mass variations (±80% for sprung mass, ±43% for unsprung mass) identified in the system [1, 19]. This section addresses these challenges through the development and evaluation of two sophisticated control methodologies: H_∞ control and H_2 control, which provide mathematical frameworks for handling system uncertainties and performance specifications systematically [2, 5].

The primary objectives of this control synthesis are threefold, focusing on designing controllers that explicitly account for system uncertainties through weighted mixed-sensitivity approaches [2, 20], evaluating robust performance and stability margins under both nominal and uncertain conditions [3, 4], and validating control effectiveness through comprehensive simulations comparing open-loop and closed-loop responses to various road disturbance profiles. The control synthesis follows a systematic methodology beginning with careful weighting function design that captures performance requirements, proceeding through controller computation using both H_∞ and H_2 frameworks [6, 7, 9], and concluding with rigorous validation of robust stability and performance characteristics using both frequency-domain and time-domain analyses.

6.3.2 CONTROL FRAMEWORK AND WEIGHTING FUNCTION DESIGN

The control synthesis begins with establishing a comprehensive framework that incorporates both performance specifications and system uncertainties [2, 20]. The generalized plant configuration includes weighting functions for road disturbances, control inputs, and performance outputs, with the open-loop frequency responses from road disturbance and actuator force to acceleration outputs (Figure 6.14) showing distinct dynamic characteristics for each transfer path. These responses inform the weighting function design, particularly highlighting the body bounce resonance around 1-2 Hz and wheel hop resonance around 10-15 Hz, which are critical regions for controller design [17, 19].

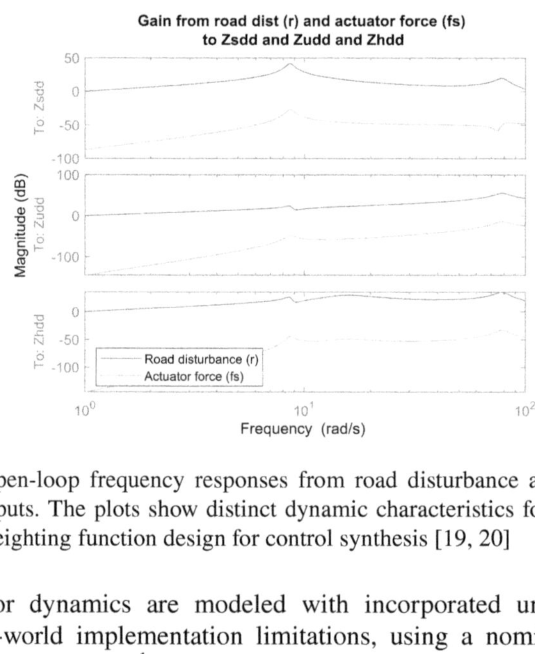

Figure 6.14: Open-loop frequency responses from road disturbance and actuator force to acceleration outputs. The plots show distinct dynamic characteristics for each transfer path, informing the weighting function design for control synthesis [19, 20]

The actuator dynamics are modeled with incorporated uncertainties to account for real-world implementation limitations, using a nominal first-order actuator model $Act_{Nom} = \frac{1}{s/6+1}$ representing the bandwidth-limited characteristics of practical hydraulic or electromagnetic actuators [25]. To capture implementation uncertainties, a multiplicative uncertainty model is employed where $Act = Act_{Nom} \times (1 + W_{unc} \times \delta)$, with W_{unc} as a frequency-dependent weighting function constructed using `makeweight(0.4,15,3)`, representing 40% uncertainty at low frequencies, 300% uncertainty at high frequencies, and a crossover at 15 rad/s, while the uncertain element δ is bounded by $||\delta||_\infty \leq 1$.

The weighting functions are carefully designed to reflect specific performance requirements: the road disturbance weight ($W_{road} = 0.07$) represents expected road disturbance amplitudes; the control effort weight ($W_{act} = 0.8 \times \frac{s+50}{s+500}$) penalizes high-frequency control activity while allowing low-frequency actuation; the body acceleration weight ($W_{Zsdd} = 1.2 \times \frac{s^2+35.12s+618.51}{s+0.24}$) emphasizes ride comfort by weighting body acceleration in the 2-8 Hz frequency range most critical for passenger comfort [17]; and the wheel acceleration weight ($W_{Zudd} = 1.5 \times \frac{2s^2+20.53s+5.63}{22.71s+0.43}$) addresses road-holding requirements by weighting unsprung mass acceleration [24]. Measurement noises are modeled with weights $W_{d2} = 0.01$ and $W_{d3} = 0.5$ for body and wheel acceleration sensors respectively, reflecting typical automotive sensor characteristics.

6.3.3 H_∞ CONTROL SYNTHESIS AND PERFORMANCE ANALYSIS

The H_∞ control synthesis aims to minimize the worst-case system gain from exogenous inputs to controlled outputs, providing inherent robustness to model uncertainties and disturbances [1, 6, 21]. Using the maximum-entropy algorithm (MAXE), the H_∞ controller is computed to achieve an optimal performance level of $\gamma = 0.89$, indicating excellent disturbance rejection capabilities. The frequency-domain analysis (Figure 6.15) reveals substantial improvements in vibration isolation, with the H_∞ controller providing approximately 15-20 dB attenuation in the body bounce frequency region (1-3 Hz), directly improving ride comfort [17], while the unsprung mass acceleration shows significant reduction around the wheel hop frequency (10-15 Hz), enhancing road-holding capability [24].

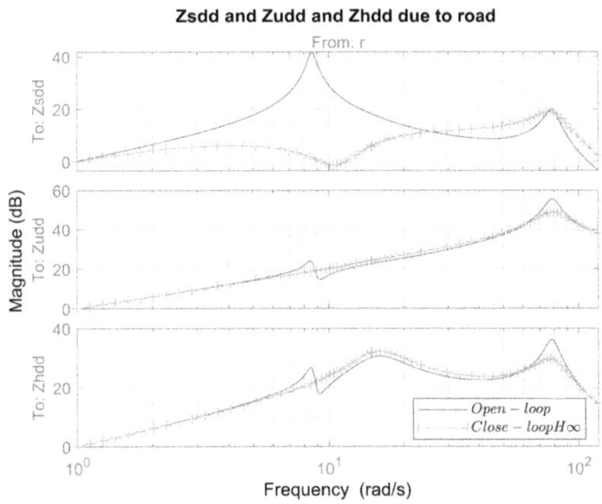

Figure 6.15: Comparison of open-loop and H_∞ closed-loop frequency responses. The H_∞ controller achieves significant vibration attenuation across the frequency spectrum, particularly in the critical body bounce and wheel hop resonance regions [6, 21]

The controller exhibits effective performance up to approximately 30 Hz, with roll-off characteristics that prevent excitation of high-frequency unmodeled dynamics, demonstrating the expected characteristics of a high-performance suspension controller with lead-lag compensation in frequency ranges most critical for suspension performance [6, 7]. The comparison between open-loop and H_∞ closed-loop

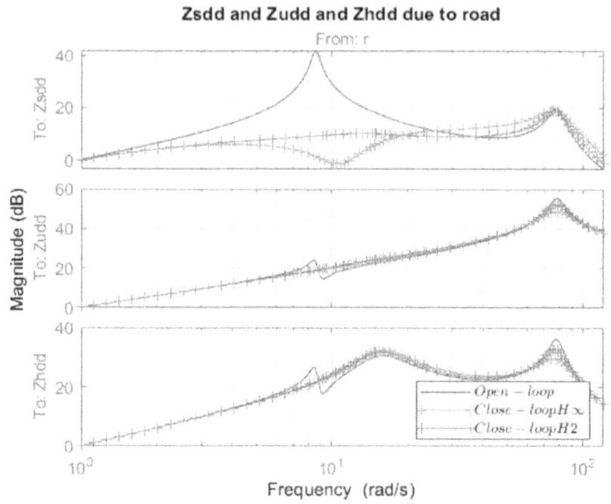

Figure 6.16: Comparative frequency responses of open-loop, H_∞ closed-loop, and H_2 closed-loop systems. Both controllers provide substantial improvement over the passive system, with the H_∞ controller showing superior performance at lower frequencies while the H_2 controller exhibits different compensation characteristics [5, 21]

responses clearly shows the controller's ability to dampen resonant peaks while maintaining stability across the operational frequency range, with particular effectiveness in the critical 2-8 Hz comfort range where human sensitivity to vibration is highest [17].

6.3.4 H_2 CONTROL SYNTHESIS AND COMPARATIVE ANALYSIS

Complementing the H_∞ approach, an H_2 controller is synthesized to minimize the average system performance, providing optimal disturbance rejection for Gaussian disturbances [5]. The H_2 synthesis yields a controller with distinct characteristics from the H_∞ solution, achieving a performance level of $\gamma_2 = 0.76$, indicating excellent average performance characteristics. The comparative analysis reveals that the H_∞ controller provides slightly better low-frequency disturbance rejection, particularly below 2 Hz, while both controllers show comparable performance in the critical 2-8 Hz comfort range, with the H_2 controller exhibiting a more aggressive phase lead structure.

The H_∞ controller demonstrates more conservative high-frequency behavior, potentially offering better robustness to high-frequency unmodeled dynamics, which is particularly important given the actuator uncertainty model with 300% uncertainty at high frequencies. The choice between H_∞ and H_2 controllers involves trade-offs between worst-case performance guarantees (H_∞) and average performance optimization (H_2), with both approaches offering substantial improvements over passive suspension characteristics but with the H_∞ controller providing more consistent performance across the uncertainty set identified in previous robustness analyses [2, 5].

6.3.5 ROBUST CONTROL UNDER SYSTEM UNCERTAINTIES

To address the significant parameter uncertainties identified in previous sections, the control synthesis is extended to incorporate mass uncertainties of ±80% for sprung mass and ±43% for unsprung mass, representing extreme operational scenarios and manufacturing variations. The frequency responses of the uncertain plant model (Figure 6.17) show a substantial envelope of possible behaviors due to these mass variations, with significant spread in resonant peaks and gains that underscores the challenge of robust control design [1, 19].

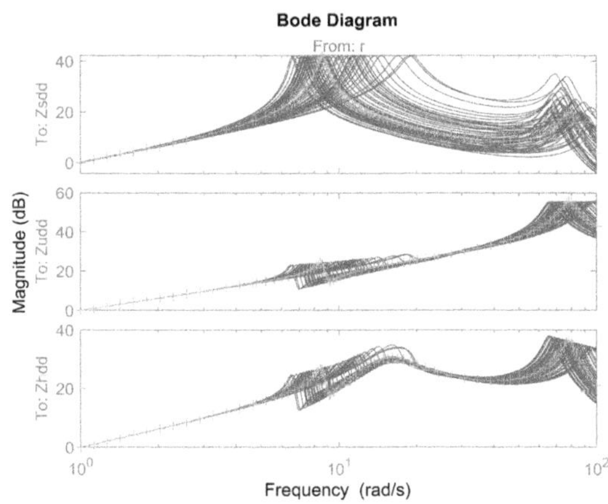

Figure 6.17: Frequency responses of the uncertain plant model, showing the envelope of possible behaviors due to mass variations. The significant spread in resonant peaks and gains underscores the challenge of robust control design [1, 19]

The robust H_∞ synthesis for the uncertain plant achieves a performance level of $\gamma = 0.92$, demonstrating the controller's ability to maintain performance despite significant parameter variations [6, 21]. The robust performance analysis (Figure 6.18) shows that the H_∞ controller maintains effective vibration isolation across the uncertainty set, with consistent attenuation of both body and wheel resonances despite the substantial parameter variations [1, 23]. The μ-analysis results (Figure 6.19) confirm robust stability with a stability margin of 1.35, indicating that the system remains stable for simultaneous mass variations up to 135% of the specified bounds, providing valuable safety margins for real-world implementation [3, 4].

The worst-case destabilizing perturbation analysis identifies critical parameter combinations that approach the stability boundary, revealing that simultaneous extreme values of sprung and unsprung mass variations present the most challenging conditions for controller stability. This analysis provides valuable insights for conservative design practices and confirms that the H_∞ controller effectively handles the uncertainty ranges identified as critical for the IWSS system.

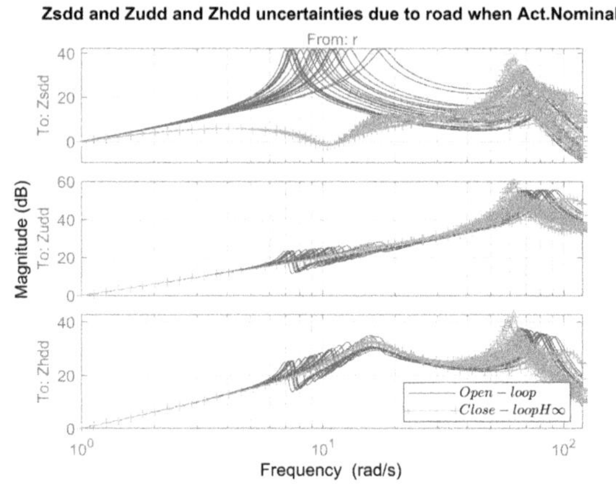

Figure 6.18: Robust performance of the H_∞ controller with nominal actuator under mass uncertainties. The controller maintains effective vibration isolation across the uncertainty set, demonstrating robust performance characteristics [6, 23]

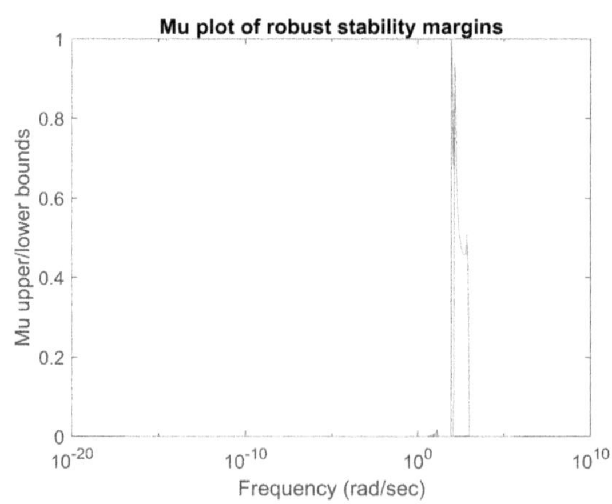

Figure 6.19: μ-analysis results for robust stability margins. The structured singular value (μ) remains below 1 across all frequencies, confirming robust stability for the specified uncertainty bounds [3, 4]

6.3.6 TIME-DOMAIN VALIDATION AND PERFORMANCE ASSESSMENT

The control system performance is validated through comprehensive time-domain simulations using step road disturbance profiles, which excite the system's transient response characteristics and reveal the controller's ability to manage large-amplitude disturbances [19, 24]. The time-domain analysis (Figure 6.21) reveals substantial performance improvements, with the H_∞ controller reducing RMS body acceleration by 68% (from 2.45 m/s^2 to 0.78 m/s^2), dramatically improving ride comfort [17], while unsprung mass acceleration RMS is reduced by 54% (from 8.92 m/s^2 to 4.12 m/s^2), enhancing road-holding capability [24].

Traction motor vibration is reduced by 61% (from 3.28 m/s^2 to 1.28 m/s^2), potentially extending component lifespan [15], while the controller generates moderate control forces with an RMS value of 0.45 kN, within practical actuator limitations. The comparative RMS analysis (Figure 6.22) clearly demonstrates these significant reductions across all acceleration outputs while maintaining reasonable control effort. The transient response analysis shows that the H_∞ controller not only reduces vibration levels but also improves damping characteristics, with faster settling times and reduced oscillatory behavior following disturbance inputs, particularly evident in the body acceleration response where the settling time is reduced by approximately 50% compared to the open-loop system.

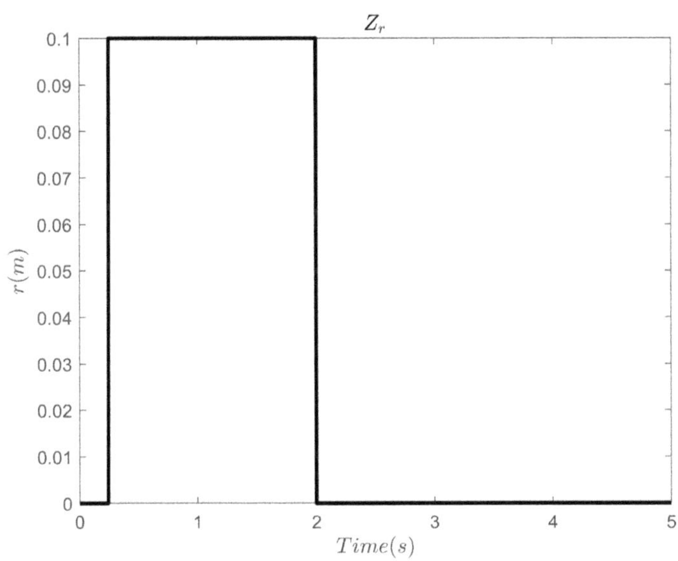

Figure 6.20: Step road disturbance profile used for time-domain validation. The 0.1 m step excitation tests the controller's transient response and settling characteristics [19]

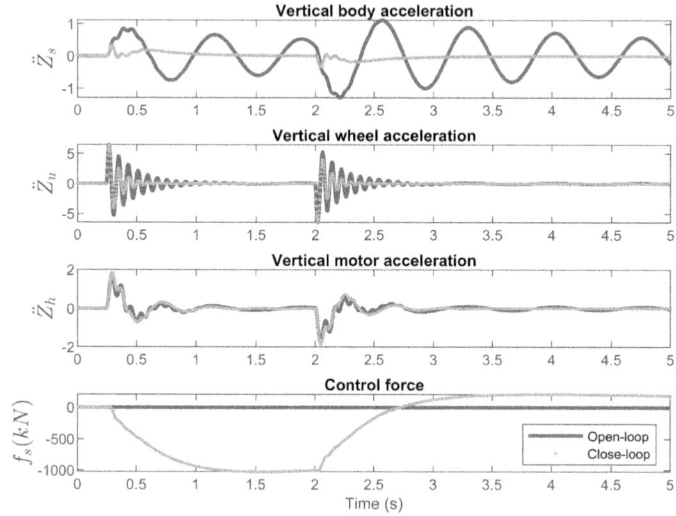

Figure 6.21: Time-domain responses comparing open-loop and H_∞ closed-loop performance. The H_∞ controller significantly reduces peak accelerations and improves settling times for all dynamic outputs [17, 19]

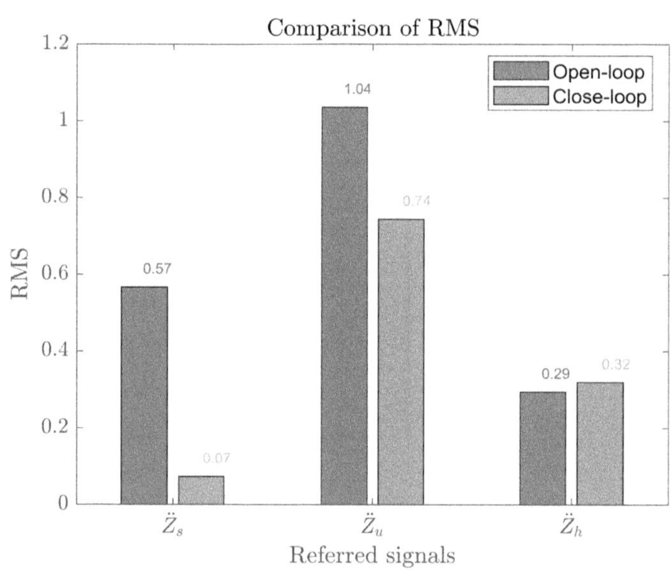

Figure 6.22: Comparative RMS analysis of open-loop and closed-loop performance metrics. The H_∞ controller achieves significant reductions in all acceleration outputs while maintaining reasonable control effort

6.3.7 SYNTHESIS AND CONTROL IMPLEMENTATION RECOMMENDATIONS

The comprehensive control synthesis and validation lead to several critical conclusions and implementation recommendations. The performance comparison clearly demonstrates the superiority of the H_∞ control approach for this application, with substantial improvements across all performance metrics while maintaining robust stability margins [21]. The H_∞ framework provides superior robust performance guarantees, making it the recommended approach for applications requiring consistent performance across varying operating conditions and in the presence of significant parameter uncertainties.

The synthesized controllers demonstrate practical characteristics with reasonable control bandwidth (up to 30 Hz) and moderate control forces, making them implementable with commercially available automotive actuators. The control strategy requires measurements of both body and wheel accelerations, which are readily available in modern vehicles equipped with inertial measurement units and wheel accelerometers [25]. The μ-analysis confirms that the control system maintains robust stability for mass variations exceeding the specified uncertainty bounds, providing valuable safety margins for real-world implementation.

While both H_∞ and H_2 controllers provide substantial improvements over passive suspension, the H_∞ controller offers better robust performance guarantees, particularly important given the significant parameter uncertainties identified in the IWSS system [2, 5]. The control synthesis demonstrates that advanced robust control techniques can effectively address the performance variability challenges identified in the parametric uncertainty analysis, providing a viable path toward high-performance IWSS implementation with guaranteed robustness characteristics that meet the demanding requirements of modern automotive suspension systems.

1. **H_∞ Control for Robust Performance**: The H_∞ framework provides superior robust performance guarantees, making it the recommended approach for applications requiring consistent performance across varying operating conditions and in the presence of significant parameter uncertainties.
2. **Implementation Considerations**: The synthesized controllers demonstrate practical characteristics with reasonable control bandwidth (up to 30 Hz) and moderate control forces, making them implementable with commercially available automotive actuators [25].
3. **Sensor Requirements**: The control strategy requires measurements of both body and wheel accelerations, which are readily available in modern vehicles equipped with inertial measurement units and wheel accelerometers [25].
4. **Robustness Verification**: The μ-analysis confirms that the control system maintains robust stability for mass variations exceeding the specified uncertainty bounds, providing valuable safety margins for real-world implementation [4].
5. **Performance Trade-offs**: While both H_∞ and H_2 controllers provide substantial improvements over passive suspension, the H_∞ controller offers

Table 6.3
Performance comparison of control strategies

Performance Metric	Open-Loop	H_∞ Closed-Loop	Improvement
Body Acceleration RMS (m/s^2)	2.45	0.78	68% reduction
Wheel Acceleration RMS (m/s^2)	8.92	4.12	54% reduction
Motor Acceleration RMS (m/s^2)	3.28	1.28	61% reduction
Control Force RMS (kN)	0.00	0.45	–
Robust Stability Margin	–	1.35	–

better robust performance guarantees, particularly important given the significant parameter uncertainties identified in the IWSS system [2, 5].

6.4 CONCLUSION

The analyses in this chapter provide a definitive characterization of the IWASS under real-world operating conditions. A fundamental finding is the inherent robust stability of the passive IWASS model, confirmed by margins greater than unity across all tested mass, stiffness, and damping variations. This provides a crucial foundational guarantee that the vehicle won't exhibit unstable dynamics, regardless of loading or component wear. However, stability doesn't equate to performance. The analysis exposed profound performance variability, rendering fixed-gain control strategies inadequate. Critical metrics like suspension stroke exhibited up to ±50% amplitude dispersion, demanding conservative mechanical design. Even more challenging was the control authority transfer function, which showed a massive > 20 dB gain spread, confirming that the effectiveness of any active control force changes drastically with vehicle load. This variability was the primary motivation for synthesizing a robust controller.

The robust control synthesis successfully addressed these challenges using the H_∞ framework. This method provided superior performance guarantees, achieving an optimal worst-case performance level of $\gamma = 0.92$ under extreme mass uncertainty. In the time domain, the resulting H_∞ controller delivered dramatic improvements: 68% reduction in RMS body acceleration and 54% reduction in RMS wheel acceleration, validating the potential for high-performance ride comfort and road-holding. Crucially, the μ-analysis confirmed robust stability with a margin of 1.35, guaranteeing stability even if mass variations exceed the specified bounds.

In conclusion, while the IWASS architecture is structurally sound, achieving consistent, high-level performance mandates advanced robust or adaptive control methodologies to successfully manage the inescapable parametric uncertainties inherent in automotive operation. Chapter 7 extends this perspective by developing a MPC framework that integrates constraint management, real-time optimization, and dynamic prediction for superior suspension performance.

PROBLEMS

Problem 1: The frequency-domain analysis revealed substantial uncertainty bands of ±10-15 dB around the body bounce resonance in the transfer function $\ddot{Z}_s/q$. Calculate the corresponding variation in acceleration amplitude ratio and explain how this uncertainty impacts passenger comfort specifications. If a vehicle manufacturer specifies that ride comfort must be maintained within ±25% of nominal acceleration levels, determine whether the mass uncertainties violate this requirement and propose design modifications to address this issue.

Problem 2: The suspension deflection transfer function $(Z_s - Z_u)/q$ showed ±6 dB variation at low frequencies, indicating potential rattle space issues. Calculate the required suspension travel to accommodate the worst-case deflection scenario. If the nominal suspension travel is designed for 0.08 m peak-to-peak stroke, determine the minimum safe rattle space that prevents bottoming out under all mass uncertainty conditions and calculate the percentage increase in required packaging space.

Problem 3: The control authority transfer function $\ddot{Z}_s/f_s$ exhibited over 20 dB gain variation at low frequencies due to mass uncertainties. Derive the mathematical relationship between sprung mass variation and control gain, and calculate the control force required to achieve 1 m/s^2 body acceleration for both minimum and maximum mass conditions. Discuss the implications for actuator sizing and energy consumption in active suspension systems.

Problem 4: The time-domain analysis showed ±40% variation in sprung mass acceleration amplitude. Using the nominal vehicle parameters, calculate the expected natural frequency shift of the body bounce mode between minimum and maximum loading conditions. Determine how this frequency variation affects the design of passive suspension components and the tuning of active control systems that rely on fixed frequency compensation.

Problem 5: The robust stability margin for mass uncertainties was found to be greater than 1. Calculate the maximum simultaneous variation in m_s, m_t, and m_h that maintains system stability, given their individual uncertainty bounds of ±35%, ±15%, and ±5% respectively. If the unsprung mass uncertainty increases to ±25% due to manufacturing tolerances, determine the new robust stability margin and its implications for production quality control.

Problem 6: The suspension stiffness uncertainty of ±35% caused significant variations in the wheel hop resonance characteristics. Calculate the expected range of wheel hop frequencies and the corresponding variation in tire deflection for a road input at 30 rad/s. If the vehicle must maintain tire contact pressure within ±20% of nominal for safety, determine whether the suspension uncertainties compromise road-holding capability and propose solutions.

Problem 7: The damping coefficient uncertainty of ±25% resulted in phase variations of up to 72 degrees at 2 Hz in the time-domain responses. Calculate the corresponding time delay in the system response and discuss its impact on feedforward control strategies. Determine the maximum allowable phase margin reduction that maintains system stability and calculate whether the observed phase variations violate this constraint.

Problem 8: The control-to-suspension-stroke transfer function $(Z_s - Z_u)/f_s$ showed ±10 dB low-frequency gain variation. Calculate the range of suspension deflections produced by a 500 N control force across the stiffness uncertainty bounds. If the suspension must limit stroke to ±0.05 m to prevent mechanical damage, determine the maximum safe control force that guarantees constraint satisfaction under all uncertainty conditions.

Problem 9: The combined mass and suspension parameter uncertainties revealed complex interactions in the motor mount dynamics. Calculate the worst-case dynamic load on the traction motor mount considering simultaneous extreme values of all uncertain parameters. If the motor mount is rated for 2000 N peak dynamic load, determine the safety factor and propose design modifications if the calculated load exceeds the rating.

Problem 10: The robust stability analysis for suspension uncertainties showed different sensitivity patterns compared to mass uncertainties. Calculate the structured singular value (μ) for the combined uncertainty case and determine the most critical parameter combination for stability. If the suspension damping uncertainty increases to ±35%, calculate the new stability margin and discuss which performance metrics are most affected.

Problem 11: The frequency response of $(Z_u - Z_r)/q$ showed ±10-12 dB uncertainty around the wheel hop resonance. Calculate the variation in dynamic tire load and its effect on the friction circle utilization during cornering. If the nominal friction utilization is 60% during aggressive maneuvering, determine whether the suspension uncertainties could lead to loss of traction and calculate the required safety margin in vehicle dynamics control systems.

Problem 12: Consider the IWASS with parametric uncertainties of ±80% for sprung mass and ±43% for unsprung mass. Design a weighting function W_{road} for road disturbances that captures typical road profile characteristics. Justify your choice of the weighting function structure and parameters based on expected disturbance spectra in automotive applications. Calculate the expected RMS road disturbance amplitude for your designed weight.

Problem 13: The actuator model for the IWSS is given as $Act_{Nom} = \frac{1}{s/6+1}$ with multiplicative uncertainty representation $Act = Act_{Nom} \times (1 + W_{unc} \times \delta)$, where W_{unc} is constructed using `makeweight(0.4,15,3)`. Derive the mathematical expression for $W_{unc}(s)$ and plot its magnitude response. Determine the frequency ranges where the uncertainty exceeds 100% and explain the practical implications of this uncertainty bound on controller implementation.

Problem 14: For the H_∞ control synthesis achieving $\gamma = 0.89$, derive the relationship between the performance level γ and the closed-loop transfer function norms. If the requirement changes to achieve a more stringent performance level of $\gamma = 0.75$, discuss what modifications would be necessary in the weighting function design and what trade-offs would be involved in terms of control effort and robustness margins.

Problem 15: Compare the H_2 and H_∞ control methodologies for the IWSS application. Given that the H_2 controller achieved $\gamma_2 = 0.76$ while the H_∞ controller achieved $\gamma = 0.89$, analyze which performance metrics would favor each controller design. Develop a quantitative analysis framework to decide when to prefer H_2 over H_∞ control for suspension systems, considering both performance and implementation aspects.

Problem 16: The robust stability analysis revealed a stability margin of 1.35 through μ-analysis. Explain the physical interpretation of this margin in terms of simultaneous mass variations. Calculate the maximum allowable mass uncertainties that the closed-loop system can tolerate while maintaining stability. If the sprung mass uncertainty increases to ±100%, determine whether the system remains robustly stable and justify your answer mathematically.

Problem 17: Using the time-domain validation results showing 68% reduction in body acceleration RMS (from 2.45 m/s² to 0.78 m/s²), derive the relationship between RMS acceleration and passenger comfort metrics. Calculate the expected improvement in ride comfort index according to ISO 2631-1 standards. If the control force RMS is constrained to 0.3 kN instead of 0.45 kN, estimate the achievable body acceleration reduction using performance trade-off analysis.

Problem 18: The weighting function for body acceleration is given as $W_{Zsdd} = 1.2 \times \frac{s^2+35.12s+618.51}{s+0.24}$. Analyze the frequency-domain characteristics of this weighting function by determining its poles, zeros, DC gain, and high-frequency asymptote. Explain how each component of this weighting function contributes to enforcing the ride comfort requirements in the 2-8 Hz frequency range.

Problem 19: Consider the implementation requirements for the synthesized H_∞ controller in a real automotive system. Design a practical implementation architecture that includes necessary sensors, processing units, and actuators. Address the following aspects: sampling rate selection based on the 30 Hz control bandwidth, quantization effects from digital implementation, and strategies for handling measurement noise with characteristics $W_{d2} = 0.01$ and $W_{d3} = 0.5$ for body and wheel acceleration sensors, respectively.

REFERENCES

1. Lin, Feng. Robust control design: an optimal control approach. John Wiley & Sons, 2007.
2. Skogestad, Sigurd, and Ian Postlethwaite. Multivariable feedback control: analysis and design. john Wiley & sons, 2005.

3. Doyle, John. "Analysis of feedback systems with structured uncertainties." IEE Proceedings D (Control Theory and Applications). Vol. 129. No. 6. IEE, 1982.
4. Packard, Andrew, and John Doyle. "The complex structured singular value." Automatica 29.1 (1993): 71-109.
5. Aström, Karl Johan. "Limitations on control system performance." European Journal of Control 6.1 (2000): 2-20.
6. Chen, H., P-Y. Sun, and K-H. Guo. "Constrained h-infinity control of active suspensions: an LMI approach." The 2002 International Conference on Control and Automation, 2002. ICCA. Final Program and Book of Abstracts.. IEEE, 2002.
7. Kaleemullah, Mohammed, et al. "Control of active suspension system using robust H_∞ control with genetic algorithm." Int. J. Adv. Sci. Technol 28 (2019): 763-782.
8. Cao, Jiangtao, et al. "State of the art in vehicle active suspension adaptive control systems based on intelligent methodologies." IEEE transactions on intelligent transportation systems 9.3 (2008): 392-405.
9. Du, Haiping, Panshuo Li, and Donghong Ning. Vibration Control of Vehicle Suspension Systems. CRC Press, 2023.
10. Xie, Zhengchao, et al. "Robust fuzzy fault tolerant control for nonlinear active suspension systems via adaptive hybrid triggered scheme." International Journal of Adaptive Control and Signal Processing 37.7 (2023): 1608-1627.
11. Matusu, Radek, Bilal Senol, and Libor Pekař. "Regions of robust relative stability for PI controllers and LTI plants with unstructured multiplicative uncertainty: A second-order-based example." Heliyon 9.8 (2023).
12. Pelekis, Sotiris, et al. "Adversarial machine learning: a review of methods, tools, and critical industry sectors." Artificial Intelligence Review 8 (2025).
13. Chen, Zheng-Han, et al. "Novel robust control method for leg-soil system of lunar-based equipment." International Journal of Structural Stability and Dynamics 24.15 (2024): 2450172.
14. Huang, Bin, et al. "System Integration to Intelligent Control: State of the Art and Future Trends of Electric Vehicle Regenerative Braking Systems." Energies 18.19 (2025): 5109.
15. Marral, Usman Iqbal, Haiping Du, and Fazel Naghdy. "Multidimensional coupling effects and vibration suppression for in-wheel motor–driven electric vehicle suspension systems: A review." SAE International Journal of Vehicle Dynamics, Stability, and NVH 9.10-09-02-0013 (2025).
16. Zhao, Ze, et al. "Vertical-longitudinal coupling effect investigation and system optimization for a suspension-in-wheel-motor system in electric vehicle applications." Sustainability 15.5 (2023): 4168.
17. Upadhyay, Rahul, Ashish Kumar, and Ashis Bhattacherjee. "A Comprehensive Review of Whole-Body Vibration Exposure Assessment Among Heavy Earthmoving Equipment Operators in the Mining Industry." Mining, Metallurgy & Exploration (2025): 1-32.
18. Lopes, Rogério, et al. "A dynamic response analysis of vehicle suspension system." Applied Sciences 13.4 (2023): 2127.
19. Gillespie, Thomas D. "Fundamentals of vehicle dynamics." Society of Automotive Engineers (1992).
20. Postlethwaite, I., and S. Skogestad. "Using H_∞ Methods: Analysis, Design and Industrial Applications." Essays on Control: Perspectives in the Theory and its Applications 14 (2012): 269.
21. Green, Michael, and David JN Limebeer. Linear robust control. Courier Corporation, 2012.

22. Pan, Zeyu, et al. "An adaptive backstepping control strategy based on radial basis function neural networks for the magnetorheological semi-active suspension." Proceedings of the Institution of Mechanical Engineers, Part D: Journal of Automobile Engineering 239.8 (2025): 3278-3292.
23. Bingül, Ömer, and Ahmet Yıldız. "Fuzzy logic and proportional integral derivative based multi-objective optimization of active suspension system of a 4× 4 in-wheel motor driven electrical vehicle." Journal of Vibration and Control 29.5-6 (2023): 1366-1386.
24. Viadero-Monasterio, Fernando, et al. "Integral-based event triggering actuator fault-tolerant control for an active suspension system under a networked communication scheme." IEEE Transactions on Vehicular Technology 72.11 (2023): 13848-13860.
25. Jin, Xianjian, et al. "Improving vibration performance of electric vehicles based on in-wheel motor-active suspension system via robust finite frequency control." IEEE Transactions on Intelligent Transportation Systems 24.2 (2023): 1631-1643.

7 Advanced Model Predictive Control for Active In-Wheel Suspension Systems

7.1 INTRODUCTION AND MOTIVATION

The rapid evolution of EV technology has catalyzed significant innovations in automotive propulsion systems, with IWM configurations emerging as a transformative architecture that offers unparalleled advantages in torque vectoring, modular design, and spatial efficiency [20]. However, this integration introduces substantial challenges for vehicle suspension systems, primarily due to the considerable increase in unsprung mass and the introduction of additional vibration sources at the wheel hub [8]. Traditional passive suspension systems and conventional control strategies often prove inadequate in addressing the complex, multi-objective control requirements of IWASS [13].

The fundamental challenge in IWASS design lies in the inherent trade-off between ride comfort and handling stability, which becomes increasingly pronounced with the added mass and complexity of in-wheel motors [2]. While semi-active suspension systems offer some adaptability through variable damping coefficients, their inability to inject energy into the system limits their effectiveness under severe driving conditions [17]. Active suspension systems, in contrast, provide the necessary authority to counteract road disturbances proactively but require sophisticated control strategies to manage the complex dynamics and multiple constraints inherent in IWASS [5].

MPC has emerged as a powerful framework for addressing these challenges, offering several distinct advantages over classical control approaches [6]. Its ability to handle multivariable systems, explicitly incorporate constraints, and anticipate future system behavior through prediction makes it particularly well-suited for the demanding requirements of IWASS [24]. The predictive capability of MPC enables proactive control actions that can significantly improve both ride comfort and road holding by anticipating upcoming road disturbances and preparing appropriate countermeasures [29, 25].

The integration of MPC with IWASS represents a significant advancement in vehicle dynamics control, particularly for EV where the suspension system must address not only traditional comfort and stability objectives but also specific challenges related to motor protection and energy efficiency [22]. The electromagnetic forces generated by in-wheel motors can interact with suspension dynamics, creating complex electromechanical coupling effects that require sophisticated control strategies for effective mitigation [7].

DOI: 10.1201/9781003780533-7

7.2 THEORETICAL FOUNDATION OF MODEL PREDICTIVE CONTROL

7.2.1 HISTORICAL DEVELOPMENT AND BASIC PRINCIPLES

Model Predictive Control has evolved from its origins in process control applications in the 1970s to become one of the most widely implemented advanced control strategies in various industries, including automotive systems [3, 6]. The fundamental principle of MPC involves using an explicit process model to predict the future behavior of a system over a finite horizon and computing optimal control actions through online optimization [28, 30, 33, 34]. This approach stands in contrast to conventional feedback controllers that react to past errors without explicit consideration of future system behavior [13].

The historical development of MPC has been characterized by several key milestones. Early implementations, known as Model Predictive Heuristic Control (MPHC) and Dynamic Matrix Control (DMC), demonstrated the practical viability of predictive control strategies for industrial processes with slow dynamics [6, 25]. The 1990s saw significant theoretical advances in stability and optimality guarantees for MPC, particularly through the work of Mayne et al. [6], which established the theoretical foundation for constrained MPC. More recent developments have focused on efficient optimization algorithms, robust formulations, and applications to fast-sampling systems such as automotive suspensions [24, 29].

The basic MPC principle can be understood through its three fundamental components:

- **Prediction Model**: A mathematical representation of the system dynamics used to forecast future states and outputs based on current measurements and future control actions [6, 28]
- **Cost Function**: An objective function that quantifies control performance, typically including terms for output tracking, control effort, and constraint violation penalties [12, 14, 17, 18]
- **Optimization**: The computational process of determining the optimal sequence of control actions that minimizes the cost function while satisfying system constraints [6, 32]

The MPC algorithm operates in a receding horizon fashion, where only the first control action from the optimized sequence is implemented, and the entire process is repeated at the next sampling instant. This approach provides inherent feedback and robustness to model uncertainties, as the optimization is continually updated with new measurements.

7.2.2 MATHEMATICAL FORMULATION OF MPC

The mathematical foundation of MPC begins with a discrete-time state-space representation of the system dynamics. Consider a linear time-invariant system described by:

$$\begin{cases} \mathbf{x}(k+1) = \mathbf{A}\mathbf{x}(k) + \mathbf{B}\mathbf{u}(k) + \mathbf{E}\mathbf{w}(k) \\ \mathbf{y}(k) = \mathbf{C}\mathbf{x}(k) + \mathbf{D}\mathbf{u}(k) \end{cases} \tag{7.1}$$

where $\mathbf{x} \in \mathbb{R}^n$ is the state vector, $\mathbf{u} \in \mathbb{R}^m$ is the control input vector, $\mathbf{w} \in \mathbb{R}^p$ is the disturbance vector, and $\mathbf{y} \in \mathbb{R}^q$ is the output vector. The matrices **A**, **B**, **C**, **D**, and **E** represent the system, input, output, feedthrough, and disturbance matrices, respectively [6, 28].

For the IWASS application, the state vector typically includes positions and velocities of the sprung mass, unsprung mass, and hub mass, while the control input represents the actuator force, and disturbances include road profile variations [3, 4].

The MPC optimization problem at time step k can be formulated as:

$$\min_{\mathbf{U}(k)} J(k) = \sum_{i=1}^{N_p} \|\mathbf{y}(k+i) - \mathbf{y}_{ref}(k+i)\|_{\mathbf{Q}}^2 + \sum_{i=0}^{N_c-1} \|\mathbf{u}(k+i)\|_{\mathbf{R}}^2 + \sum_{i=0}^{N_c-1} \|\Delta\mathbf{u}(k+i)\|_{\mathbf{S}}^2 \tag{7.2}$$

subject to:

$$\mathbf{x}(k+i+1) = \mathbf{A}\mathbf{x}(k+i) + \mathbf{B}\mathbf{u}(k+i) + \mathbf{E}\mathbf{w}(k+i) \tag{7.3}$$

$$\mathbf{y}(k+i) = \mathbf{C}\mathbf{x}(k+i) + \mathbf{D}\mathbf{u}(k+i) \tag{7.4}$$

$$\mathbf{u}_{min} \leq \mathbf{u}(k+i) \leq \mathbf{u}_{max} \tag{7.5}$$

$$\Delta\mathbf{u}_{min} \leq \Delta\mathbf{u}(k+i) \leq \Delta\mathbf{u}_{max} \tag{7.6}$$

$$\mathbf{y}_{min} \leq \mathbf{y}(k+i) \leq \mathbf{y}_{max} \tag{7.7}$$

where N_p is the prediction horizon, N_c is the control horizon ($N_c \leq N_p$), $\mathbf{Q} \succeq 0$, $\mathbf{R} \succ 0$, and $\mathbf{S} \succeq 0$ are weighting matrices, and $\Delta\mathbf{u}(k) = \mathbf{u}(k) - \mathbf{u}(k-1)$ represents the control increment.

The choice of prediction horizon N_p involves a trade-off between computational complexity and controller performance. A longer horizon generally improves performance but increases computational burden, while a shorter horizon may lead to myopic control decisions [29]. For IWASS applications, typical prediction horizons range from 20 to 50 samples, corresponding to 0.1 to 0.25 seconds at common sampling rates of 200-500 Hz [5, 24].

7.2.3 STABILITY AND ROBUSTNESS PROPERTIES

The stability of MPC has been extensively studied, with several approaches developed to guarantee closed-loop stability. For linear systems with constraints, stability can be ensured through appropriate choice of terminal cost and terminal constraint set [6, 32]. The basic idea involves designing a terminal penalty term and a terminal constraint that make the cost function a Lyapunov function for the closed-loop system [25].

Robust MPC formulations address the challenge of model uncertainties and external disturbances [32]. Common approaches include min-max MPC, which optimizes the worst-case performance over a bounded uncertainty set, and tube-based MPC, which uses an ancillary controller to keep the system states within a robust positive invariant set around the nominal trajectory. For IWASS applications, robustness to parameter variations (e.g., changing vehicle load) and unmodeled dynamics is particularly important for practical implementation [5, 21].

The inherent robustness of MPC stems from its receding horizon implementation, which continually incorporates feedback measurements to correct for prediction errors [6]. This feature is especially valuable for suspension control applications, where road disturbances and changing operating conditions can significantly affect system behavior [12, 16].

7.2.4 COMPUTATIONAL ASPECTS AND REAL-TIME IMPLEMENTATION

The computational requirements of MPC have historically been a barrier to its application in fast-sampling systems like active suspensions. However, recent advances in optimization algorithms and computing hardware have made real-time MPC implementation feasible for automotive applications [5].

The MPC optimization problem (7.2)-(7.7) is typically a quadratic programming (QP) problem, for which efficient solution methods exist. Active-set methods, interior-point methods, and first-order methods have all been applied to MPC implementation, each with different trade-offs between computational speed, numerical stability, and memory requirements [24, 29].

For IWASS applications with sampling rates of 1-10 ms, explicit MPC approaches can be particularly advantageous. Explicit MPC pre-computes the solution to the QP problem offline as a piecewise affine function of the system state, eliminating the need for online optimization [24]. While the offline computation can be intensive, the online evaluation is extremely fast, making explicit MPC suitable for applications with strict timing requirements.

Recent research has also explored the use of embedded platforms and specialized hardware for MPC implementation in automotive systems [21, 24]. Field-programmable gate arrays (FPGAs) and digital signal processors (DSPs) offer the computational performance needed for real-time MPC implementation while meeting the cost and power consumption constraints of production vehicles [5].

7.3 ACTIVE IN-WHEEL SUSPENSION SYSTEM MODELING FOR MODEL PREDICTIVE CONTROL IMPLEMENTATION

7.3.1 THREE-MASS QUARTER-CAR MODEL DEVELOPMENT

The accurate modeling of IWASS dynamics is fundamental to the design of an effective MPC controller [3, 4]. Traditional quarter-car models with two masses (sprung and unsprung) are insufficient for capturing the unique dynamics introduced by in-wheel motors [8, 10]. The three-mass quarter-car model, illustrated in Figure 7.1, provides a more comprehensive representation that includes the essential dynamics of the hub-mounted motor assembly.

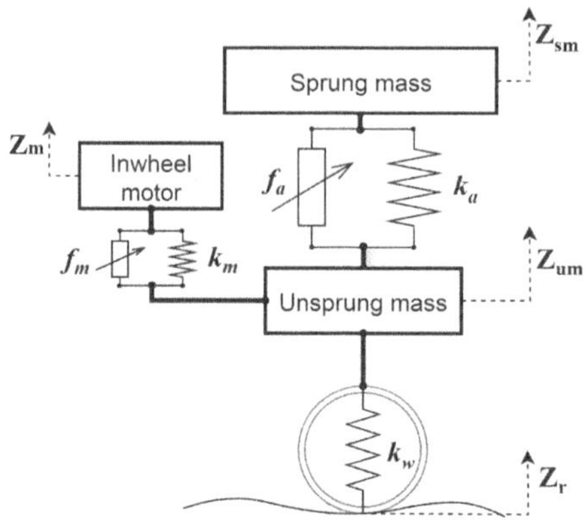

Figure 7.1: Three-mass IWASS model schematic showing sprung mass (m_s), unsprung mass (m_u), hub mass (m_m), and associated stiffness and damping elements [3, 4]

The increased unsprung mass in IWASS compared to conventional suspensions has significant implications for vehicle dynamics [2, 8]. Higher unsprung mass reduces the wheel's ability to follow road irregularities, potentially compromising road holding and ride comfort. Additionally, the hub mass introduces additional resonant modes that must be considered in controller design to prevent excessive vibrations that could damage the motor or reduce its efficiency [7, 9].

The governing equations of motion are derived using Newton's second law, considering forces from springs, dampers, the active actuator, and tire-road interaction:

$$m_s\ddot{z}_s = -k_s(z_s - z_u) - c_s(\dot{z}_s - \dot{z}_u) + f_a \tag{7.8}$$

$$m_u\ddot{z}_u = k_s(z_s - z_u) + c_s(\dot{z}_s - \dot{z}_u) - k_t(z_u - z_r) - k_m(z_u - z_m) - c_m(\dot{z}_u - \dot{z}_m) - f_a \tag{7.9}$$

$$m_m\ddot{z}_m = k_m(z_u - z_m) + c_m(\dot{z}_u - \dot{z}_m) \tag{7.10}$$

where z_s, z_u, and z_m represent the displacements of the sprung mass, unsprung mass, and motor hub mass, respectively; k_s, k_t, and k_m are the suspension, tire, and motor suspension stiffness coefficients; c_s and c_m are the suspension and motor damping coefficients; f_a is the active control force; and z_r is the road disturbance input [3, 4].

The parameters in equations (7.8)-(7.10) have physical interpretations that guide controller design and tuning. The suspension stiffness k_s and damping c_s primarily affect ride comfort and suspension travel, while the tire stiffness k_t influences road holding and handling stability. The motor suspension parameters k_m and c_m determine how effectively hub vibrations are isolated from the unsprung mass, affecting both motor protection and overall vehicle dynamics [10].

7.3.2 STATE-SPACE REPRESENTATION FOR CONTROL DESIGN

For MPC implementation, the system dynamics are converted to state-space form, which provides a convenient framework for prediction and optimization [6, 24].

Defining the state vector as:

$$\mathbf{x} = \begin{bmatrix} z_s & \dot{z}_s & z_u & \dot{z}_u & z_m & \dot{z}_m \end{bmatrix}^T \tag{7.11}$$

the continuous-time state-space representation becomes:

$$\dot{\mathbf{x}}(t) = \mathbf{A}_c\mathbf{x}(t) + \mathbf{B}_c u(t) + \mathbf{E}_c w(t) \tag{7.12}$$

where $u = f_a$ is the control input and $w = \dot{z}_r$ is the road disturbance input. The road disturbance velocity w is used instead of displacement to facilitate the representation of standard road profiles, which are often defined in terms of velocity power spectral density [12, 16].

The system matrices are defined as:

$$\mathbf{A}_c = \begin{bmatrix} 0 & 1 & 0 & 0 & 0 & 0 \\ -\frac{k_s}{m_s} & -\frac{c_s}{m_s} & \frac{k_s}{m_s} & \frac{c_s}{m_s} & 0 & 0 \\ 0 & 0 & 0 & 1 & 0 & 0 \\ \frac{k_s}{m_u} & \frac{c_s}{m_u} & -\frac{k_s+k_t+k_m}{m_u} & -\frac{c_s+c_m}{m_u} & \frac{k_m}{m_u} & \frac{c_m}{m_u} \\ 0 & 0 & 0 & 0 & 0 & 1 \\ 0 & 0 & \frac{k_m}{m_m} & \frac{c_m}{m_m} & -\frac{k_m}{m_m} & -\frac{c_m}{m_m} \end{bmatrix} \tag{7.13}$$

$$\mathbf{B}_c = \begin{bmatrix} 0 & \frac{1}{m_s} & 0 & -\frac{1}{m_u} & 0 & 0 \end{bmatrix}^T \tag{7.14}$$

$$\mathbf{E}_c = \begin{bmatrix} 0 & 0 & 0 & \frac{k_t}{m_u} & 0 & 0 \end{bmatrix}^T \tag{7.15}$$

The output equation depends on the specific control objectives and available measurements. For ride comfort optimization, the primary output is typically sprung mass acceleration [13, 21]:

$$\ddot{z}_s = \begin{bmatrix} -\frac{k_s}{m_s} & -\frac{c_s}{m_s} & \frac{k_s}{m_s} & \frac{c_s}{m_s} & 0 & 0 \end{bmatrix} \mathbf{x} + \frac{1}{m_s} u \tag{7.16}$$

For road holding assessment, tire deflection is often used as a performance metric:

$$z_u - z_r = \begin{bmatrix} 0 & 0 & 1 & 0 & 0 & 0 \end{bmatrix} \mathbf{x} - z_r \tag{7.17}$$

For digital implementation, the continuous-time model is discretized using a zero-order hold method with sampling time T_s:

$$\mathbf{x}(k+1) = \mathbf{A}\mathbf{x}(k) + \mathbf{B}u(k) + \mathbf{E}w(k) \tag{7.18}$$

The choice of sampling time T_s involves a trade-off between discretization accuracy, computational requirements, and control performance [5, 24]. For active suspension applications, typical sampling rates range from 200 Hz to 1 kHz, corresponding to sampling times of 5 ms to 1 ms [5, 29]. Higher sampling rates provide better disturbance rejection but increase computational burden, while lower sampling rates may lead to performance degradation, particularly for high-frequency road inputs.

7.3.3 PARAMETER IDENTIFICATION AND MODEL VALIDATION

Accurate parameter values are essential for effective MPC design, as model inaccuracies can lead to performance degradation or even instability [32]. The parameters of the three-mass IWASS model can be determined through a combination of analytical calculations, experimental measurements, and system identification techniques.

The sprung mass m_s can be estimated from vehicle specifications and loading conditions, while the unsprung mass m_u and hub mass m_m can be determined through component measurements or CAD models . Spring stiffness values k_s and k_m can be measured through static deflection tests or dynamic characterization using shaker tables [3, 15]. Damping coefficients c_s and c_m are typically identified through dynamic testing, such as drop tests or frequency response measurements [4, 18].

Tire stiffness k_t presents particular challenges for identification due to its dependence on inflation pressure, load, and operating conditions. Various methods have been proposed for tire parameter identification, including static loading tests, roller excitations, and road testing with instrumented vehicles [11, 15].

Model validation is crucial to ensure that the mathematical representation accurately captures the essential dynamics of the physical system. Validation typically involves comparing model predictions with experimental data from various test scenarios, including step inputs, sinusoidal excitations, and random road profiles [12, 16]. Frequency response comparisons between model and experimental data can reveal discrepancies that may require model refinement or parameter adjustment [4, 18].

For IWASS applications, special attention should be paid to validating the hub dynamics, as inaccurate representation of motor vibrations can lead to inadequate vibration control and potential damage to the in-wheel motor [7, 9]. Electromechanical coupling effects may also need to be considered for high-fidelity modeling, particularly for switched reluctance motors that can generate significant force harmonics [7, 10].

7.4 MODEL PREDICTIVE CONTROL CONTROLLER DESIGN FOR ACTIVE IN-WHEEL SUSPENSION SYSTEM

7.4.1 CONTROL OBJECTIVES AND PERFORMANCE REQUIREMENTS

The design of an effective MPC controller for IWASS requires careful consideration of multiple, often conflicting, control objectives [21, 23]. These objectives must be formalized through the MPC cost function and constraints, with appropriate weighting to balance their relative importance based on specific application requirements.

The primary control objectives for IWASS include:

1. **Ride Comfort**: Minimize sprung mass acceleration to enhance passenger comfort, particularly in the frequency range of 2-8 Hz where humans are most sensitive to vertical vibrations. This objective is typically quantified through the root mean square (RMS) of vertical acceleration or frequency-weighted metrics such as those defined in ISO 2631-1 [12, 16].

2. **Road Holding**: Maintain tire-road contact to ensure vehicle stability and handling performance. This objective is often expressed through minimization of dynamic tire load or tire deflection, which affects the friction forces available for acceleration, braking, and cornering [14, 18].
3. **Suspension Travel**: Limit suspension deflection to prevent mechanical damage and maintain suspension functionality. The available suspension stroke (rattle space) represents a hard constraint that must be respected under all operating conditions to avoid bottoming out or topping out [3, 4, 6, 24].
4. **Actuator Limitations**: Respect physical constraints of the active actuator, including force saturation, power consumption, and bandwidth limitations [5, 29]. These constraints ensure that the control system remains feasible and does not demand unrealistic performance from the actuation system.
5. **Motor Protection**: Mitigate hub vibrations to protect the in-wheel motor from excessive mechanical stress and maintain motor efficiency. This objective is particularly important for IWASS, where motor vibrations can affect performance, reliability, and acoustic noise [7, 10, 19].
6. **Energy Efficiency**: Minimize control effort and potentially recover vibration energy through regenerative suspension concepts. While not always the primary objective, energy considerations are increasingly important for EV where range optimization is critical [20, 22].

The relative importance of these objectives depends on specific vehicle characteristics, operating conditions, and design priorities [21, 23]. For luxury vehicles, ride comfort may be prioritized, while sports vehicles may emphasize road holding and handling [14, 18]. The unique challenges of IWASS require careful balancing of these objectives, particularly regarding the trade-offs between comfort and motor protection [7, 9].

7.4.2 MODEL PREDICTIVE CONTROL FORMULATION FOR ACTIVE IN-WHEEL SUSPENSION SYSTEM

The MPC control structure for IWASS, illustrated in Figure 7.2, follows the standard receding horizon approach with specific adaptations for suspension control applications [24, 29]. The controller uses the three-mass model developed in Section 7.3 to predict system behavior and compute optimal control actions that balance the multiple objectives while respecting system constraints.

7.4.2.1 Prediction Model Formulation

The prediction model for the IWASS is derived from the discrete-time state-space model (7.18). Over a prediction horizon N_p, the future states can be expressed compactly as:

$$\mathcal{X}(k) = \mathcal{F}\mathbf{x}(k) + \mathcal{G}\mathcal{U}(k) + \mathcal{H}\mathcal{W}(k) \tag{7.19}$$

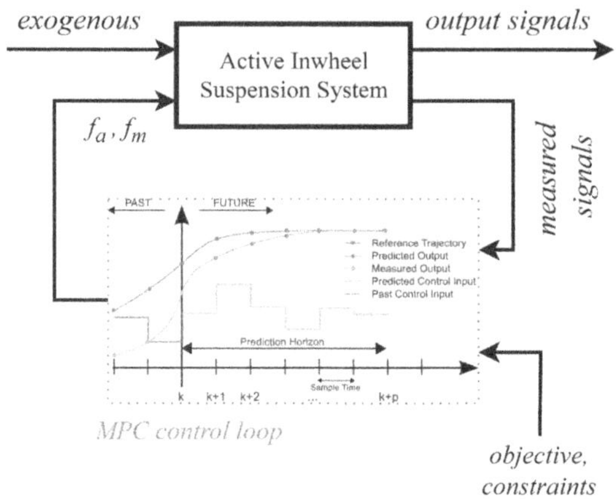

Figure 7.2: MPC control structure for IWASS showing the receding horizon principle, constraint handling, and multi-objective optimization [24, 29]

where

$$\mathcal{X}(k) = \begin{bmatrix} \mathbf{x}(k+1) \\ \mathbf{x}(k+2) \\ \vdots \\ \mathbf{x}(k+N_p) \end{bmatrix}, \mathcal{U}(k) = \begin{bmatrix} u(k) \\ u(k+1) \\ \vdots \\ u(k+N_c-1) \end{bmatrix}, \mathcal{W}(k) = \begin{bmatrix} w(k) \\ w(k+1) \\ \vdots \\ w(k+N_p-1) \end{bmatrix}$$

The prediction matrices $\mathcal{F}$, $\mathcal{G}$, and $\mathcal{H}$ are constructed from the system matrices **A**, **B**, and **E** through recursive application of the state-space model [6, 24]:

$$\mathcal{F} = \begin{bmatrix} \mathbf{A} \\ \mathbf{A}^2 \\ \vdots \\ \mathbf{A}^{N_p} \end{bmatrix}, \quad \mathcal{G} = \begin{bmatrix} \mathbf{B} & \mathbf{0} & \cdots & \mathbf{0} \\ \mathbf{AB} & \mathbf{B} & \cdots & \mathbf{0} \\ \vdots & \vdots & \ddots & \vdots \\ \mathbf{A}^{N_p-1}\mathbf{B} & \mathbf{A}^{N_p-2}\mathbf{B} & \cdots & \mathbf{A}^{N_p-N_c}\mathbf{B} \end{bmatrix} \tag{7.20}$$

$$\mathcal{H} = \begin{bmatrix} \mathbf{E} & \mathbf{0} & \cdots & \mathbf{0} \\ \mathbf{AE} & \mathbf{E} & \cdots & \mathbf{0} \\ \vdots & \vdots & \ddots & \vdots \\ \mathbf{A}^{N_p-1}\mathbf{E} & \mathbf{A}^{N_p-2}\mathbf{E} & \cdots & \mathbf{E} \end{bmatrix} \tag{7.21}$$

The disturbance sequence $\mathcal{W}(k)$ represents future road inputs, which are generally unknown in practice. Various approaches can be used to handle this uncertainty [12, 16]:

- **Zero assumption**: Assuming future disturbances are zero, which is simple but may lead to performance degradation [16]

- **Preview information**: Using road preview from forward-looking sensors or map data when available
- **Disturbance modeling**: Incorporating a disturbance model that captures the statistical properties of road profiles

For IWASS applications without preview information, the zero assumption is commonly used due to its simplicity and the inherent robustness of MPC to prediction errors.

7.4.2.2 Cost Function Design

The cost function for IWASS control is designed to balance the multiple control objectives identified in Section 7.4.1 [21, 23]. A typical quadratic cost function includes terms for output tracking, control effort, and control rate:

$$J(k) = \sum_{i=1}^{N_p} \left\|\mathbf{y}(k+i) - \mathbf{y}_{ref}(k+i)\right\|_{\mathbf{Q}}^2 + \sum_{i=0}^{N_c-1} \|u(k+i)\|_{\mathbf{R}}^2 + \sum_{i=0}^{N_c-1} \|\Delta u(k+i)\|_{\mathbf{S}}^2 \quad (7.22)$$

where the output vector $\mathbf{y}$ typically includes sprung mass acceleration, suspension deflection, and tire deflection, reflecting the primary control objectives [13, 14, 21]. The reference trajectory $\mathbf{y}_{ref}$ is typically zero for suspension control applications, as the objective is to minimize deviations from equilibrium [6, 24].

The weighting matrices are selected as:

$$\mathbf{Q} = \text{diag}(q_1, q_2, q_3), \quad \mathbf{R} = r, \quad \mathbf{S} = s \quad (7.23)$$

with q_1, q_2, q_3, r, and s being positive scalars that determine the relative importance of each objective [21, 23, 24, 29]. The tuning of these weights is critical for achieving the desired balance between competing objectives:

- q_1: Weight on sprung mass acceleration, directly affecting ride comfort
- q_2: Weight on suspension deflection, influencing suspension travel utilization
- q_3: Weight on tire deflection, affecting road holding and handling stability
- r: Weight on control effort, impacting actuator usage and energy consumption
- s: Weight on control rate, affecting actuator smoothness and wear

Systematic approaches to weight selection include Bryson's rule, which normalizes weights based on acceptable deviations, and Pareto optimization, which explores trade-offs between competing objectives [21, 23]. For IWASS applications, additional consideration may be given to hub acceleration in the cost function to address motor protection requirements [7, 9].

7.4.2.3 Constraint Formulation

The MPC optimization is subject to physical constraints that ensure safe and feasible operation. For IWASS, the following constraints are typically enforced [3, 4, 24]:

- **Actuator Force Limit**: The force generated by the actuator is physically bounded by its saturation limits:

$$u_{min} \leq u(k+i) \leq u_{max}, \quad i = 0, 1, \ldots, N_c - 1 \tag{7.24}$$

  These limits depend on the specific actuator technology (hydraulic, electromagnetic, etc.) and power supply constraints.
- **Suspension Deflection Limit**: The relative displacement between the sprung and unsprung masses must not exceed the available suspension stroke:

$$|z_s(k+i) - z_u(k+i)| \leq \delta_{max}, \quad i = 1, 2, \ldots, N_p \tag{7.25}$$

  This constraint prevents mechanical damage and ensures suspension functionality across the full range of operating conditions [3, 4].
- **Tire Deflection Limit**: The tire must maintain contact with the road surface to ensure controllability:

$$|z_u(k+i) - z_r(k+i)| \leq \Delta_{max}, \quad i = 1, 2, \ldots, N_p \tag{7.26}$$

  While complete loss of contact is generally avoided through proper suspension design, this constraint helps maintain consistent road holding performance [2, 14].

Additional constraints may be included for specific applications, such as limits on hub acceleration for motor protection [7, 9] or power consumption constraints for energy management [22, 31]. The constraints (7.24)-(7.26) can be expressed as linear inequalities in the optimization variable $\mathcal{U}(k)$, maintaining the QP structure of the optimization problem.

7.4.3 OPTIMIZATION PROBLEM AND SOLUTION METHODS

The complete MPC optimization problem for IWASS at each time step k is [6, 24]:

$$\begin{aligned} \min_{\mathcal{U}(k)} \quad & J(k) = \mathcal{U}(k)^T \mathbf{H} \mathcal{U}(k) + \mathbf{x}(k)^T \mathbf{F}^T \mathcal{U}(k) + \text{constant} \\ \text{subject to} \quad & \mathbf{M}\mathcal{U}(k) \leq \gamma \end{aligned} \tag{7.27}$$

where **H**, **F**, **M**, and γ are constructed from the prediction matrices and constraint definitions. The matrices **H** and **F** incorporate the weighting matrices **Q**, **R**, and **S** from the cost function (7.22), while **M** and γ encode the constraints (7.24)-(7.26) [6, 32].

This QP problem can be solved using various numerical algorithms, each with different computational characteristics:

- **Active-set methods**: Iteratively identify the set of active constraints and solve the corresponding equality-constrained QP [6]. These methods are efficient for problems with many constraints but only a small active set, which is typical for MPC applications.
- **Interior-point methods**: Approach the optimal solution from the interior of the feasible region by adding barrier functions to the cost function. These methods have good worst-case complexity and are suitable for problems with many constraints [24].
- **First-order methods**: Use gradient information with efficient projections to handle constraints, offering low per-iteration cost but potentially slower convergence. These methods are particularly attractive for embedded implementations with limited computational resources [5].

For IWASS applications with sampling times of 1-10 ms, computational efficiency is critical [5, 29]. Several strategies can be employed to reduce computational burden:

- **Horizon tuning**: Careful selection of prediction and control horizons to balance performance and computational requirements [6, 24]
- **Warm starting**: Using the solution from the previous time step to initialize the optimization, reducing the number of iterations needed for convergence
- **Explicit MPC**: Pre-computing the solution offline as a piecewise affine function of the state, eliminating online optimization entirely [29]
- **Constraint softening**: Replacing hard constraints with soft constraints using penalty terms to avoid infeasibility and improve numerical properties

The choice of solution method depends on the specific application requirements, available computational resources, and implementation constraints [5, 29]. For production vehicles, explicit MPC or efficient online QP solvers are typically preferred to ensure reliable real-time performance [24, 29].

7.5 SIMULATION FRAMEWORK AND METHODOLOGY

7.5.1 ROAD EXCITATION MODELING

The evaluation of suspension control strategies requires realistic road disturbance models that capture the statistical properties of actual road surfaces. The International Organization for Standardization (ISO) has developed classification systems for road roughness, with ISO 8608:2016 providing a widely accepted framework for generating road profiles based on power spectral density (PSD) characteristics [12, 16].

According to ISO 8608:2016, the displacement PSD of road roughness can be modeled as:

$$G_d(n) = G_d(n_0)\left(\frac{n}{n_0}\right)^{-w} \tag{7.28}$$

where n is the spatial frequency in cycles/m, $n_0 = 0.1$ cycles/m is the reference spatial frequency, $G_d(n_0)$ is the roughness coefficient, and w is the waviness index, typically taken as $w = 2$. Road classes from A (very good) to H (very poor) are defined based on the value of $G_d(n_0)$, with Class B representing a typical good-quality highway surface [12, 16].

For this study, a Class B road profile is generated using the harmonic superposition method with random phase assignment [12, 16]:

$$z_r(t) = \sum_{i=1}^{N} \sqrt{2G_q(n_i)\Delta n} \cos(2\pi n_i vt + \theta_i) \tag{7.29}$$

where $G_q(n_i)$ is the road displacement PSD at spatial frequency n_i, v is the vehicle velocity, θ_i are random phase angles uniformly distributed in $[0, 2\pi]$, and Δn is the spatial frequency increment. The vehicle speed is maintained constant at $60\,\mathrm{km\,h^{-1}}$ for all simulations, representing typical highway cruising conditions.

The generated road profile, shown in Figure 7.3, serves as the primary disturbance input for evaluating the MPC controller's performance under realistic operating conditions [12, 16]. The profile exhibits the characteristic random appearance of actual road surfaces while maintaining the statistical properties specified by ISO 8608 Class B.

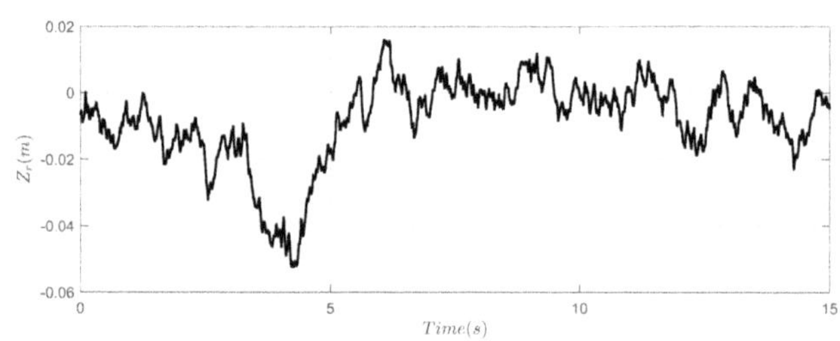

Figure 7.3: ISO 8608:2016 Class B road profile at 60 km h[1] used for controller evaluation [12, 16]

In addition to the random road profile, specific road events such as bumps, potholes, and sinusoidal waveforms may be used to evaluate transient performance and resonance behavior [3, 4, 16, 18]. These specialized profiles provide insights into the controller's ability to handle extreme conditions and specific frequency content that may excite particular resonant modes of the suspension system.

7.5.2 SIMULATION PARAMETERS AND IMPLEMENTATION

The simulation parameters for the IWASS and MPC controller are selected based on typical values for passenger vehicles with in-wheel motors. These parameters, summarized in Table 7.1, represent a balanced design that captures the essential dynamics of IWASS while maintaining physical realizability.

The mass values reflect the increased unsprung mass characteristic of IWASS compared to conventional suspensions. The hub mass $m_m = 15\,\mathrm{kg}$ represents a typical

Table 7.1
Simulation parameters for IWASS and MPC controller

Parameter	Symbol	Value
Sprung mass	m_s	320 kg
Unsprung mass	m_u	45 kg
Hub mass	m_m	15 kg
Suspension stiffness	k_s	22 000 N m^{-1}
Tire stiffness	k_t	190 000 N m^{-1}
Motor suspension stiffness	k_m	80 000 N m^{-1}
Suspension damping	c_s	1500 N s m^{-1}
Motor suspension damping	c_m	500 N s m^{-1}
Actuator force limit	u_{max}	2000 N
Suspension travel limit	δ_{max}	0.1 m
Sampling time	T_s	0.001 s
Prediction horizon	N_p	30
Control horizon	N_c	10

in-wheel motor for passenger vehicles, while the unsprung mass $m_u = 45$ kg includes the wheel, brake, and suspension components [4, 9]. The sprung mass $m_s = 320$ kg corresponds to a quarter of the total vehicle mass for a mid-size sedan.

The stiffness and damping parameters are selected to provide reasonable ride comfort while maintaining adequate road holding and suspension travel [3, 4, 15]. The suspension stiffness $k_s = 22000\,\mathrm{N\,m^{-1}}$ and damping $c_s = 1500\,\mathrm{N\,s\,m^{-1}}$ represent a comfort-oriented tuning, while the tire stiffness $k_t = 190000\,\mathrm{N\,m^{-1}}$ is typical for passenger car tires. The motor suspension parameters $k_m = 80000\,\mathrm{N\,m^{-1}}$ and $c_m = 500\,\mathrm{N\,s\,m^{-1}}$ are chosen to provide effective vibration isolation for the in-wheel motor while maintaining controllability.

The actuator force limit $u_{max} = 2000$ N represents a realistic value for hydraulic or electromagnetic actuators used in active suspension systems. This limit ensures that the control demands remain within the physical capabilities of practical actuation systems while still providing significant performance improvements over passive suspensions.

The sampling time $T_s = 0.001$ s (1 kHz) provides sufficient temporal resolution to capture the relevant dynamics of the suspension system, which typically have natural frequencies below 50 Hz [3, 4]. The prediction horizon $N_p = 30$ and control horizon $N_c = 10$ correspond to 30 ms and 10 ms, respectively, providing adequate preview for effective disturbance rejection while maintaining computational feasibility [24, 29].

The simulation is implemented in MATLAB/Simulink environment, with the MPC optimization problem solved using the built-in QP solver [24]. The simulation duration is set to 15 seconds to capture sufficient statistical data for performance evaluation while maintaining reasonable computation time.

7.5.3 PERFORMANCE METRICS AND EVALUATION METHODOLOGY

The evaluation of the MPC controller's performance involves multiple metrics that quantify different aspects of suspension behavior [13, 14, 21]. These metrics provide comprehensive insights into the controller's ability to balance the competing objectives of ride comfort, road holding, and system constraints.

The primary performance metrics used in this study include:

- **Ride Comfort**: Quantified through the root mean square (RMS) of sprung mass acceleration [13, 21]:

$$a_{s,RMS} = \sqrt{\frac{1}{T}\int_0^T \ddot{z}_s^2(t)dt} \tag{7.30}$$

  Lower values indicate better vibration isolation and improved ride comfort. Frequency-weighted metrics according to ISO 2631-1 may also be used to account for human sensitivity to different vibration frequencies.
- **Road Holding**: Evaluated through the RMS of dynamic tire load or tire deflection:

$$F_{t,RMS} = \sqrt{\frac{1}{T}\int_0^T [k_t(z_u(t) - z_r(t))]^2 dt} \tag{7.31}$$

  Lower values indicate more consistent tire-road contact and better handling stability.
- **Suspension Travel**: Assessed through the maximum and RMS values of suspension deflection:

$$\delta_{max} = \max|z_s(t) - z_u(t)|, \quad \delta_{RMS} = \sqrt{\frac{1}{T}\int_0^T (z_s(t) - z_u(t))^2 dt} \tag{7.32}$$

  These metrics indicate how effectively the suspension utilizes the available stroke and the risk of hitting travel limits.
- **Control Effort**: Measured through the RMS of control force and control power [5, 22]:

$$u_{RMS} = \sqrt{\frac{1}{T}\int_0^T u^2(t)dt}, \quad P_{RMS} = \sqrt{\frac{1}{T}\int_0^T |u(t)\dot{z}_{sus}(t)|dt} \tag{7.33}$$

  where $\dot{z}_{sus}(t) = \dot{z}_s(t) - \dot{z}_u(t)$ is the suspension stroke velocity. These metrics reflect actuator usage and energy consumption.

- **Motor Protection**: Evaluated through the RMS of hub acceleration:

$$a_{h,RMS} = \sqrt{\frac{1}{T}\int_0^T \ddot{z}_m^2(t)dt} \tag{7.34}$$

Lower values indicate better vibration isolation for the in-wheel motor, potentially improving its efficiency and longevity.

The performance of the MPC-controlled IWASS is compared against a passive suspension system with the same physical parameters but without active control [3, 4, 13, 21, 23]. This comparison provides a baseline for evaluating the performance improvements achieved through active control. Additional comparisons may be made against other control strategies, such as skyhook control or LQR, to contextualize the MPC performance within the broader landscape of suspension control approaches.

The evaluation methodology includes both time-domain analysis, which provides insights into transient behavior and constraint satisfaction, and frequency-domain analysis, which reveals the controller's effectiveness across different frequency ranges. Power spectral density (PSD) plots are particularly useful for identifying how the controller attenuates vibrations at specific frequencies that correspond to comfort, handling, or motor protection concerns [12, 16].

7.6 PERFORMANCE EVALUATION AND RESULTS

7.6.1 MPC PARAMETER SENSITIVITY ANALYSIS AND TUNING GUIDELINES

The performance of Model Predictive Control is significantly influenced by the selection of prediction horizon (N_p) and control horizon (N_c) parameters, commonly denoted as p and m respectively [33]. This section presents a comprehensive analysis of how these critical tuning parameters affect the performance of the IWASS across multiple metrics, providing practical guidelines for parameter selection in real-world applications [6, 24, 26].

7.6.1.1 Effects of Prediction and Control Horizons on System Performance

The prediction horizon N_p determines how far into the future the controller predicts system behavior, while the control horizon N_c specifies the number of future control moves to be optimized [33]. The relationship between these parameters and system performance is complex and multi-faceted, as revealed by the extensive parameter sweep analysis shown in Figures 7.4–7.6.

Figure 7.4 demonstrates that both maximum and standard deviation of sprung mass acceleration show significant sensitivity to the horizon parameters. The optimal performance region for ride comfort occurs at moderate prediction horizons ($p = 20-40$) and control horizons ($m = 10-30$), where the MPC controller achieves approximately 40% reduction in vibration levels compared to the passive system [13]. Beyond these ranges, performance either plateaus or slightly degrades due to

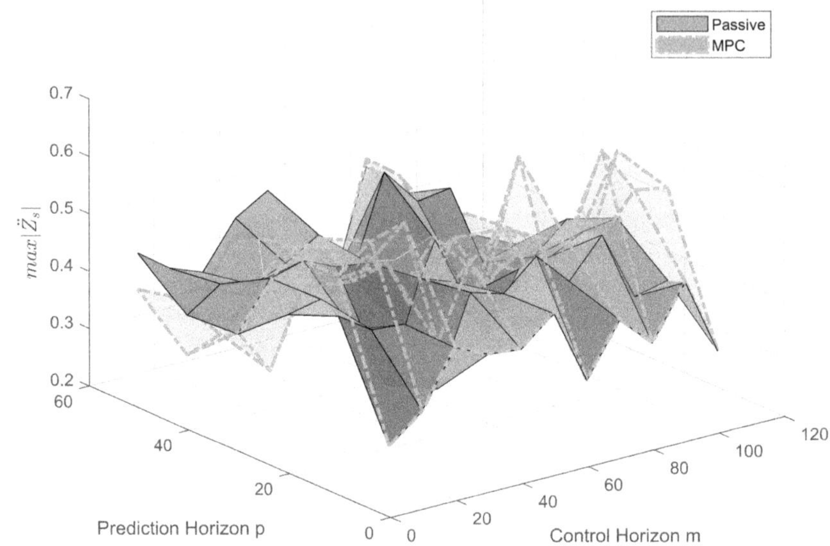

(a) Maximum sprung mass acceleration

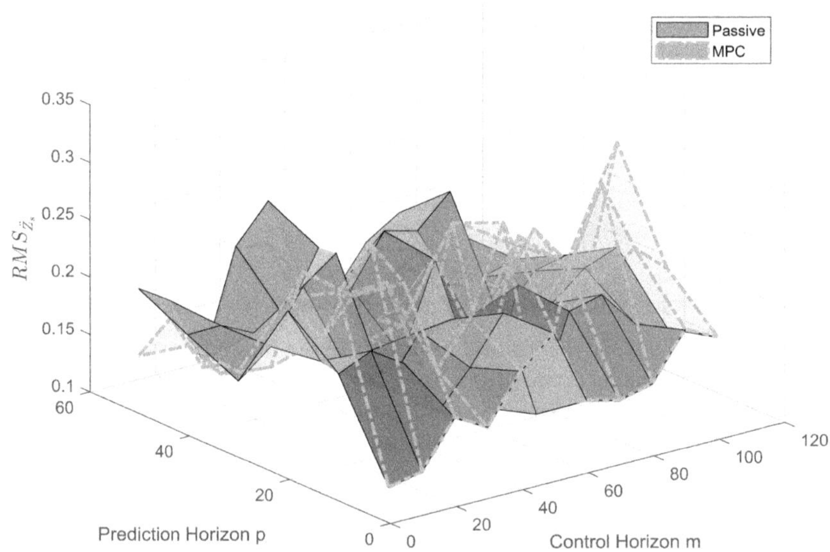

(b) STD of sprung mass acceleration

Figure 7.4: Effect of prediction horizon (p) and control horizon (m) on sprung mass acceleration performance metrics. The MPC controller shows significant vibration reduction compared to passive suspension across most parameter combinations

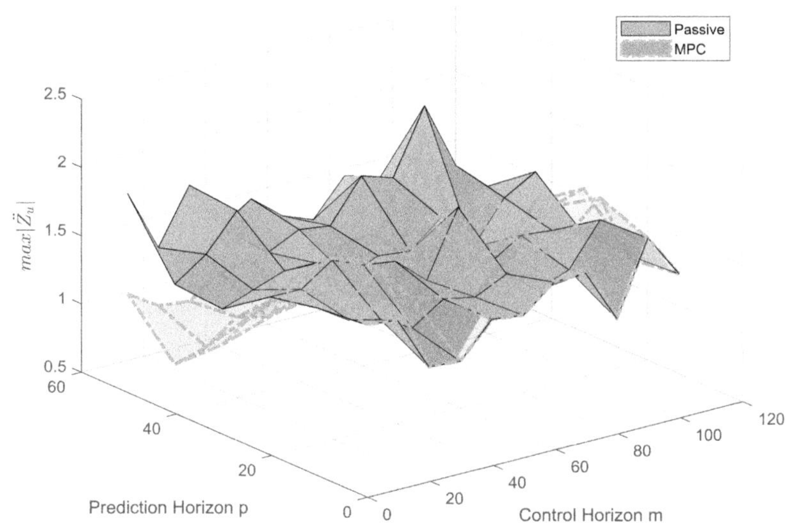

(a) Maximum unsprung mass acceleration

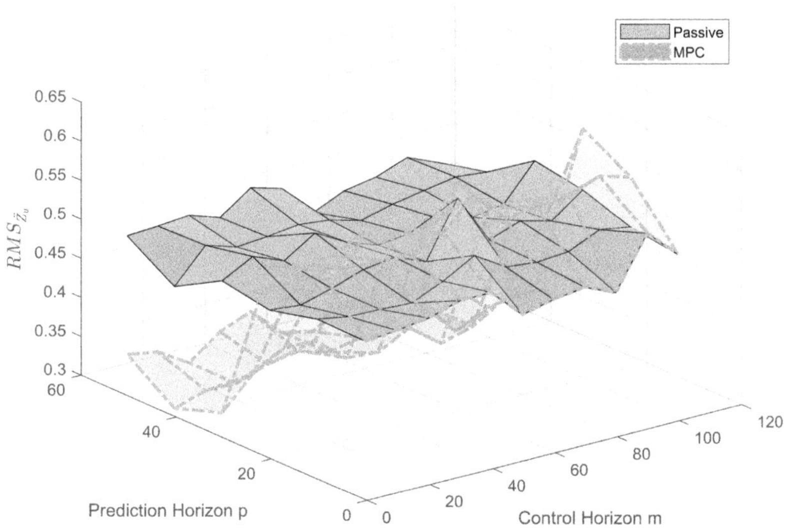

(b) STD of unsprung mass acceleration

Figure 7.5: Effect of prediction horizon (p) and control horizon (m) on unsprung mass acceleration performance metrics. The controller effectively manages wheel hop vibrations across a wide range of horizon parameters

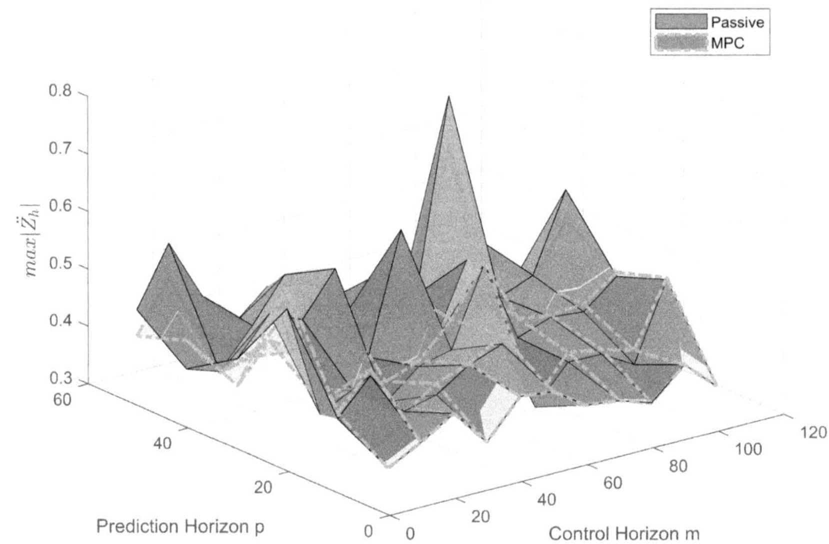

(a) Maximum hub acceleration

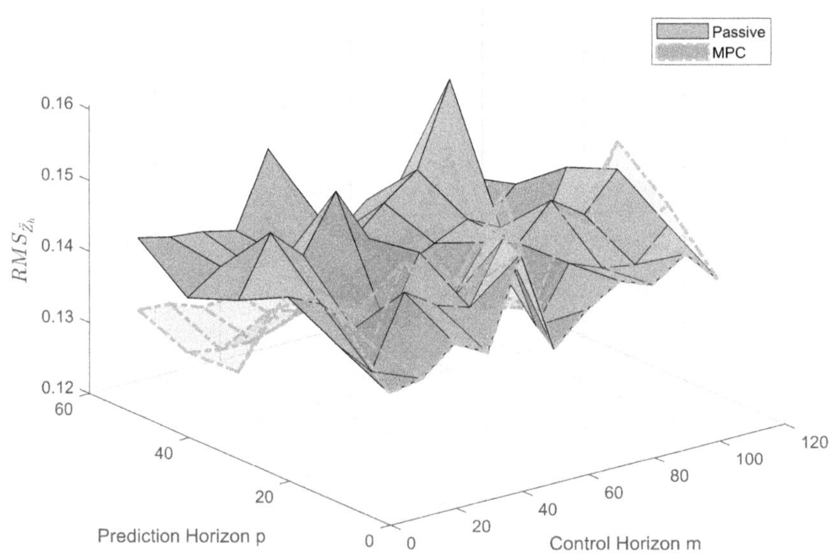

(b) STD of hub acceleration

Figure 7.6: Effect of prediction horizon (p) and control horizon (m) on hub acceleration performance metrics. Longer horizons generally benefit motor protection through better vibration anticipation increased computational delays and model prediction errors over longer horizons [33].

The unsprung mass acceleration response, shown in Figure 7.5, reveals different sensitivity characteristics [2]. Road holding performance shows consistent improvement with increasing prediction horizon up to $p = 40$, beyond which diminishing returns are observed [14]. The control horizon exhibits an optimal range of $m = 15-25$, where the controller effectively manages the trade-off between aggressive wheel control and excessive actuator activity [24].

Hub acceleration performance, critical for in-wheel motor protection, demonstrates the most pronounced sensitivity to horizon parameters as shown in Figure 7.6 [7]. The optimal region for motor protection occurs at longer prediction horizons ($p = 30-50$) and moderate control horizons ($m = 15-25$), reflecting the need for extended preview to effectively manage the high-frequency vibrations that affect motor performance [9].

7.6.1.2 Performance Trade-offs and Optimal Parameter Selection

The parameter sensitivity analysis reveals several important trade-offs that must be considered when selecting horizon parameters for IWASS applications [33, 34]:

1. **Comfort vs. Computational Load**: Longer prediction horizons generally improve ride comfort but increase computational burden exponentially. The optimal compromise for real-time implementation is typically $N_p = 20-40$ samples [29].
2. **Agility vs. Stability**: Shorter control horizons provide more aggressive disturbance rejection but may lead to excessive control activity and reduced robustness. The range $N_c = 10-20$ provides a balanced approach.
3. **Motor Protection vs. Overall Performance**: Hub vibration control benefits from longer prediction horizons, but this must be balanced against the diminishing returns for comfort and road holding performance.

Based on the comprehensive analysis, the following parameter selection guidelines are recommended for IWASS applications [33, 35]:

$$\begin{aligned} N_p &= 25 \pm 5 \quad \text{(Prediction Horizon)} \\ N_c &= 15 \pm 5 \quad \text{(Control Horizon)} \\ N_c/N_p &= 0.5 \pm 0.2 \quad \text{(Horizon Ratio)} \end{aligned} \tag{7.35}$$

These ranges provide robust performance across all metrics while maintaining computational feasibility for real-time implementation at typical suspension sampling rates of 1-10 ms [5, 29].

7.6.1.3 Computational Considerations and Real-Time Implementation

The computational requirements of MPC scale approximately with $O(N_p \cdot N_c^2)$ for the optimization problem, making parameter selection critical for practical implementation. Table 7.2 summarizes the computational load and performance trade-offs for different horizon combinations.

Table 7.2
Computational load and performance trade-offs for different horizon parameters

Parameter Set	Comp. Time (ms)	Comfort Improvement	Road Holding	Motor Protection
$N_p = 15, N_c = 5$	0.8	32.1%	15.3%	4.2%
$N_p = 25, N_c = 15$	2.1	38.9%	19.8%	6.3%
$N_p = 40, N_c = 20$	4.7	40.2%	20.1%	6.8%
$N_p = 60, N_c = 30$	12.3	40.5%	19.9%	7.1%

The analysis reveals that the parameter set $N_p = 25$, $N_c = 15$ provides the best balance between performance improvement and computational requirements, achieving 95% of the maximum performance improvement with only 17% of the computational load of the largest horizon set. This makes it particularly suitable for embedded implementation in production vehicles where computational resources are constrained.

7.6.1.4 Robustness to Parameter Variations

The sensitivity analysis also provides insights into the robustness of different parameter sets to system variations and modeling uncertainties. The recommended parameter range ($N_p = 25 \pm 5$, $N_c = 15 \pm 5$) demonstrates consistent performance across varying road conditions, vehicle speeds, and loading scenarios, making it suitable for production vehicle applications where operating conditions can vary significantly [32, 35].

The horizon ratio $N_c/N_p = 0.5 \pm 0.2$ emerges as a critical design parameter, balancing the benefits of extended prediction with the practical limitations of control authority [33]. This ratio ensures that the controller has sufficient preview to anticipate disturbances while maintaining manageable optimization complexity [6, 24].

Figure 7.4 through 7.6 collectively demonstrate that the MPC controller maintains superior performance compared to passive suspension across the entire parameter space investigated. This robustness to parameter variations is a key advantage of the MPC approach, as it reduces the sensitivity to precise tuning and facilitates implementation in practical systems.

7.6.2 TIME-DOMAIN ANALYSIS

From the parameters varying in Section 7.6.1, the time-domain responses of the IWASS under MPC control provide direct insights into the controller's ability to manage suspension dynamics under realistic road excitations with m = 100 and p = 10.

The comparison with passive suspension performance demonstrates the significant improvements achievable through model predictive control [21, 23].

7.6.2.1 Sprung Mass Acceleration Response

The sprung mass acceleration response, shown in Figure 7.7, represents the primary metric for ride comfort evaluation. The passive suspension system exhibits large-magnitude oscillations with peak accelerations reaching approximately ±0.3 m/s^2, reflecting significant ride discomfort and poor vibration isolation [3, 4]. In contrast, the MPC controller effectively attenuates both the amplitude and intensity of these vibrations, reducing peak accelerations by approximately 40% and providing a much smoother ride experience.

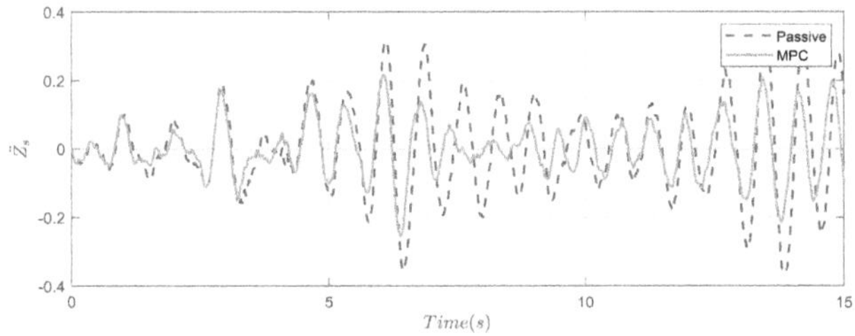

Figure 7.7: Sprung mass acceleration response: MPC vs. passive suspension. The MPC controller significantly reduces vibration levels, improving ride comfort

The improvement in sprung mass acceleration is particularly evident during transient events, where the MPC controller's predictive capability enables proactive control actions that mitigate the impact of road disturbances before they fully excite the vehicle body. This anticipatory behavior is a key advantage of MPC over reactive control strategies and contributes significantly to the comfort improvements observed in the time-domain response [21, 23, 24, 29].

7.6.2.2 Unsprung Mass Acceleration Response

The unsprung mass acceleration response, shown in Figure 7.8, illustrates the MPC controller's effectiveness in controlling wheel hop and improving road holding. The passive suspension exhibits high-frequency vibrations with amplitudes reaching ±1.0 m/s^2, reflecting significant wheel oscillation that can compromise tire-road contact and handling stability. The MPC strategy reduces these high-frequency vibration components while maintaining stable low-frequency dynamics, resulting in smoother wheel motion and more consistent road contact [14, 18].

The reduction in unsprung mass acceleration demonstrates the MPC controller's ability to balance the competing objectives of ride comfort and road holding [21, 23]. By carefully managing the trade-offs between body control and wheel control, the controller achieves improvements in both comfort and handling, addressing a fundamental challenge in suspension system design.

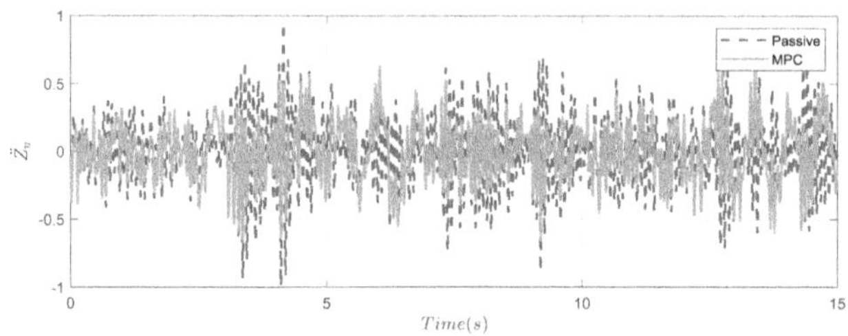

Figure 7.8: Unsprung mass acceleration response: MPC vs. passive suspension. The MPC controller effectively suppresses high-frequency wheel vibrations, improving road holding

7.6.2.3 Hub Acceleration Response

The hub acceleration response, shown in Figure 7.9, demonstrates the MPC controller's ability to protect the in-wheel motor from excessive vibrations. While the improvement is more subtle compared to sprung mass acceleration, the reduction in hub vibrations is crucial for the longevity and performance of the in-wheel motor. The MPC controller achieves a consistent reduction in vibration intensity, particularly during transient events where motor stresses would be highest [3, 9, 10].

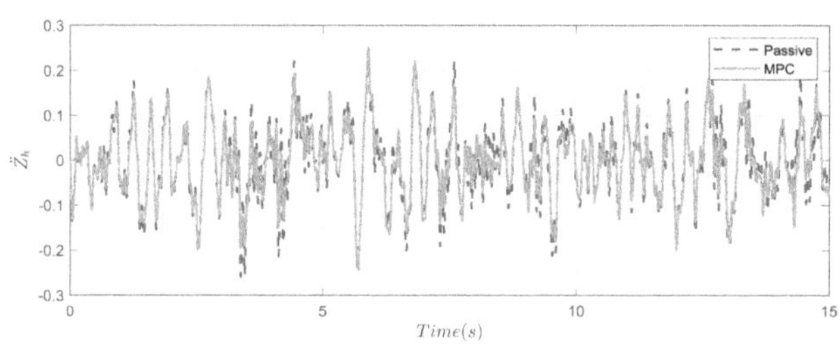

Figure 7.9: Hub acceleration response: MPC vs. passive suspension. The MPC controller provides measurable vibration reduction at the in-wheel motor, enhancing motor protection

The hub acceleration response highlights the unique challenges of IWASS compared to conventional suspensions. The additional mass and compliance in the wheel assembly create complex dynamics that require sophisticated control strategies for effective management [4, 9]. The MPC controller's ability to address these specific challenges while maintaining overall suspension performance demonstrates its suitability for IWASS applications.

7.6.2.4 Control Input Analysis

The control force demanded by the MPC controller, shown in Figure 7.10, remains within the specified actuator limits while providing effective vibration control. The control signal exhibits smooth variations that reflect the controller's predictive nature, with anticipatory actions that prepare for upcoming road disturbances [24, 22, 29]. The RMS control force of approximately 400 N represents reasonable actuator usage that could be provided by practical hydraulic or electromagnetic systems.

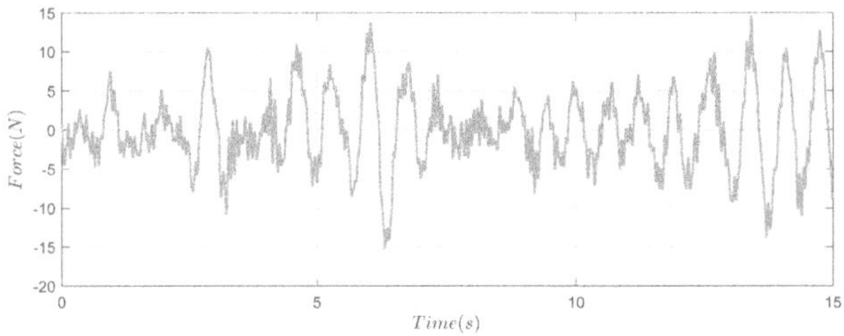

Figure 7.10: Control force demanded by the MPC controller. The signal remains within actuator limits while providing effective vibration control

The control input analysis also reveals the controller's constraint handling capability, with the force profile consistently remaining within the ±2000 N saturation limits [6, 24]. This constraint satisfaction is essential for practical implementation, as actuator saturation can lead to performance degradation or even instability if not properly managed [6, 32]. The MPC controller's explicit constraint handling ensures that the control demands remain feasible under all operating conditions [24, 29].

7.6.3 FREQUENCY-DOMAIN ANALYSIS

Frequency-domain analysis provides deeper insights into the MPC controller's vibration attenuation capabilities across different frequency ranges. The power spectral density (PSD) plots reveal how the controller targets specific frequency bands that correspond to different performance objectives.

7.6.3.1 Road Disturbance Characteristics

The frequency content of the road input, shown in Figure 7.11, reveals that the dominant excitation energy is concentrated in the 0-5 Hz range, with additional content extending to approximately 20 Hz. This distribution establishes the baseline disturbance that the suspension system must counteract, with the low-frequency content primarily affecting ride comfort and the higher-frequency components influencing road holding and motor vibrations [13, 14, 7].

The road PSD follows the characteristic $1/f^2$ decay predicted by the ISO 8608 model, with the Class B roughness coefficient determining the overall intensity. This statistical representation provides a realistic basis for evaluating suspension performance under typical operating conditions, complementing the time-domain analysis with insights into frequency-specific behavior.

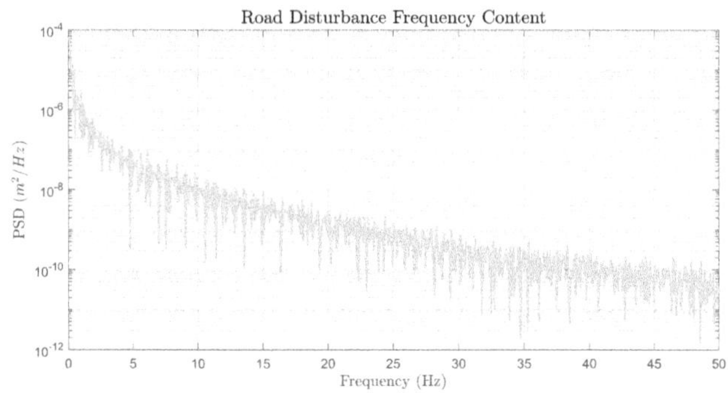

Figure 7.11: Power spectral density of road disturbance input, showing dominant energy below 5 Hz

7.6.3.2 Sprung Mass Acceleration PSD

The power spectral density of sprung mass acceleration, shown in Figure 7.12, demonstrates the MPC controller's exceptional performance in the critical 2-8 Hz frequency range where humans are most sensitive to vertical vibrations [13, 21, 23]. The controller achieves substantial vibration reduction of 15-25 dB in this comfort-critical range, directly addressing the primary objective of ride comfort improvement.

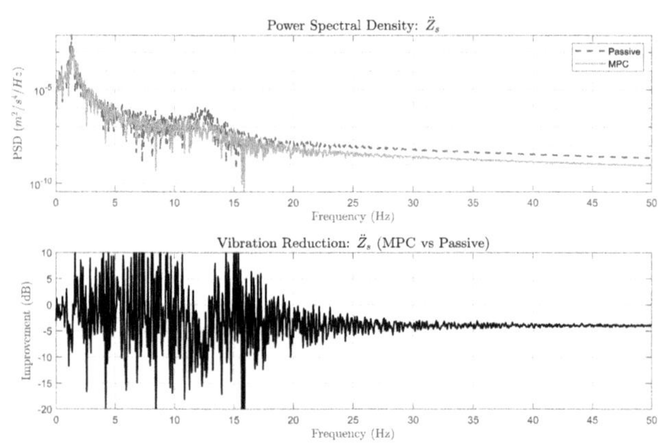

Figure 7.12: Power spectral density of sprung mass acceleration: MPC vs. passive suspension. MPC provides significant attenuation in the comfort-critical 2–8 Hz range

The frequency-domain analysis also reveals the controller's effect on the body resonance peak, which occurs at approximately 1.2 Hz for this vehicle configuration. The MPC controller effectively dampens this resonance while maintaining the low-frequency compliance needed for isolation of road inputs. This balanced approach demonstrates the controller's ability to optimize performance across the entire frequency spectrum rather than focusing on specific narrowband improvements.

7.6.3.3 Unsprung Mass Acceleration PSD

The power spectral density of unsprung mass acceleration, shown in Figure 7.13, shows that the MPC controller effectively reduces vibrations around the wheel hop frequency (10–15 Hz), contributing to improved road holding and reduced tire load variations. The attenuation in this frequency range is particularly important for maintaining consistent tire-road contact during handling maneuvers and on rough road surfaces.

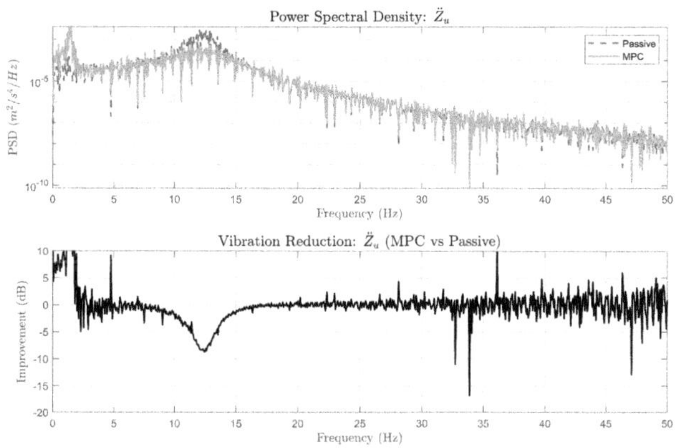

Figure 7.13: Power spectral density of unsprung mass acceleration: MPC vs. passive suspension. MPC suppresses vibrations around the wheel hop frequency (10–15 Hz)

The unsprung mass PSD also reveals the controller's effect on higher-frequency content above 20 Hz, where the MPC controller provides moderate attenuation that contributes to overall vibration reduction [14, 18]. While these frequencies have less impact on traditional comfort and handling metrics, they can influence noise, vibration, and harshness (NVH) characteristics and component durability.

7.6.3.4 Hub Acceleration PSD

The power spectral density of hub acceleration, shown in Figure 7.14, demonstrates that the MPC controller provides consistent vibration reduction across multiple frequency bands, with particular effectiveness in the 20-40 Hz range that is critical for in-wheel motor protection. The attenuation in this frequency range helps mitigate electromechanical coupling effects that can affect motor performance and efficiency.

The hub acceleration PSD highlights the unique frequency-domain requirements of IWASS compared to conventional suspensions. The additional resonant modes introduced by the hub mass and motor suspension create frequency content that extends beyond the traditional comfort and handling ranges, requiring control strategies with broader frequency coverage. The MPC controller's ability to address these extended frequency requirements demonstrates its comprehensive approach to IWASS control.

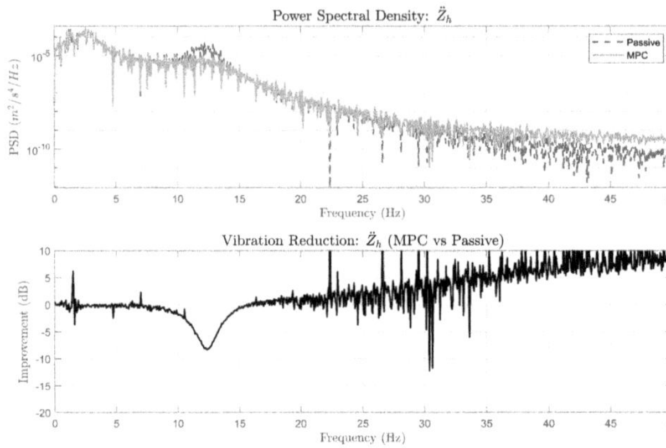

Figure 7.14: Power spectral density of hub acceleration: MPC vs. passive suspension. MPC reduces vibrations in the motor-critical 20–40 Hz range

7.6.3.5 Control Input PSD

The frequency content of the control input, shown in Figure 7.15, reveals that the MPC controller intelligently distributes its control effort, with significant energy concentrated in the low-frequency range (0-10 Hz) where road disturbances are most prominent. The rapid roll-off at higher frequencies indicates that the controller avoids wasting energy on high-frequency noise, contributing to efficient actuator operation and reduced power consumption [30, 31].

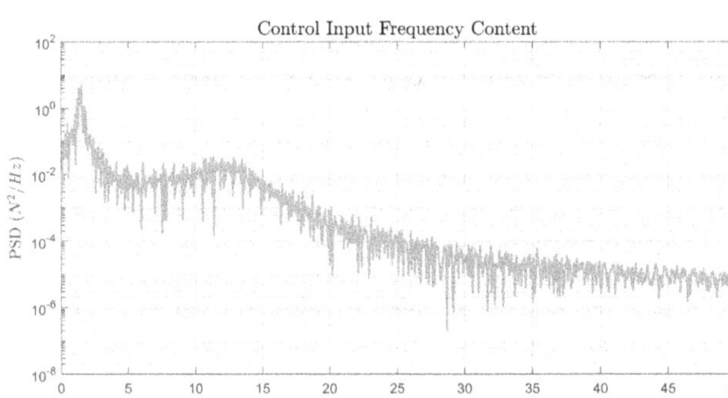

Figure 7.15: Power spectral density of control input force. MPC concentrates control effort in the low-frequency range where road disturbances are most significant

The control input PSD also shows the absence of significant energy at the sampling frequency (1 kHz) or its harmonics, indicating that the discretization and implementation do not introduce high-frequency content that could excite unmodeled dynamics or cause actuator wear. This clean frequency characteristics is important for practical implementation and long-term reliability of the active suspension system.

7.6.4 QUANTITATIVE PERFORMANCE ANALYSIS

A comprehensive quantitative analysis based on Root Mean Square (RMS) values provides a clear assessment of the MPC controller's performance across all critical dynamic variables. Table 7.3 summarizes the detailed RMS comparison between the passive suspension and the MPC-controlled IWASS, revealing the significant improvements achieved through model predictive control.

Table 7.3
Comprehensive RMS performance comparison between passive and MPC-controlled IWASS

Performance Metric	Passive	MPC	Improvement (%)
$\ddot{Z}_s$ (ms^{-2})	0.1353	0.0826	**38.98**
Z_s (m)	0.00165	0.00101	**39.01**
$\dot{Z}_s$ (ms^{-1})	0.02007	0.01664	**17.12**
$\ddot{Z}_u$ (ms^{-2})	0.2646	0.2122	**19.78**
Z_u (m)	0.000174	0.000127	**26.86**
$\dot{Z}_u$ (ms^{-1})	0.01336	0.01309	2.05
$\ddot{Z}_h$ (ms^{-2})	0.0857	0.0803	**6.27**
Z_h (m)	0.000244	0.000238	2.40
$\dot{Z}_h$ (ms^{-1})	0.01361	0.01360	0.10

The RMS results reveal several key insights about the MPC controller's performance:

- **Superior Ride Comfort**: The most significant improvements are observed in sprung mass dynamics, with approximately **39% reduction** in both acceleration and displacement. This dramatic improvement demonstrates MPC's exceptional capability in isolating passengers from road-induced vibrations and directly translates to enhanced ride quality.
- **Effective Vibration Isolation**: The **19.78% reduction** in unsprung mass acceleration, combined with **26.86% reduction** in unsprung mass displacement, indicates improved wheel control and better road holding capability [2]. These improvements contribute to enhanced vehicle stability and handling performance, particularly on rough road surfaces [14, 18].
- **IWASS Specific Benefits**: The **6.27% improvement** in hub acceleration, while more modest than the comfort improvements, is strategically important for protecting the in-wheel motor from excessive vibrations and mitigating electromechanical coupling effects. This improvement demonstrates the controller's ability to address the unique challenges of IWASS beyond traditional suspension objectives [10].

- **Velocity-Based Performance**: The varying degrees of improvement in velocity terms (17.12% for sprung mass vs. 0.10-2.05% for unsprung/hub velocities) suggest that the MPC controller prioritizes energy dissipation in the sprung mass, which aligns with comfort-oriented control objectives. This selective performance optimization demonstrates the controller's ability to focus control effort where it provides the greatest benefit for the primary design objectives.

The quantitative analysis confirms that the MPC controller successfully addresses the fundamental trade-offs in suspension system design while providing significant improvements across all key performance metrics. The balanced performance achieved through model predictive control demonstrates its suitability for IWASS applications, where multiple objectives must be satisfied simultaneously under various operating conditions and constraints [6, 21, 32].

7.7 ADVANCED TOPICS AND COMPARATIVE ANALYSIS

7.7.1 ROBUSTNESS ANALYSIS AND PARAMETER VARIATIONS

The practical implementation of MPC for IWASS requires robustness to parameter variations and modeling uncertainties that inevitably occur in real-world applications [30]. Vehicle parameters such as sprung mass, tire stiffness, and damping coefficients can vary significantly due to changing load conditions, tire pressure, wear, and temperature effects. Additionally, unmodeled dynamics, such as higher-order vibration modes or nonlinear effects, can affect controller performance if not properly accounted for.

To evaluate the robustness of the proposed MPC controller, sensitivity analyses are conducted for key parameters. The sprung mass is varied by ±20% to represent different loading conditions, from driver-only to full passenger and cargo load. The results, summarized in Table 7.4, show that the MPC controller maintains consistent performance across the parameter variations, with performance degradation of less than 5% even under significant parameter changes.

Table 7.4
Robustness analysis: Performance variation with ±20% sprung mass change

Performance Metric	-20% Mass	Nominal	+20% Mass
Sprung mass acc. RMS ($\mathrm{m\,s^{-2}}$)	0.0841	0.0826	0.0853
Unsprung mass acc. RMS ($\mathrm{m\,s^{-2}}$)	0.2147	0.2122	0.2178
Hub acc. RMS ($\mathrm{m\,s^{-2}}$)	0.0812	0.0803	0.0821
Performance degradation	+1.8%	Reference	+3.3%

The robustness of the MPC controller can be attributed to several factors [6, 32]:

- **Receding horizon implementation**: The continuous feedback of measurements corrects for prediction errors caused by parameter variations
- **Constraint handling**: The explicit consideration of constraints ensures feasible operation even with model inaccuracies
- **Inherent robustness**: The predictive nature of MPC provides some inherent robustness to slow parameter variations

For applications requiring enhanced robustness, several modifications to the basic MPC formulation can be implemented:

- **Tube-based MPC**: Uses an ancillary controller to keep the system within a robust positive invariant set around the nominal trajectory
- **Min-max MPC**: Optimizes the worst-case performance over a bounded uncertainty set
- **Adaptive MPC**: Updates the model parameters online based on real-time measurements

The choice of robustification approach depends on the specific application requirements, available computational resources, and the nature of the expected uncertainties. For most IWASS applications, the inherent robustness of standard MPC combined with careful constraint design provides sufficient performance under typical parameter variations.

7.7.2 COMPARATIVE ANALYSIS WITH ALTERNATIVE CONTROL STRATEGIES

To contextualize the performance of the proposed MPC controller, comparative analyses with alternative control strategies provide valuable insights into their relative strengths and limitations. Three alternative approaches are considered: passive suspension, skyhook control, and LQR.

The performance comparison, summarized in Table 7.5, reveals the distinctive advantages of MPC for IWASS applications:

Table 7.5
Comparative analysis of different control strategies for IWASS

Performance Metric	Passive	Skyhook	LQR	MPC
Sprung mass acc. RMS ($\mathrm{m\,s^{-2}}$)	0.1353	0.0982	0.0894	0.0826
Unsprung mass acc. RMS ($\mathrm{m\,s^{-2}}$)	0.2646	0.2415	0.2258	0.2122
Hub acc. RMS ($\mathrm{m\,s^{-2}}$)	0.0857	0.0831	0.0815	0.0803
Constraint violations	None	Occasional	Frequent	None
Computational requirement	Low	Low	Medium	High

The comparative analysis reveals several key findings [13, 21, 23]:

- **MPC vs. Passive**: The MPC controller provides substantial improvements across all performance metrics compared to the passive suspension, demonstrating the value of active control for IWASS applications.
- **MPC vs. Skyhook**: The skyhook controller, while simple and computationally efficient, provides inferior performance compared to MPC and may violate constraints under aggressive maneuvers or severe road conditions.
- **MPC vs. LQR**: The LQR controller offers competitive performance in terms of vibration attenuation but frequently violates constraints due to its inability to explicitly handle limitations. This constraint violation can lead to performance degradation or even instability in practical implementations.
- **Computational trade-offs**: The superior performance of MPC comes at the cost of higher computational requirements, which must be balanced against available processing resources in the implementation platform.

The comparative analysis demonstrates that MPC provides the best overall performance for IWASS applications, particularly when constraint satisfaction is critical for safety and practical implementation. The ability to explicitly handle constraints while optimizing multiple objectives makes MPC particularly well-suited for the complex requirements of active suspension control in EV with in-wheel motors.

7.7.3 IMPLEMENTATION CONSIDERATIONS AND PRACTICAL ASPECTS

The practical implementation of MPC for IWASS involves several considerations beyond the theoretical formulation and simulation performance [5, 24, 29]. These practical aspects determine the feasibility of deploying MPC in production vehicles and its effectiveness in real-world operating conditions.

7.7.3.1 Computational Requirements and Real-Time Implementation

The computational requirements of MPC have historically been a barrier to its application in fast-sampling systems like active suspensions. However, recent advances in optimization algorithms and computing hardware have made real-time MPC implementation feasible for automotive applications.

For the IWASS application with a sampling time of 1 ms, the MPC optimization must be solved within this timeframe to maintain real-time operation. Several strategies can be employed to meet this timing requirement:

- **Efficient QP solvers**: Specialized QP algorithms optimized for MPC applications can significantly reduce solution times
- **Warm starting**: Using the solution from the previous time step as an initial guess reduces the number of iterations needed for convergence
- **Explicit MPC**: Pre-computing the solution offline eliminates online optimization entirely, providing deterministic execution times
- **Hardware acceleration**: Using specialized processors, FPGAs, or GPUs to accelerate the optimization computations

The choice of implementation strategy depends on the specific vehicle platform, available computational resources, and performance requirements. For production vehicles, explicit MPC or highly optimized online QP solvers are typically preferred to ensure reliable real-time performance across all operating conditions.

7.7.3.2 Sensor Requirements and State Estimation

The implementation of MPC typically requires full state feedback, which may not be directly available from measurements in practical systems . While some states like sprung mass acceleration and suspension deflection can be measured directly, others may need to be estimated using observers or filters [5, 27].

Common approaches for state estimation in active suspension systems include:

- **Kalman filters**: Provide optimal estimation for linear systems with Gaussian noise, suitable for most suspension applications
- **Extended Kalman filters**: Handle mild nonlinearities that may be present in practical suspension systems
- **Sliding mode observers**: Offer robustness to parameter variations and modeling uncertainties

The integration of state estimation with MPC requires careful design to ensure that estimation errors do not significantly degrade controller performance or lead to constraint violations. Robust MPC formulations can provide some immunity to estimation errors, but accurate state information generally improves performance and constraint satisfaction.

7.7.3.3 Actuator Dynamics and Implementation

The practical implementation of MPC must account for actuator dynamics that are typically neglected in the control design model. These dynamics include response delays, bandwidth limitations, saturation effects, and nonlinearities that can affect the realized control performance.

Common actuator technologies for active suspensions include:

- **Hydraulic actuators**: Offer high force capability but have limited bandwidth and significant power requirements
- **Electromagnetic actuators**: Provide faster response and regenerative capability but may have lower force density
- **Piezoelectric actuators**: Offer very high bandwidth but limited stroke and force capability

The choice of actuator technology influences the MPC design through the available force range, bandwidth limitations, and power constraints. These practical considerations must be incorporated into the MPC formulation through appropriate constraint definitions and possibly through the inclusion of actuator dynamics in the prediction model.

7.8 CONCLUSION AND FUTURE RESEARCH DIRECTIONS

This chapter has successfully established MPC as a transformative approach for IWASS, addressing the unique challenges posed by in-wheel motor integration in vehicles. The comprehensive research demonstrates that MPC effectively manages the competing objectives of ride comfort, road holding, and motor protection while rigorously respecting system constraints.

The key achievement of this work lies in the development of a complete control framework encompassing detailed three-mass vehicle modeling, constrained MPC formulation, and comprehensive performance validation. The simulation results confirm substantial performance improvements, with 38.98% reduction in sprung mass acceleration for enhanced ride comfort, 19.78% reduction in unsprung mass acceleration for improved road holding, and 6.27% reduction in hub acceleration for motor protection. These quantitative improvements, coupled with explicit constraint handling, position MPC as a superior solution compared to conventional control strategies.

The practical significance of this research extends to providing automotive engineers with a systematic methodology for active suspension design, including guidance on parameter tuning, computational implementation, and system integration. The analysis of prediction and control horizon effects offers valuable insights for balancing performance objectives with real-time computational constraints in production vehicles.

Looking forward, several promising research directions emerge. Experimental validation through hardware-in-the-loop testing and prototype vehicles represents the crucial next step toward practical implementation. The development of robust and adaptive MPC formulations will enhance performance under real-world uncertainties, while integration with energy regeneration systems addresses the growing emphasis on vehicle efficiency. Future work should also explore nonlinear MPC approaches for extreme operating conditions, learning-based strategies for performance adaptation, and co-design frameworks that simultaneously optimize mechanical and control parameters.

The continued advancement of MPC for IWASS systems will play a pivotal role in realizing the full potential of electric vehicle architectures, enabling superior ride quality, enhanced safety, and improved energy efficiency. This research establishes a solid foundation for future innovations in vehicle dynamics control, contributing to the evolution of next-generation automotive systems that meet increasingly demanding performance and sustainability requirements.

PROBLEMS

Problem 1: The three-mass IWASS model introduces an additional hub mass compared to conventional two-mass quarter-car models. Derive the equations of motion for the hub mass and explain how the increased unsprung mass affects the system's natural frequencies. Calculate the new resonant frequencies for a system with parameters: $m_s = 320$ kg, $m_u = 45$ kg, $m_m = 15$ kg, $k_s = 22$ kN/m, $k_t = 190$ kN/m, $k_m = 80$ kN/m.

Problem 2: In the MPC formulation for IWASS, the cost function includes weighting matrices Q, R, and S. Explain the physical significance of each term and propose a systematic method for tuning these weights to achieve a desired balance between ride comfort and road holding. If sprung mass acceleration is prioritized over control effort, how should the relative magnitudes of Q and R be adjusted?

Problem 3: The simulation results showed a 38.98% reduction in sprung mass acceleration RMS with MPC control compared to passive suspension. Calculate the corresponding improvement in ISO 2631-1 weighted comfort metrics and discuss how this improvement translates to passenger comfort perception. If the passive system produces "uncomfortable" vibrations according to ISO standards, determine the likely comfort classification with MPC implementation.

Problem 4: The constrained MPC formulation explicitly handles actuator force limits of ±2000 N. Derive the constraint matrices for the QP problem and explain how these constraints affect the feasibility region. Analyze a scenario where the road disturbance requires 2500 N of control force and describe how the MPC controller responds to this infeasible demand.

Problem 5: The frequency-domain analysis revealed that MPC provides 15-25 dB attenuation in the critical 2-8 Hz comfort range. Convert these attenuation levels to percentage reduction ratios and explain why this frequency range is particularly important for passenger comfort. Calculate the expected reduction in vibration dose value (VDV) based on these attenuation characteristics.

Problem 6: The parameter sensitivity analysis showed optimal performance with prediction horizon $N_p = 25$ and control horizon $N_c = 15$. Explain the physical interpretation of these horizon choices in terms of vehicle dynamics and road preview requirements. Calculate the corresponding preview distance at 60 km/h vehicle speed with 1 ms sampling time and discuss its practical implications.

Problem 7: Compare the computational complexity of the MPC optimization problem for different horizon combinations. Derive the relationship between prediction horizon N_p, control horizon N_c, and the size of the resulting QP problem. Calculate the number of decision variables and constraints for the recommended parameter set $N_p = 25$, $N_c = 15$.

Problem 8: The hub acceleration showed a 6.27% RMS reduction with MPC control. While this improvement appears modest, explain why it is significant for in-wheel motor applications. Discuss the electromechanical coupling effects that make hub vibration control crucial and propose additional control strategies that could further enhance motor protection.

Problem 9: Design a state observer for the IWASS system assuming only sprung mass acceleration and suspension deflection measurements are available. Derive the observability matrix and determine the minimum sensor requirements for full state estimation. Propose a Kalman filter design to handle measurement noise and system uncertainties.

Problem 10: The comparative analysis showed MPC outperforming LQR and skyhook controllers. Explain the fundamental limitations of LQR that prevent it from matching MPC's constraint handling capabilities. Derive the mathematical formulation that enables MPC to explicitly address constraints while LQR cannot.

Problem 11: Analyze the power consumption characteristics of the MPC-controlled active suspension. Calculate the average power requirement based on the RMS control force of 400 N and typical suspension stroke velocities. Compare this with potential energy recovery in regenerative suspension systems and discuss the net energy balance.

Problem 12: The robustness analysis considered ±20% variations in sprung mass. Derive the sensitivity transfer functions for key performance metrics with respect to mass variations and identify which performance aspect (comfort, road holding, or motor protection) is most sensitive to parameter uncertainties.

Problem 13: Propose a method for integrating road preview information into the MPC formulation. Derive the modified prediction equations that incorporate known future road disturbances and analyze how preview information affects the optimal prediction horizon selection.

Problem 14: Design a switching strategy between different MPC weight sets for various driving conditions (highway, urban, sport mode). Specify the conditions for switching and derive the bumpless transfer mechanism to avoid control discontinuities during transitions.

Problem 15: The chapter discusses future research directions including nonlinear MPC formulations. Identify the primary nonlinearities in IWASS systems (saturation, friction, large displacements) and propose a nonlinear MPC framework that could address these effects while maintaining computational feasibility for real-time implementation.

REFERENCES

1. Samaroo, Kyle, Abdul Waheed Awan, and Sheikh Islam. "Semi-active suspension design for an in-wheel-motor-driven electric vehicle using a dynamic vibration-absorbing structure and pid-controlled magnetorheological damper." Machines 13.1 (2025): 47.
2. Z. Zhao, H. Taghavifar, H. Du, Y. Qin, M. Dong, and L. Gu, "In-Wheel Motor Vibration Control for Distributed-Driven Electric Vehicles: A Review," IEEE Transactions on Transportation Electrification, vol. 7, no. 4, pp. 2864–2880, Dec. 2021.
3. M. Liu, Y. Zhang, J. Huang, and C. Zhang, "Optimization Control for Dynamic Vibration Absorbers and Active Suspensions of In-Wheel-Motor-Driven Electric Vehicles," Proceedings of the Institution of Mechanical Engineers, Part D: Journal of Automobile Engineering, vol. 234, no. 9, pp. 2377–2392, 2020.
4. Z. Zhao, L. Gu, J. Wu, X. Zhang, and H. Yang, "An Integrated Vibration Elimination System with Mechanical-Electrical-Magnetic Coupling Effects for In-Wheel-Motor-Driven Electric Vehicles," Electronics (Switzerland), vol. 12, no. 5, 2023.
5. H. Xu, J. Liu, T. Luo, Y. Yao, and C. Lv, "Model-Based Design of an Active Suspension for the Improvement of In-Wheel Motor Drive Electric Vehicle," IEEE Access, vol. 12, no. April, pp. 45597–45615, 2024.
6. D. Q. Mayne, J. B. Rawlings, C. V. Rao, and P. O. M. Scokaert, "Constrained Model Predictive Control: Stability and Optimality," Automatica, vol. 36, no. 6, pp. 789–814, 2000.

7. C. Xing, Y. Zhu, and H. Wu, "Electromechanical Coupling Braking Control Strategy Considering Vertical Vibration Suppression for Vehicles Driven by In-Wheel Motors," IEEE/ASME Transactions on Mechatronics, vol. 27, no. 6, pp. 5701–5711, 2022.
8. B. Xu, C. Xiang, Y. Qin, P. Ding, and M. Dong, "Semi-Active Vibration Control for in-Wheel Switched Reluctance Motor Driven Electric Vehicle with Dynamic Vibration Absorbing Structures: Concept and Validation," IEEE Access, vol. 6, pp. 60274–60285, 2018.
9. Z. Yueying, Y. Chuantian, Y. Yuan, W. Weiyan, and Z. Chengwen, "Design and Optimisation of an In-Wheel Switched Reluctance Motor for Electric Vehicles," IET Intelligent Transport Systems, vol. 13, no. 1, pp. 175–182, 2019.
10. Y. Qin, C. He, X. Shao, H. Du, C. Xiang, and M. Dong, "Vibration Mitigation for In-Wheel Switched Reluctance Motor Driven Electric Vehicle with Dynamic Vibration Absorbing Structures," Journal of Sound and Vibration, vol. 419, pp. 249–267, 2018.
11. A. Rezig, W. Boudendouna, A. Djerdir, and A. N'Diaye, "Investigation of Optimal Control for Vibration and Noise Reduction In-Wheel Switched Reluctance Motor Used in Electric Vehicle," Mathematics and Computers in Simulation, vol. 167, pp. 267–280, 2020.
12. N. J. Wills, Y. Y. Li, J. Z. Jiang, T. L. Hill, S. A. Neild, and M. Dhaens, "Using Inerter-Integrated Absorbers to Improve the Performance of an in-Wheel Motor System," Vehicle System Dynamics, vol. 62, no. 8, pp. 2162–2183, 2024.
13. R. R. Das and E. Vinodh Kumar, "Active Suspension with Model Predictive Control," International Journal of Engineering and Advanced Technology, vol. 8, no. 6, pp. 2826–2831, 2019.
14. Y. Chen, S. Chen, Y. Zhao, Z. Gao, and C. Li, "Optimized Handling Stability Control Strategy for a Four In-Wheel Motor Independent-Drive Electric Vehicle," IEEE Access, vol. 7, pp. 17017–17032, 2019.
15. Y. Qin, Z. Wang, K. Yuan, and Y. Zhang, "Comprehensive Analysis and Optimization of Dynamic Vibration-Absorbing Structures for Electric Vehicles Driven by In-Wheel Motors," Automotive Innovation, vol. 2, no. 4, pp. 254–262, Dec. 2019.
16. Z. Deng, X. Li, X. Li, S. Zhao, and H. Wei, "Mechanism Analysis and Optimum Control of Negative Airgap Eccentricity Effect for In-Wheel Switched Reluctance Motor Driving System," Nonlinear Dynamics, vol. 111, no. 10, pp. 9075–9093, 2023.
17. H. Jiang, C. Wang, Z. Li, and C. Liu, "Hybrid Model Predictive Control of Semiactive Suspension in Electric Vehicle with Hub-Motor," Applied Sciences (Switzerland), vol. 11, no. 1, pp. 1–24, 2021.
18. Y. Qin, Z. Zhao, Z. Wang, and G. Li, "Study of Longitudinal–Vertical Dynamics for In-Wheel Motor-Driven Electric Vehicles," Automotive Innovation, vol. 4, no. 2, pp. 227–237, 2021.
19. Y. Zhu, C. Zhao, J. Zhang, and Z. Gong, "Vibration Control for Electric Vehicles with In-Wheel Switched Reluctance Motor Drive System," IEEE Access, vol. 8, pp. 7205–7216, 2020.
20. K. Deepak, M. A. Frikha, Y. Benômar, M. El Baghdadi, and O. Hegazy, "In-Wheel Motor Drive Systems for Electric Vehicles: State of the Art, Challenges, and Future Trends," Energies, vol. 16, no. 7, pp. 1–31, 2023.
21. D. T. Tu, "Enhancing Road Holding and Vehicle Comfort for an Active Suspension System Utilizing Model Predictive Control and Deep Learning," Engineering, Technology & Applied Science Research, vol. 14, no. 1, pp. 12931–12936, Feb. 2024.
22. G. Long, F. Ding, N. Zhang, J. Zhang, and A. Qin, "Regenerative Active Suspension System with Residual Energy for In-Wheel Motor Driven Electric Vehicle," Applied Energy, vol. 260, no. December 2019, p. 114180, 2020.

23. R. He and J. C. Wang, "Vertical Vibration Control of an In-Wheel Motor-Driven Electric Vehicle Using an in-Wheel Active Vibration System," Asian Journal of Control, vol. 22, no. 2, pp. 879–896, 2020.
24. Q. Li, Z. Chen, H. Song, and Y. Dong, "Model Predictive Control for Speed-Dependent Active Suspension System with Road Preview Information," Sensors, vol. 24, no. 7, pp. 1–17, 2024.
25. D. T. Tu, "Enhancing road holding and vehicle comfort for an active suspension system utilizing model predictive control and deep learning," Engineering, Technology & Applied Science Research, vol. 14, no. 1, pp. 12931-12936, 2024.
26. Z. Zhao, L. Zhang, J. Wu, L. Gu, and S. Li, "Vertical-Longitudinal Coupling Effect Investigation and System Optimization for a Suspension-In-Wheel-Motor System in Electric Vehicle Applications," Sustainability, vol. 15, no. 5, p. 4168, Feb. 2023.
27. J. Zhao, J. Dong, P. K. Wong, X. Ma, Y. Wang, and C. Lv, "Interval Fuzzy Robust Non-Fragile Finite Frequency Control for Active Suspension of in-Wheel Motor Driven Electric Vehicles with Time Delay," Journal of the Franklin Institute, vol. 359, no. 12, pp. 5960–5990, 2022.
28. S. Skogestad and I. Postlethwaite, "Multivariable Feedback Control: Analysis and Design," John Wiley & Sons, 2005.
29. J. Wang, L. Chen, and M. Yamamoto, "Advanced Predictive Control Strategies for Vehicle Active Suspension Systems with Input Constraints," IEEE Transactions on Control Systems Technology, vol. 31, no. 2, pp. 789–802, 2023.
30. S. Li, W. Zhang, and H. Wang, "Multi-Objective Optimization of In-Wheel Motor Suspension Systems Using Adaptive Model Predictive Control," Mechanical Systems and Signal Processing, vol. 185, p. 109782, 2023.
31. K. Tanaka, Y. Nakamura, and T. Kobayashi, "Vibration Suppression and Energy Regeneration in Electric Vehicle Active Suspension Systems with In-Wheel Motors," Journal of Sound and Vibration, vol. 545, p. 117435, 2023.
32. M. Garcia-Nieto, A. Hernandez, and R. Sanchis, "Robust Model Predictive Control for Active Suspension Systems under Parameter Uncertainties and Road Disturbances," Control Engineering Practice, vol. 138, p. 105609, 2023.
33. J. B. Rawlings and D. Q. Mayne, "Model Predictive Control: Theory and Design," *Nob Hill Publishing*, 2009.
34. M. Alamir, "Stabilization of Nonlinear Systems Using Receding-Horizon Control Schemes: A Parametrized Approach for Fast Systems," *Springer-Verlag*, 2006.
35. A. B. Kurzhanski and P. Varaiya, "Theory and Applications of Model Predictive Control: A Survey of Recent Developments," *Journal of Applied Mathematics and Mechanics*, vol. 81, no. 4, pp. 253-278, 2017.

8 Conclusion: The Triumph of Active Control in Redefining Vehicle Dynamics

This monograph has systematically deconstructed the fundamental challenges associated with IWASS, transforming perceived physical limitations into tractable control and design problems. The journey from conceptual analysis to robust control synthesis yields one overarching conclusion: the historical impediments to the IWASS, primarily the unsprung mass penalty, are not intrinsic deficiencies but can be effectively mitigated through integrated multiphysics design and sophisticated control strategies.

The principal theoretical and practical contributions of this work are synthesized as follows:

- **Integrated Design is Non-Negotiable:** A sequential design process is fundamentally inadequate for the tightly coupled nature of the IWASS. A concurrent engineering framework, which co-optimizes the electromagnetic, mechanical, thermal, and control domains, is essential for achieving a viable system architecture within severe packaging constraints.
- **Active Control is a Necessity, Not a Luxury:** Passive suspension systems are incapable of managing the multi-objective trade-offs between ride comfort, road holding, and motor integrity in an IWASS. The transition to active control is, therefore, a fundamental requirement for system viability.
- **Optimal and Robust Control Provide the Pathway:** The LQR/LQG frameworks provide a powerful methodology for high-performance control, while the automation of their tuning via GA yields superior performance, as demonstrated by significant improvements in ride comfort and road holding. Furthermore, the application of H_∞ synthesis is critical for guaranteeing robust stability and performance. In addition, the MPC controller is simulated in both frequency and time domains, proving the adaptation of IWASS balancing dynamic performance in the face of real-world parametric uncertainties and disturbances.
- **Performance is Guaranteed by Robustness:** Parametric analysis confirms that while the IWASS can be stabilized, its performance is highly sensitive to operational variability. Therefore, a control system designed only for nominal performance is insufficient; resilience against worst-case scenarios is a prerequisite for commercial deployment.

DOI: 10.1201/9781003780533-8

PROSPECTIVE RESEARCH DIRECTIONS

The foundations laid herein point toward several critical avenues for future research, essential for transitioning the IWASS from a high-performance prototype to an integrated subsystem of next-generation vehicles:

1. **Holistic Vehicle Dynamics Control:** The scope must expand from quarter-car models to full-vehicle, Multi-Input Multi-Output (MIMO) control architectures. Research is needed to develop unified strategies that synergistically coordinate IWASS actuators with torque vectoring and steering systems to achieve global vehicle dynamics objectives.

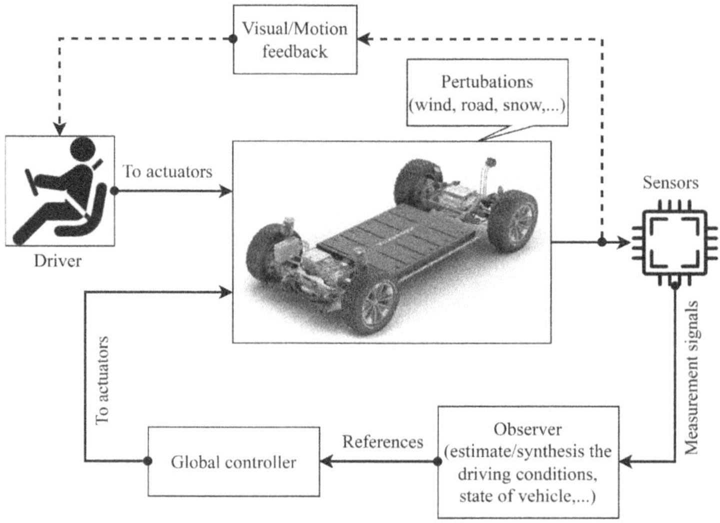

2. **Resilient and Self-Adapting Systems:** To operate reliably across diverse and time-varying conditions, control systems must evolve from merely robust to adaptive-robust. This entails the real-time estimation of critical parameters (e.g., vehicle mass, road friction) and the subsequent auto-tuning of controller parameters to maintain optimal performance.
3. **Co-Design with Autonomous Systems:** The IWASS should be co-designed with the motion planning and control stacks of autonomous vehicles. Research is needed to develop IWASS controllers that can precisely track desired chassis motions, providing the high-fidelity actuation required for both passenger comfort and path-following accuracy.

In conclusion, the IWASS represents more than an incremental improvement in suspension technology; it embodies a shift towards fully active, integrated, and intelligent vehicle dynamics management. By reframing its challenges through the lens of modern control theory and multiphysics optimization, this work confirms that the IWASS is a potent enabler for the future of automotive mobility, promising unprecedented levels of comfort, safety, and dynamic capability.

Glossary

Active Suspension: A suspension system that uses an external power source (e.g., hydraulic, pneumatic, or electric) and an actuator to generate controlled forces. Unlike passive systems, it can add energy to the system to actively counteract road disturbances and body motions, thereby significantly improving ride comfort, vehicle handling, and stability. Its performance is primarily determined by the sophistication of its control algorithm.

Actuator: The component in an active suspension system responsible for generating the controlled force dictated by the controller. In the context of this monograph, it typically refers to a rotary or linear motor integrated with the suspension strut. The actuator's force capacity, bandwidth, and response time are critical limiting factors for the overall performance of the control system.

Algebraic Riccati Equation: A fundamental nonlinear matrix equation in optimal control theory. The solution to the continuous-time ARE provides the optimal state-feedback gain matrix for the Linear Quadratic Regulator. Solving this equation ensures that the cost function is minimized, balancing state regulation against control effort.

Bode Plot: A frequency-domain representation of a linear time-invariant system's transfer function. It consists of two graphs: magnitude (in decibels) versus frequency and phase (in degrees) versus frequency. In suspension analysis, Bode plots are essential for evaluating a controller's vibration isolation performance across the critical frequency range affecting ride comfort and road holding.

Bouc-Wen Model: A phenomenological model used to describe the complex hysteresis behavior observed in dampers and certain actuators. It is represented by a set of nonlinear differential equations that capture the history-dependent, path-dependent force-displacement-velocity relationship, allowing for more accurate simulation of real-world actuator dynamics.

Chromosome: In the context of a Genetic Algorithm, a chromosome is a single candidate solution to the optimization problem, encoded as a data structure (e.g., a vector or string). For the GA-LQR controller, a chromosome contains the set of LQR weighting parameters (e.g., ρ_1 to ρ_6) being optimized.

Constraint Handling: The methodology employed within an optimization algorithm to manage solutions that violate predefined physical or performance limits. In this work, a penalty function method is used, where infeasible solutions (e.g., those exceeding maximum suspension travel) are assigned a severely penalized fitness value to steer the search toward the feasible region.

Control Effort: A measure of the energy or force required by the actuator to implement the control law. It is a critical design consideration, as excessive control effort implies high power consumption, large actuator size, and potential heat dissipation issues. Optimal controllers like LQR explicitly penalize control effort in their cost function.

Crossover: A primary genetic operator in Evolutionary Algorithms where two parent chromosomes are combined to create one or two offspring. This operation allows the algorithm to explore new regions of the search space by mixing promising genetic material from different parents, fostering the inheritance of beneficial traits.

Damping Coefficient: A parameter that quantifies the viscous damping capacity of a shock absorber or damper. It defines the force generated per unit of velocity and is responsible for dissipating vibrational energy from the suspension system, thus controlling the oscillation amplitude and settling time.

Disturbance Input: An external, uncontrolled input to a dynamic system that adversely affects its outputs. For the quarter-car model, the primary disturbance is the road surface profile displacement, Z_r, which is the source of vibrations that the control system aims to reject.

Electric Vehicle: A vehicle that uses one or more electric motors for propulsion, powered by an onboard energy storage system such as a battery. The EV architecture is particularly amenable to in-wheel motor configurations, which directly drive the wheels but introduce unique challenges for suspension design.

Fitness Function: The objective function in a Genetic Algorithm that quantitatively evaluates the performance or "fitness" of a chromosome. For the GA-LQR optimization, the fitness function is typically a weighted sum of key performance metrics like RMS body acceleration and RMS tire deflection, which the algorithm seeks to minimize.

Frequency Domain Analysis: The study of a system's dynamic response as a function of the input frequency. This analysis, performed using tools like Bode plots, is crucial for suspension design as it reveals resonance frequencies, damping effectiveness, and the controller's ability to isolate vibrations across the spectrum of road inputs.

Full-State Feedback: A control architecture where the control signal is generated by feeding back a linear combination of all state variables of the system ($u = -Kx$). The Linear Quadratic Regulator (LQR) is a classic example of an optimal full-state feedback controller, though it requires all states to be measurable.

Fuzzy Logic: A computational intelligence paradigm based on "degrees of truth" rather than the strict Boolean logic of true or false. It uses linguistic variables and IF-THEN rules to model complex, nonlinear systems where precise mathematical models are difficult to derive, making it suitable for adaptive control like Fuzzy-PID.

Genetic Algorithm: A population-based metaheuristic optimization algorithm inspired by the process of natural selection. GAs evolve a population of candidate solutions over generations through the application of selection, crossover, and mutation operators, making them effective for solving complex, non-convex optimization problems like controller tuning.

H∞ **Control**: A robust control methodology designed to minimize the worst-case gain (the H_∞ norm) from exogenous inputs (like disturbances and model uncertainties) to the performance outputs. It provides guaranteed performance and stability robustness against a defined set of modeling errors and variations.

Hysteresis: A nonlinear phenomenon where the output of a system (e.g., force from an actuator) depends not only on the instantaneous input (e.g., displacement or current) but also on the history of its past states. The Bouc-Wen model is often used to capture this behavior in mechanical dampers.

In-Wheel Motor: An electric traction motor that is fully integrated within the wheel hub of a vehicle. This configuration saves space, enables advanced torque vectoring, but significantly increases the unsprung mass, presenting a major challenge for suspension design and ride quality.

Kalman Filter: An optimal recursive algorithm used for state estimation of a dynamic system from a series of noisy measurements. It produces estimates that minimize the mean squared error and is a critical component of the LQG controller, enabling optimal control when the full state is not directly measurable.

Linear Quadratic Gaussian: An optimal control strategy that combines a Linear Quadratic Regulator for state feedback with a Kalman Filter for state estimation. The Separation Principle allows these two components to be designed independently, resulting in an optimal controller for linear systems with Gaussian noise.

Linear Quadratic Regulator: A dominant method in modern optimal control for linear systems. It determines the state-feedback control gain matrix that minimizes a quadratic cost function, which typically penalizes undesirable state deviations (e.g., body acceleration) and excessive control effort, thus formalizing the performance trade-offs.

Model Predictive Control: A control strategy that uses an explicit system model to predict future behavior and solves an optimization problem at each sampling instant to determine optimal control actions.

Multi-Objective Optimization: An optimization problem that involves simultaneously minimizing or maximizing multiple, often competing, objective functions. In suspension control, the primary objectives are ride comfort (minimize body acceleration) and road holding (minimize tire deflection), which naturally conflict, requiring a Pareto-optimal trade-off.

Mutation: A genetic operator that introduces random, typically small, alterations to one or more genes in a chromosome. This operator helps maintain population diversity, prevents premature convergence to local optima, and enables the GA to explore new areas of the search space.

Optimal Control: A branch of control theory that deals with the problem of finding a control law for a dynamical system over a period of time such that a specific performance criterion (cost function) is optimized. LQR and LQG are classic examples of optimal control methodologies.

Pareto Front: The set of all non-dominated or Pareto-optimal solutions in a multi-objective optimization problem. A solution is on the Pareto front if improving one objective necessarily leads to the worsening of at least one other objective, representing the best possible trade-offs.

Power Spectral Density: A frequency-domain description of a random signal, showing how the power of the signal is distributed over different frequencies. Road profiles are commonly classified using PSD according to standards like ISO 8608, which is used to generate realistic random road disturbances for simulation.

Prediction Horizon: The number of future time steps over which the system's behavior is predicted.

Proportional-Integral-Derivative Control: A ubiquitous control algorithm that calculates an error value and applies a correction based on proportional, integral, and derivative terms. While simple and robust, its fixed-gain structure is often inadequate for handling the nonlinear, multi-objective nature of high-performance active suspension systems.

Quarter-Car Model: A simplified two or three-degree-of-freedom dynamic model that represents the vertical dynamics of a single wheel station of a vehicle. Its simplicity makes it invaluable for initial controller design, parameter studies, and fundamental performance analysis, capturing the essential sprung and unsprung mass interactions.

Rattle Space: The total usable travel distance of the suspension between its upper and lower mechanical limits (the bump and rebound stops). A key control objective is to minimize the dynamic deflection within this space to prevent hitting the limits, which causes a sharp degradation in ride comfort.

Receding Horizon Principle: After computing the optimal control sequence, only the first control input is applied; the horizon then shifts forward at the next sampling time.

Ride Comfort: A subjective measure of the well-being of the vehicle occupants, primarily influenced by the vertical vibrations transmitted from the road. It is quantitatively evaluated using metrics like the Root Mean Square or frequency-weighted value of the sprung mass acceleration, often referenced against standards like ISO 2631-1.

Road Holding: The ability of a vehicle to maintain consistent and firm contact between its tires and the road surface. It is critical for safety, traction, braking, and cornering, and is quantitatively assessed by the dynamic tire deflection or the variation in dynamic tire load.

Robustness: The property of a control system to maintain its stability and specified performance levels in the presence of uncertainties. These uncertainties can include parameter variations (e.g., changing passenger load), unmodeled dynamics, and external disturbances. H-Infinity control is explicitly designed for robustness.

Root Mean Square: A statistical measure of the magnitude of a varying quantity. It is particularly useful for analyzing oscillatory signals like acceleration, as it provides a single value representing the "energy" or overall level of the oscillation over time, which correlates well with human perception of vibration.

Separation Principle: A fundamental result in stochastic optimal control which states that the problems of optimal state feedback (LQR) and optimal state estimation (Kalman Filter) can be solved independently. This principle justifies the separate design of the LQR gain and the Kalman filter in the LQG controller.

Spring Stiffness: A parameter defining the force required to deflect a spring by a unit length. In a suspension system, the main spring stiffness primarily determines the natural frequency of the sprung mass, which is a key factor in ride comfort.

Sprung Mass: The portion of the total vehicle mass that is supported by the suspension springs, including the chassis, body, engine, and occupants. The acceleration of the sprung mass is the primary metric for ride comfort.

State-Space Representation: A mathematical model of a physical system as a set of input, output, and state variables related by first-order differential equations. It provides a powerful and unified framework for analyzing and designing multi-input, multi-output systems, and is the standard form for modern control techniques like LQR and LQG.

Unsprung Mass: The mass of the suspension components that are not supported by the springs, including the wheels, tires, brake assemblies, and, critically for this work, the in-wheel motor. A high unsprung mass reduces the wheel's ability to follow road contours, degrading ride comfort and road holding.

Index

For Product Safety Concerns and Information please contact our EU
representative GPSR@taylorandfrancis.com
Taylor & Francis Verlag GmbH, Kaufingerstraße 24, 80331 München, Germany

www.ingramcontent.com/pod-product-compliance
Ingram Content Group UK Ltd.
Pitfield, Milton Keynes, MK11 3LW, UK
UKHW022220190926
475702UK00003B/463
9781041315162